Encyclopedia of
Animal Rights and
Animal Welfare

ENCYCLOPEDIA OF ANIMAL RIGHTS AND ANIMAL WELFARE

Edited by **Marc Bekoff**
with **Carron A. Meaney**

Foreword by **Jane Goodall**

Greenwood Press
Westport, Connecticut

Library of Congress Cataloging-in-Publication Data

Encyclopedia of animal rights and animal welfare / edited by Marc
 Bekoff with Carron A. Meaney ; foreword by Jane Goodall.
   p. cm.
   Includes bibliographical references and index.
   ISBN 0–313–29977–3 (alk. paper)
     1. Animal rights—Encyclopedias. 2. Animal welfare—
 Encyclopedias. I. Bekoff, Marc. II. Meaney, Carron A., 1950– .
 HV4708.E53   1998
 179'.3—dc21   97–35098

British Library Cataloguing in Publication Data is available.

Library of Congress Catalog Card Number: 97–35098
ISBN: 0–313–29977–3

First published in 1998

Greenwood Press, 88 Post Road West, Westport, CT 06881
An imprint of Greenwood Publishing Group, Inc.

Printed in the United States of America

The paper used in this book complies with the
Permanent Paper Standard issued by the National
Information Standards Organization (Z39.48–1984).

10 9 8 7 6 5 4 3 2

Cover Acknowledgments:

Photo of chickens courtesy of Joy Mench. Photo of *Macaca experimentalis* courtesy
of Viktor Reinhardt. Photo of Lyndon B. Johnson courtesy of the Lyndon Baines
Johnson Presidential Library Archives.

Contents

Foreword

It is an honor for me to contribute a foreword to this unique, informative, and exciting volume. Never before has an attempt been made to gather together, between two covers, comprehensive information about the use and abuse of nonhuman animals by our own human species, along with the complex issues that must be understood by those who are concerned with animal welfare and animal rights, and some of the ways in which different groups are tackling these issues. Because human beings are animals, this book could have been expanded to include the horrible abuse and torture to which we subject other humans—theoretically, there could be a whole section on human rights. But that is not the purpose of the editors. This book is concerned with the essential dignity of the wondrous nonhuman beings with whom we share this planet, and our human responsibilities towards them: the beings known in common parlance as "animals"—which is how I shall refer to them here.

Of course, we humans are much more like other animals than was once thought, much more so than many people like to, or are prepared to, believe. I have been privileged to spend 35 years learning about and from the chimpanzees, our closest living relatives. A detailed understanding of chimpanzee nature has helped, perhaps more than anything else, to blur the line, once thought to be so clear and sharp, dividing humans from the rest of the animal kingdom. Once we are prepared to accept that it is not only humans who have personalities, not only humans who are capable of rational thought and simple problem solving, and above all, not only humans who can experience emotions such as joy, sorrow, fear, despair, and mental as well as physical suffering, then we are surely compelled to have new respect not only for chimpanzees but also for so many other amazing animal species. (In fact, I received my first lessons about the amazing capabilities of nonhumans from my dog, Rusty, before I was 10 years old.)

The only thing that we humans do that no other animals do in the same way is to communicate by means of a sophisticated spoken—and written—language, and this, I believe, lays on us certain responsibilities towards the rest of the animal kingdom. (It might be mentioned that in English translations of the Old Testament—Psalm 8—"dominion" is often used, but this is somewhat misleading. "Dominion" is not the best translation of the original Hebrew word, which is actually a verb meaning "made to rule over," as a wise king rules over his subjects with care and respect. Whatever English word is chosen, it is clear that the original Hebrew phrasing implies a respectful and caring attitude towards creation and suggests a sense of responsibility. This, of course, gives the text a completely different meaning than some of the narrower meanings, such as domination, which are often read into the English translation "dominion.")

I have been fortunate. I have been able to spend many years observing chimpanzees and other animals in their own natural environments, thereby gaining unique insights into their true nature. For this reason, I believe it is my particular responsibility to share my knowledge with as large an audience as possible for the benefit of the animals themselves. Chimpanzees have given me as much, and I am haunted at the thought of those who are imprisoned in the name of entertainment or science. As I have written elsewhere, "The least I can do is to speak out for the hundreds of chimpanzees who right now, sit hunched, miserable and without hope, staring out with dead eyes from their metal prisons. They cannot speak for themselves."

This is why I am so very glad that this encyclopedia has been put together—for it speaks out for animals, for all kinds of animals. It broadcasts a simple message, a plea, that needs desperately to be heard as we head into the 21st century. Give animals the respect that, as sentient beings, is their due. And this simple message is delivered here by a multitude of voices from many different disciplines: from biology, including ethology (the study of behavior) and ecology, anthropology, psychology, philosophy, sociology, education, law, ethnology, history, politics, theology, veterinary science, and public administration. This multidisciplinary collection of contributors means that the essays discuss the central theme from different perspectives: collectively they provide an astonishingly rich overview of the extent of animal suffering in our modern society and the various steps that have been taken by those fighting for animal welfare and animal rights. And, importantly, the material is presented in a straightforward way intended to appeal to the general public as well as the scientists. Once this encyclopedia reaches the shelves of libraries in schools and universities, many young people, as well as their teachers, will have access to this valuable information.

The encyclopedia provides the reader with an opportunity to acquire in-depth understanding of complex issues. And because different contributors voice differing opinions, the reader will also be able to develop his or her own carefully reasoned arguments to use when discussing controversial issues with people who hold different views. This is important. The more passionate one feels about animal abuse, the more important it becomes to try to understand what is behind it. However distasteful it may seem, it really is necessary to become fully informed about a given issue. Dogmatism, a refusal to listen to any point of view differing from one's own, results in moral and intellectual arrogance. This is far from helpful and is most unlikely to lead to any kind of progress. The "us" v. "them" attitude brings useful dialogue to an end. In fact, most issues are quite complex and can seldom be described in simple terms of black and white. And until we become fully cognizant of all that is involved, we had better not start arguing, let alone throwing bricks at anyone.

Let me give an example. Recently, during a semi-official visit to South Korea, a press conference was set up by my host organization. The subject of cruelty came up. I said that I would like to discuss their habit of eating dogs. My interpreter blanched. Quite clearly she felt that this was politically insensitive and would embarrass my hosts! I explained that in the country where I grew up (England), people typically ate cows and pigs and chickens, and that pigs at least are quite as intelligent as dogs and, in fact, make wonderful pets. Yet only too often they are kept in horrendous conditions. I suggested that the most important issue, if one was going to eat an animal at all (which I did not), was not so much the species as how it was treated in life. At this point one of the journalists assured me that the dogs they ate were bred for eating. This led to discussions about whether or not this made any difference, the ways in which dogs—and pigs—were kept, and a variety of other issues. The point was that an almost taboo subject was aired in public, and this led, for a number of people, to new ways of thinking about animals in general.

Perhaps the bitterest pill that we who care about animals have to swallow is that only too often, it is through a series of compromises that progress is actually made, and this seems agonizingly slow. There are, of course, situations when the cruelty inflicted is so great that no compromise is possible. Then it is equally important, if not more so, to know as much as possible about the situation: this encyclopedia may provide the animal activist with information about how similar situations have been successfully tackled.

The essays in the volume are necessarily brief, summarizing information which in some cases is extensive. Each essay can serve to stimulate the reader to pursue a particular issue in greater depth, guided by the extensive lists of references and key organizations that have been compiled for the encyclopedia. These lists will be a goldmine for all those who care about animal issues.

Albert Schweitzer once said, "We need a boundless ethic that includes animals too." At the present our ethic concerning animals is limited and confused. For me, cruelty, in any shape or form, whether it be directed towards humans or sentient nonhumans, is the very worst of human sins. To fight cruelty brings us into direct conflict with that unfortunate streak of inhumanity that lurks in all of us. For all who are like I am, committed to joining this particular battle, this encyclopedia will prove invaluable. A great deal of the behavior that we deem cruel is not deliberate but due to a lack of understanding. It is that lack of understanding that we must overcome. And every time cruelty is overcome by compassion, we are moving towards that new and boundless ethic that will respect all living beings. Then indeed we shall stand at the threshold of a new era in human evolution—the realization of our most unique quality: humanity.

—*Jane Goodall*

Preface

The preparation of this encyclopedia was a difficult and time-consuming task. Emily Birch first contacted me in April 1995 and asked whether I was interested in undertaking this project. I hesitated and then said "yes," later wishing on more than one occasion that I had not shown such weakness. Contacting authors, developing a working index, preventing and putting out fires, and editing consumed me daily (and in and of itself, there is an interesting sociological story that can be told at another time). Carron Meaney helped primarily with editorial matters. Interestingly, only about five people said "no" to the invitation that was extended to them. Three thought that they could not write a substantial essay, and two were uneasy about having their names associated with a book whose title included the word "rights." This was unfortunate, for the final product deals with much more than animal rights.

Many people were extremely helpful in making this project grow, bloom, and mature into the finished product. First, I thank all contributors for their efforts on our and other animals' behalf. I appreciate their patience and understanding during the long process of organizing and seeing this volume through to completion. Their entries were written and rewritten and finally edited for length, audience, consistency in style, and overlap. Most authors did not see the final edited versions of their essays. My editor, Emily Birch, was a pleasure to work with and always supported me; she provided comic relief when it was sorely needed and expert editorial assistance. Emily also was a source of inspiration for the daily grind of downloading, reading, editing, and sending entries back to contributors. Charles Eberline did an outstanding job of copyediting. Andrew Linzey and Bernard Unti wrote about 200 biographical essays from which I chose a representative handful;

their efforts and their help in making these difficult choices went well beyond the call of duty. Andrew Rowan and David Morton provided extensive advice on the development of the chronological list, as did Christine Stevens, who also provided useful historical insights. David Anderson offered his long list of organizations and worked closely with me to come up with a representative final product. Colin Allen was always there to help in downloading and translating entries that I could not handle, and Linda Bowden and Randy Abrams in the office of Environmental, Population, and Organismic Biology, University of Colorado, Boulder, retyped a number of essays.

—*Marc Bekoff*

Introduction

Interest in the nature of human–nonhuman-animal (hereafter animal) interactions is growing as we head into the 21st century, for it is clear that there
are many important associated issues that demand immediate and careful
attention. Basically, while most people agree that animals are important to
humans and that we must pay attention to their well-being, there also is a
good deal of disagreement about the types, if any, of obligations that humans
have toward other animals. People who have thought a lot about these issues
often use the same information to come to vastly different conclusions or
use very different information to come to the same conclusions. Because so
many people come to these issues from very different walks of life (academic
and nonacademic) and many different areas of interest, most of which are
represented in this volume (for example, social, political, educational, philosophical, psychological, legal, zoological, ethological, ecological, theological, anthropological, sociological, historical, biographical, veterinary science,
ethnological, and public health), I thought it important to collect as much
information as possible in one easy-to-read reference book.

The issues with which humans need to deal to develop informed views
about human-animal interactions require that people from many different
disciplines be involved in the discussions. Of course, these exchanges of ideas
must be open and people must be sensitive to all different views if we are
to make progress. I hope that I have been successful in having all sides
presented; balance is essential, for there are many difficult and contentious
issues. "Us-them" interactions are not very helpful and tend to alienate,
rather than to unite, individuals. It is important for all people to listen to
one another and for all of us to listen to the animals with whom we are
privileged to share the planet and interact. Respect for the dignity of all

animals' lives needs to underlie consideration of how humans interact with other animals. Thus I hope that I and my authors have covered the issues from all sides, including theoretical matters and practical applications, using information gathered from animals living in highly controlled laboratory environments and those living in the wild. All types of data are important, and much useful information about the complexity, diversity, and richness of animals' lives has come from the study of free-living animals.

It also is important to stress that there is a long, rich, and diverse history of events that center on how animals have been used by human animals in various sorts of activities. Thus I have included a representative sample of biographical sketches to show just how rich and diversified the tradition is. Some difficult choices had to be made about whom to include, and I decided not to include any among the living (those living persons who have made major contributions are mentioned in many entries).

In a nutshell, this encyclopedia offers, for the first time, a discussion of just about all of the major issues between its two widely separated covers. I hope that it becomes clear that humans have unique responsibilities to the world and that these need to be taken seriously. We and the animals whom we use should be viewed as partners in a joint venture. We can teach one another respect and trust, and animals can facilitate contact with ourselves and help us learn about our place in this complex but awe-inspiring world. If we forget that humans and other animals are all part of the same world, and if we forget that humans and animals are deeply connected at many levels of interaction, when things go amiss in our interactions with animals and animals are set apart from and inevitably below humans, it is certain that we will miss the animals more than the animal survivors will miss us. The interconnectivity and spirit of the world will be lost forever, and these losses will make for a severely impoverished universe. As Paul Shepard wrote:

> There is a profound, inescapable need for animals that is in all people every-where, an urgent requirement for which no substitute exists. This need is no vague, romantic, or intangible yearning, no simple sop to our loneliness or nostalgia for Paradise. . . . Animals have a critical role in the shaping of per-sonal identity and social consciousness. . . . Because of their participation in each stage of the growth of consciousness, they are indispensable to our be-coming human in the fullest sense.[1]

Entries for this encyclopedia were mainly chosen by going through nu-merous books and essays and listing the topics that were covered in these works. In many instances, indexes to various books provided good alpha-betical listings of important topics. The entries in this volume were listed in numerous indexes or were included as major topics in a large number of books and papers, indicating their importance in debates concerning the subjects of animal rights and animal welfare.

Entries are arranged in alphabetical order. They are cross-referenced through the use of the asterisk (*). "See" references and "see also" references are meant to lead the reader to other relevant topics. There are also separate sections on sources and organizations, and a chronological listing of historical events. Readers can use all of these tools to further their research and to gather more information on a specific topic. Entries should not be read as being complete works, nor should the selected bibliography after each entry be thought of as complete. Rather, each entry and the summary of resources should be viewed as points of departure for further investigations, rather like kindling wood that can be used to ignite larger fires.

NOTE

1. Paul Shepard, *Traces of an Omnivore* (Washington, DC: Island Press, 1996), 3.

Chronology

This is a chronology of some historical events (in the United States if not otherwise indicated; UK stands for United Kingdom) related to the use of animals and to animal rights and animal welfare. For more information see the sources chapter, including Rowan (1984), Ritvo (1987), Ryder (1989), Animal Welfare Institute (1990), Orlans (1993), Finsen and Finsen (1994), Salisbury (1994), Zurlo, Rudacille, and Goldberg (1994), Cohen (1995), Sherry (1995), and Francione (1995, 1996). The Animal Welfare Information Center (AWIC) Newsletter updates information in its "Congress in Action" section.

1822	Ill-Treatment of Cattle Act
1822	Martin's Anticruelty Act (UK)
1824	Society for the Prevention of Cruelty to Animals (SPCA) (UK) founded
1826	Bill to Prevent the Cruel and Improper Treatment of Dogs
1832	Warburton Anatomy Act (UK)
1840	SPCA becomes the Royal Society for the Prevention of Cruelty to Animals (RSPCA) with patronage of Queen Victoria (UK)
1866	American Society for the Prevention of Cruelty to Animals (ASPCA) founded
1868	Massachusetts Society for the Prevention of Cruelty to Animals (MSPCA) founded
1875	Victoria Street Society for the Protection of Animals from Vivisection (UK) founded
1876	Cruelty to Animals Act (UK)
1877	American Humane Association founded

1883	American Anti-Vivisection Society founded
1889	American Humane Education Society (AHES) founded
1891	The Humanitarian League founded
1895	New England Anti-Vivisection Society founded
1898	British Union for the Abolition of Vivisection (UK)
1906	Animal Defence and Anti-Vivesection Society (UK) founded
1911	Protection of Animals Act (England, UK)
1912	Millennium Guild founded
1912	Protection of Animals Act (Scotland, UK)
1925	The Performing Animals (Regulations) Act (UK)
1926	University of London Animal Welfare Society founded (name changed to Universities Federation for Animal Welfare [UFAW] in 1938) (UK)
1929	National Anti-Vivisection Society (UK) founded (formerly Victoria Street Society for the Protection of Animals from Vivisection)
1946	National Society for Medical Research founded
1948	Morris Animal Foundation founded
1949	The Docking and Nicking of Animals Act (UK)
1950	Animal Protection Law (covers farm animals and bans battery cages) (Denmark)
1951	Animal Welfare Institute founded
1952	Institute for Animal Laboratory Resources founded
1954	Humane Society of the United States (HSUS) founded
1954	The Protection of Animals (Anaesthetics) Act (UK)
1955	Society for Animal Protective Legislation founded
1957	Friends of Animals founded
1958	Humane Slaughter Act
1959	Beauty without Cruelty (UK) founded
1959	Wild Horses Act
1959	Catholic Society for Animal Welfare (now International Society for Animal Rights) founded
1960	The Abandonment of Animals Act (UK)
1961	Lawson-Tait Trust (UK) founded
1962	Bald and Golden Eagle Protection Act
1962	The Animals (Cruel Poisons) Act (UK)
1963	British Hunt Saboteurs Association (UK) founded
1965	Brambell Report on Farm Animal Welfare (UK)
1965	Littlewood Report (UK)
1965	American Association for Accreditation of Laboratory Animal Care founded

1966	Laboratory Animal Welfare Act
1967	Fund for Animals (UK) founded
1967	Farm Animal Welfare Advisory Committee (UK) founded
1968	Animal Protection Institute founded
1969	Council of Europe Convention on Animals in Transport
1969	International Fund for Animal Welfare (IFAW) founded
1969	Endangered Species Act
1969	Fund for the Replacement of Animals in Medical Experiments (FRAME) (UK) founded
1969	International Association against Painful Experiments on Animals (UK) founded
1970	Laboratory Animal Welfare Act broadened and renamed Animal Welfare Act; legislation extended to include all warm-blooded animals (including pet and exhibition trades)
1970	Dr. Hadwen Trust for Humane Research (UK) founded
1971	Greenpeace (now International) founded
1971	Wild Free-roaming Horse and Burro Act
1971	Law requiring approval of new buildings for animal protection (Sweden)
1972	American Zoo and Aquarium Association accreditation standards and code of professional ethics
1972	Marine Mammal Protection Act
1972	Animal Protection Act (Germany)
1973	International Primate Protection League founded
1973	National Antivivisection Society founded
1973	Endangered Species Act strengthened
1973	Convention on International Trade in Endangered Species (CITES) of wild fauna and flora (international)
1976	Animal Rights International (Henry Spira) founded
1976	Animal Welfare Act broadened to cover, among other things, transportation and prohibitions against dogfighting and cockfighting
1976	Horse Protection Act
1976	Fur Seal Act
1976	Protest at American Museum of Natural History (Henry Spira)
1976	The Dangerous Wild Animals Act (UK)
1977	First International Conference on the Rights of Animals, Trinity College, Cambridge, England (organized by Andrew Linzey and Richard Ryder)
1978	Humane Slaughter Act broadened
1978	Scientists Center for Animal Welfare (SCAW) founded
1978	Animal Legal Defense Fund founded

1978   Swiss Animal Welfare Act

1979   Association for Biomedical Research (founded as Research Animal Alliance) founded

1979   Coalition to Abolish the Draize Test (Henry Spira) founded

1979   First European Conference on Farm Animal Welfare, the Netherlands

1979   Packwood-Magnuson Amendment to the International Fishery Conservation Act

1980   People for the Ethical Treatment of Animals (PETA) founded

1980   Psychologists for the Ethical Treatment of Animals (PsyETA) founded

1981   Association of Veterinarians for Animal Rights (AVAR) founded

1981   Johns Hopkins Center for Alternatives to Animal Testing founded

1981   Silver Spring monkeys case

1981   The Zoo Licensing Act (UK)

1981   Foundation for Biomedical Research founded

1982   Marine Mammal Protection Act reauthorized

1982   World Women for Animal Rights/Empowerment Vegetarian Activist Collective founded

1982   Canadian Council on Animal Care founded

1983   In Defense of Animals founded

1984   Humane Farming Association founded

1984   Performing Animal Welfare Society founded

1984   Break-in, Head Injury Clinical Research Laboratory, University of Pennsylvania

1985   Improved Standards for Laboratory Animals Act (an amendment of the Animal Welfare Act)

1985   Head Injury Clinical Research Laboratory closed

1985   National Association for Biomedical Research (merger of National Society for Medical Research, Association for Biomedical Research, and Foundation for Biomedical Research) founded

1985   Jews for Animal Rights founded

1986   Farm Animal Reform Movement (FARM) founded

1986   Animals (Scientific Procedures) Act (UK)

1986   Animal Welfare Information Center founded

1986   European Directive Regarding the Protection of Animals Used for Experimental and Other Scientific Purposes (European Communities)

1986   European Convention for the Protection of Vertebrate Animals Used for Experimental and Other Scientific Purposes (Council of Europe)

1988   Swedish Animal Welfare Act

1989   Veal Calf Protection Bill hearings (U.S. Congress)

1990   Veal Crate Ban (UK)

1990   Pet Theft Act, amendment to the Animal Welfare Act
1990   Rutgers Animal Rights Law Center founded
1991   The Ark Trust, Incorporated, founded
1991   Americans for Medical Progress founded
1991   European Union Regulation against Leghold Traps
1992   Czechoslovakian Law against Cruelty on Animals (first welfare legisla-
        tion in the former Communist countries)
1992   Wild Bird Conservation Act
1992   International Dolphin Conservation Act
1992   Driftnet Fishery Conservation Act
1992   Protection of Animal Facilities Act
1992   Animal Enterprise Protection Act
1993   National Health Revitalization Act
1993   First World Congress on Alternatives and Animals in the Life Sciences,
        Baltimore, Maryland
1993   European Centre for the Validation of Alternative Methods (ECVAM)
1995   Second World Congress on Alternatives and Animals in the Life Sci-
        ences, Utrecht, Netherlands

Encyclopedia of
Animal Rights and
Animal Welfare

A

ACTIVISM FOR ANIMALS

Animal protection as a social movement is a modern development, arising in England early in the 18th century. From the beginning, activists working to protect animals have enlisted the support of wealthy and powerful individuals whose political influence and economic privilege have greatly advanced the animal-protection agenda. At the same time, a high degree of tension has always existed between those promoting gradual improvement and proponents of revolutionary change. Societies for the protection of animals were formed in both England and the United States in connection with the passing of the first animal protection legislation (*see* AMERICAN SOCIETY FOR THE PREVENTION OF CRUELTY TO ANIMALS; ROYAL SOCIETY FOR THE PREVENTION OF CRUELTY TO ANIMALS). Those who fought for this legislation were often the same individuals who formed the societies.

Some authors draw a sharp distinction between the humane movement and activism opposing the use of animals in science (the antivivisection movement; *see* ANTIVIVISECTIONISM), which arose decades later, pointing to both ideological and class differences between the two. However, ideology and class divided individuals within the antivivisection movement as well, as demonstrated most acutely in the rivalry between Anna Kingsford and Frances Power Cobbe,* the two most important figures in 19th-century British antivivisection. While Kingsford, a physician, linked the suffering of laboratory animals with the suffering of "animals in the meat-trade, the fur-trade, in the hunting field, and in the barnyard," Cobbe retained a single-minded focus on vivisection, continuing to wear furs and eat meat. Nonetheless, the two activists were equally intense in their opposition to the scientific use of animals and both refused to compromise or consider anything other than the immediate ending of the practice.

Victorian antivivisectionists tended to use the same methods of protest developed by other groups advocating social change. Foreshadowing contemporary "celebrity activism," Cobbe enlisted the support of individuals prominent in law, government, and the church to lobby for the cause. Antivivisection and animal welfare organizations produced a huge volume of literature in the 19th century, including periodicals, advertisements, and tracts. Five antivivisection congresses drawing activists from all over Europe were held from 1898 to 1909, with the last culminating in a demonstration in London that included seven marching bands.

Louise Lind-af-Hageby* and Leisa Schartau, two Swedish medical students, anticipated the undercover investigations of 20th-century animal rights* groups by attending physiology demonstrations at University College, Kings College, and the University of London and then writing a book about their observations titled *The Shambles of Science*, which created an enormous outpouring of public revulsion. Nonetheless, the increasingly successful record of experimental medicine in developing vaccines and treating infectious diseases effectively killed public support for antivivisection until late in the 20th century.

Interest in animal protection began to peak once again following the publication of philosopher Peter Singer's book *Animal Liberation* in 1975. Singer's critical analysis of human exploitation of other animals, which he termed "speciesism,"* found a receptive audience and instigated an upsurge in animal-related activism that rivaled 19th-century efforts. Enormous sums of money were donated to existing organizations, and a number of new groups were soon founded, most notably People for the Ethical Treatment of Animals (PETA), which grew from 25 to 250,000 members during the 1980s.

Unlike their 19th-century predecessors, 20th-century activists could claim some clear victories. Henry Spira, head of the New York–based Animal Rights International, achieved antivivisection's first major success by forcing the cessation of experiments on cats at the Museum of Natural History in New York City after over a year of protest in 1977. Spira's Coalition to Abolish the Draize Test fought for and eventually achieved radical changes in product safety testing worldwide. In the standard Draize test, a liquid or solid substance is placed in one eye each of a group of rabbits, and changes in the cornea, conjunctiva, and iris are then observed and scored. Both injury and potential for recovery are noted. Consumer protests against widespread use of the Draize test created the momentum that led to the development of alternatives to many types of whole-animal testing. Campaigns against fur wearing led by PETA and other organizations resulted in significant drops in fur sales by the mid-1990s.

Despite its philosophical basis in ethics and its emphasis on compassion, animal protection also displayed a violent face in the activities of the Animal Liberation Front (ALF) and other radical groups. Arson, vandalism, and malicious destruction of property by animal rights activists from 1977 to 1993 resulted in damages of $7.75 million, leading the biomedical research community to press for the Animal Enterprise Protection Act, passed by Congress in August 1992. This legislation makes theft and destruction of property at a research facility a federal crime. In the final years of the 20th century, the focus of animal rights activism is shifting to factory farming* and the environmental, ethical, and health costs of a meat-based diet.

Selected Bibliography. French, R. D., *Antivivisection and Medical Science in Victorian Society* (Princeton: Princeton University Press, 1975); Hume, E. D., *The Mind-Changers* (London: Michael Joseph, 1939); Rowan, A. N., F. M. Loew, and J. C. Weer, *The Animal Research Controversy: Protest, Process, and Public Policy* (Boston: Center for Animals and Public Policy, 1995); Sperling, S., *Animal Liberators: Research and Morality* (Berkeley: University of California Press, 1988); Turner, J., *Reckoning with the Beast: Animals, Pain, and Humanity in the Victorian Mind* (Baltimore: Johns Hopkins University Press, 1980).

DEBORAH RUDACILLE

ADVERTISING, USE OF ANIMALS IN

The use of live animals in advertising takes many different forms. Domestic animals and wild animals are often trained for use in television commercials. While the advertising industry purports to adhere to standards set by the American Humane Association in regard to the treatment of animal "actors," some would argue that the manipulation (i.e., training) of an animal for use in advertising is unethical. The use of wild animals in commercials is particularly controversial. Animal rights* advocates maintain that when an animal is shown in a setting that is completely unrelated to its natural environment, a message about that animal's nature is conveyed that is both false and damaging to an accurate public understanding of the particular animal's nature. Even when domestic animals are used in advertising in ways that portray them more accurately, such as domestic dogs* or cats* in some animal food commercials, many proponents of animal rights believe that the individual animals used are being exploited. Often, dogs or cats are dressed in human clothing, and cinematographic technology is used to make them appear to be dancing or performing other humanlike behaviors. This use of animals is considered to be demeaning and trivializing to individual animals and to animals in general.

Live animals have also been kept in cages and other enclosures for advertising purposes. Considerable attention has been given to the imprisonment of great apes such as gorillas in small cages in stores and shopping malls.

The only argument in favor of such a use of animals is one that disclaims the fact that animals have any inherent rights at all and considers humans to have the right to employ an animal for any purpose that benefits a human being. In other words, this view argues that humans have the right to complete control and dominion over animals. From an animal rights perspective, this practice is abusive and unethical because it causes harm to an animal by restricting his or her freedom, places him or her in an unnatural setting, and isolates the animal from others of his or her kind.

The effect and implications of using images of animals in advertising are more subtle. Animals used to sell products and services that are aimed at children are usually shown as silly or "cute." "Tony the Tiger" is just one example of an animal image with which we are all familiar and that has come to be closely associated with a particular food product marketed to children. Tigers, many would argue, should be valued as the wild and independent creatures that they are in nature and should not be portrayed as friendly purveyors of breakfast cereal. Although most people would view the use of animal images as harmless, many advocates of animal rights argue that these images exploit animals, contribute to the perpetuation of a view of animals that is paternalistic and trivializing, and ultimately contribute to a lack of respect for members of other species.

ANN B. WOLFE

ALTERNATIVES TO ANIMAL EXPERIMENTS

In the early 1970s, British antivivisectionists (*see* ANTIVIVISECTION-ISM) established humane research charities with twin aims: to advance medical progress and to replace animal experiments with solely nonanimal methods. This was the first coordinated effort anywhere in the world to identify and develop alternative, nonanimal research as a serious scientific enterprise. Despite initial resistance from the scientific community, progress with alternative techniques has been dramatic. Animal experiments are being replaced by alternative methods, called nonanimal techniques, that range from the inanimate, such as computer systems and chemical tests, through research at the molecular and the cellular level to clinical research and population studies at the other end of the spectrum. Computer programs can offer insights into the action of new medicines on the basis of their molecular structures, even when they exist only in the chemist's imagination. On a systems level, complex aspects of physiology and drug metabolism can also be modeled with computers. For example, there are computer programs that can predict, with 80% accuracy, whether or not a chemical may be metabolized by the liver into a cancer-causing substance.

Understanding basic processes of health and disease through use of human cells and tissues grown outside the body in laboratory cultures leads to better

diagnosis and treatments. Rabies diagnosis used to require infecting mice*
with the disease, inevitably causing suffering and death. A tissue-culture test
has now saved many tens of thousands of mice and produces results in 4
rather than 35 days.

Human cell and tissue cultures, sometimes combined with silicon-chip
technology and fluorescent dyes, are replacing animals in medical research
and vaccine production. Cancer, Parkinson's disease, diabetes, asthma, co-
litis, spinal injury, and multiple sclerosis are all being researched in the "test
tube." In the Netherlands, scientists have replaced lethal vaccine tests on
guinea pigs with cell-culture alternatives. Sometimes, microscopic organisms
such as bacteria and yeasts are simple analogues of a human system. For
example, tests with bacterial cultures have partly replaced the use of rats and
mice to determine whether chemicals cause cancer. As a result, many
thousands of animals have been spared from chemical-induced tumors. Volunteer studies provide direct information about human health and disease.
Cancer, heart disease, muscle disorders, epilepsy, arthritis, and psychiatric
illness can be researched with new scanning and imaging techniques. Lasers
and ultrasound probes can safely monitor the internal effects of some novel
treatments.

Population studies of diet, lifestyle, and occupation have revealed causes
of heart disease, stroke, cancer, osteoporosis, and birth defects. Diabetes,
arthritis, and multiple sclerosis are among other major health problems for
which population research is providing breakthroughs.

Today, nonanimal methods of research, testing, and teaching are widely
accepted and increasingly implemented. Medical students can learn physiology and pharmacology from interactive computer models and self-
experimentation, instead of using dogs* and rabbits; cell-culture tests are
replacing experiments on mice and guinea pigs; studies of the brain are pursued safely in volunteers instead of through invasive research on monkeys.
Nonanimal techniques allow us to save lives tomorrow without taking lives
today.

Selected Bibliography. Langley, G. R. (Ed.), *Animal Experimentation: The Consensus Changes* (Basingstoke, England: Macmillan, 1989); Orlans, F. B., *In the Name of
Science: Issues in Responsible Animal Experimentation* (Oxford: Oxford University Press,
1993); Sharpe, R., *Let's Liberate Science: Humane Research for All Our Futures* (Jenkintown, PA: American Anti-Vivisection Society, 1992).

GILL LANGLEY

Reduction, Refinement, and Replacement (the Three Rs)

The concept of alternatives or the Three Rs, reduction, refinement, and
replacement of laboratory animal use,* first appeared in a book by two British scientists, William M. S. Russell and Rex Burch, published in 1959 entitled *The Principles of Humane Experimental Technique*. The book was the

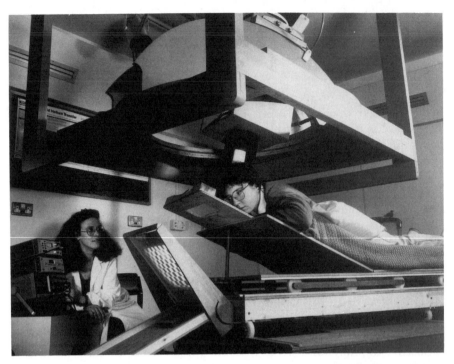

Alternatives to Animal Experiments. The MEG brain scanner is entirely noninvasive and is here being used to study photosensitive epilepsy in a volunteer. This is an alternative to distressing experiments on baboons. *Source*: Dr. Hadwen Trust for Humane Research, Clinical Neurophysiology Unit, Aston University, Birmingham, England.

report of their scientific study of humane techniques in laboratory animal experiments, commissioned by the Universities Federation for Animal Welfare (UFAW). Russell and Burch maintained that scientific excellence and the humane use of laboratory animals were inextricably linked and proceeded to define in detail how both of these goals could be achieved through reduction, refinement, and replacement of animal use. In 1978, physiologist David Smyth used the term "alternatives" to refer to the Three Rs. Since then, the Three Rs have become interchangeable with the word "alternatives." In some circles, however, the word "alternatives" is understood to signify only replacement. Hence, in order to avoid possible misinterpretations, one of the Three Rs should precede the term "alternatives" when discussing specific methods (reduction alternative, refinement alternative, or replacement alternative).

A reduction alternative is a method that uses fewer animals to obtain the same amount of data or that allows more information to be obtained from a given number of animals. The goal of reduction alternatives is to decrease the total number of animals that must be used. In research, scientists can decrease the number of animals they use by more efficient planning of experiments and by more precise use of statistics to analyze their results. Researchers can also reduce the number of experimental animals by using ever-evolving cellular and molecular biological methods. These systems are sometimes more suitable for testing hypotheses and for gaining substantial information prior to conducting an animal experiment.

Refinement alternatives are methods that minimize animal pain* and distress* or that enhance animal well-being.* An important consideration in developing refinement alternatives is being able to assess the level of pain an animal is experiencing. In the absence of good objective measures of pain, it is appropriate to assume that if a procedure is painful to humans, it will also be painful to animals. Refinement alternatives include the use of analgesics and/or anesthetics to alleviate any potential pain. They also include the use of proper handling techniques and environment enrichment.* Such enrichment ranges from placing species-appropriate objects for play and exploration in animal cages to group housing of social species.

Replacement alternatives are methods that do not use live animals, such as in vitro systems. The term "in vitro" literally means "in glass" and refers to studies carried out on living material or components of living material cultured in petri dishes or in test tubes under defined conditions. These may be contrasted to "in vivo" studies, or those carried out "in the living animal." Certain tests that were done in live animals, such as pregnancy tests, have been completely replaced by in vitro tests. Other examples of replacement alternatives are mathematical and computer models; use of organisms with limited sentience such as invertebrates, plants, and microorganisms; and human studies, including the use of human volunteers, postmarketing surveillance, and epidemiology.

The Three Rs of reduction alternatives, refinement alternatives, and replacement alternatives are considered by many to be the middle ground where scientists and animal welfare* advocates can meet to reconcile the interests of human health and animal well-being. Those interested in promoting the Three Rs have begun a series of World Congresses on Alternatives and Animals in the Life Sciences, the first of which took place in Baltimore, Maryland, in 1993 and the second in Utrecht, the Netherlands, in 1995. These meetings provide a forum for scientists to participate in dialogues with the animal-protection community to focus not on the differences between the two groups, but on opportunities for collaborative efforts and shared concerns.

Selected Bibliography. Animal Welfare Information Center and Universities Federation for Animal Welfare, *Environmental Enrichment Information Resources for Laboratory Animals, 1965–1995: Birds, Cats, Dogs, Farm Animals, Ferrets, Rabbits, and Rodents* (Washington, DC: U.S. Department of Agriculture, 1995); Balls, M., A. M.

Goldberg, J. H. Fentem, C. L. Broadhead, R. L. Burch, M. F. W. Festing, J. M. Frazier, C. F. M. Hendriksen, M. Jennings, M. D. O. van der Kamp, D. B. Morton, A. N. Rowan, C. Russell, W. M. S. Russell, H. Spielmann, M. L. Stephens, W. S. Stokes, D. W. Straughan, J. D. Yager, J. Zurlo, and B. F. M. van Zutphen, The Three Rs: The Way Forward, *Alternatives to Laboratory Animals* 23 (1995): 838–866; Russell, W. M. S., and R. L. Burch, *The Principles of Humane Experimental Technique* London: Methuen, 1959; reprint, Potters Bar, Herts, UK: Universities Federation for Animal Welfare, 1992); Smyth, D., *Alternatives to Animal Experiments* (London: Scolar Press, 1978); Zurlo, J., D. Rudacille, and A. M. Goldberg, *Animals and Alternatives in Testing: History, Science, and Ethics* (New York: Mary Ann Liebert, 1994).

JOANNE ZURLO AND ALAN M. GOLDBERG

Refinement Alternatives

Refinement is one of the Three Rs that are the cornerstone in providing alternatives: refine, reduce, replace. Both replacement and reduction focus on the alternatives of lowering the numbers of animals used. In contrast, refinement considers the quality of life (*see* WELL-BEING OF ANIMALS) for animals in laboratory or teaching situations. It addresses the current situation by asking how the lives of laboratory animals can be improved. Refinement requires improving handling procedures and husbandry of the animals.

In the past, efforts at refinement were focused primarily on reducing animal pain* and suffering.* Recent legislation reflects a broader view of refinement as the general well-being of the animals. The Laboratory Animal Welfare Act,* as amended in 1985 in Public Law 99–198, and the revised 1991 regulations emphasize training of the animal care staff in providing comfort, good husbandry and housing, and gentle handling. They require environmental enrichment* for primates. Engineering standards specify certain cage size and structure requirements for animal well-being. Performance standards focus on the functional and mental state of the animals, as indicated by their behavioral repertoires and stress* indicators.

Environmental enrichment is an aspect of refinement that has been pursued in particular for primates in laboratories and a variety of wild mammals in zoos.* Animal laboratories may house several hundred individuals and often require sterile environments. Even within these constraints, economic refinements such as caging illumination, sound quality, nesting material, and social environment can often be made, once the species' preferences are well understood.

Assessing the quality of life for the animal requires some understanding of the animal's point of view of its world. The discipline of animal behavior provides tools to evaluate an animal's well-being. With the awareness of the importance of the human caregiver and the potential stress of various restraining techniques that limit movement, animals can be trained, by positive reinforcement, to cooperate with medical examination procedures so that restraint is not required.

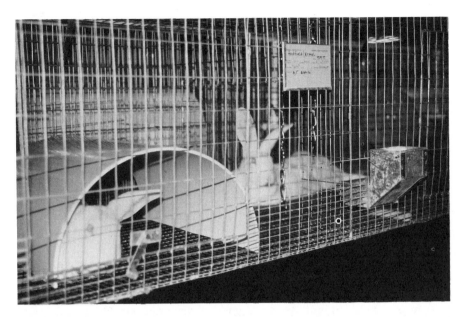

Alternatives to Animal Experiments (Refinement Alternatives): Animal caregivers in this laboratory conducted systematic observations of rabbits and concluded they preferred social housing, as seen here. Photo by Lynette A. Hart.

Selected Bibliography. Dawkins, M. S., *Animal Suffering: The Science of Animal Welfare* (London: Chapman and Hall, 1980); Hart, L., Improving Implementation of "The Third R," Refinement, *Humane Innovations and Alternatives* 6: (1992): 385–387; Hart, L., Opportunities for Environmental Enrichment in the Laboratory, *Lab Animal* 23 (2) (1994): 24–27; Russell, W. M. S., and R. L. Burch, *The Principles of Humane Experimental Technique* (London: Methuen, 1959); U.S. Government, Title 9 Code of Federal Regulations (9 CFR), Part 3, Animal Welfare, Standards, Final Rule, *Federal Register* 56(32) (February 15, 1991): 6426–6505.

LYNETTE A. HART

AMERICAN HUMANE EDUCATION SOCIETY. *See* HUMANE EDUCATION MOVEMENT.

AMERICAN INDIANS. *See* NATIVE PEOPLES AND ANIMALS.

AMERICAN SOCIETY FOR THE PREVENTION OF CRUELTY TO ANIMALS (ASPCA)

The American Society for the Prevention of Cruelty to Animals (ASPCA), the United States' first humane society, was founded by Henry Bergh* on

April 10, 1866. Shortly after its founding, it served as the inspiration and model for the formation of SPCAs and humane societies across the country. Bergh organized a meeting of influential business and political leaders at Clinton Hall on February 8, 1866. He gave a speech enumerating the many terrible deeds done to animals, the important role that animals played, and the need for a society to protect them. Just nine days after the charter was granted by the New York State legislature, Bergh convinced the legislature to pass an anticruelty law that gave the new society the authority to enforce it.

From the very start the ASPCA was active in publicizing the plight of animals and intervening on their behalf. One of the first cases that Bergh and the new ASPCA brought before the court was that of a cart driver beating his fallen horse with a spoke from one of the cart's wheels. This event was eventually depicted in the seal adopted by the ASPCA, showing an avenging angel rising up to protect a fallen horse.

Within its first year Bergh and the ASPCA addressed many of the same questions that would occupy the efforts of his successors at the ASPCA and other humane societies, including the treatment of farm animals, dogfighting, horses used to pull trolleys, and turtles transported for food and vivisection. Recognizing the difficulty of coordinating the efforts of a far-ranging national organization, Bergh encouraged and helped others to start independent SPCAs across the country. The ASPCA became the model for hundreds of other societies, many of which used a variation of the SPCA name, the charter, and even the seal.

The issues in the society's early years were frequently played out in the pages of the newspapers. Stories about the ASPCA's arrests, court cases, and rescues of animals were given great attention. In addition, Bergh wrote many letters to the papers to explain the actions of the ASPCA and to point out problems that needed to be addressed. The newspapers were soon in the middle of a long feud between two of America's most famous men, Henry Bergh and P. T. Barnum. Bergh attacked Barnum over the care provided for the animals in his menagerie or performing in his shows. Barnum defended his practices and used the publicity from the dispute to attract even larger crowds. Over time, Barnum became a grudging admirer of Bergh and the work of the ASPCA and eventually helped to form an SPCA in Connecticut.

The ASPCA helped to change the way that Americans thought about animals. The organization also helped to introduce a number of innovations that provided for their care and protection. Bergh helped to design and introduce an ambulance for horses and promoted an early version of the "clay pigeon" as a target for shooters instead of live pigeons. This innovation continued in the 1950s when the ASPCA helped with the design and implementation of equipment for the humane slaughter of animals for food (*see* TRANSPORTATION AND SLAUGHTER).

Seal of the American Society for the Prevention of Cruelty to Animals. *Source*: "ASPCA History," the American Society for the Prevention of Cruelty to Animals.

The ASPCA is one of the world's largest humane societies. It maintains animal hospitals and shelters* in New York City, and its humane law-enforcement agents enforce the anticruelty laws in New York State. The ASPCA also promotes education and legislative activities that fulfill the original mission described for the organization by its founder Henry Bergh, "to provide effective means for the prevention of cruelty to animals throughout the United States."

Selected Bibliography. Franz, William C., Bergh's War: The First Crusade for Animal Rights, *Elks Magazine* (October 1980); Loeper, John J., *Crusade for Kindness: Henry Bergh and the ASPCA* (New York: Atheneum, 1991); Pace, Mildred Mastin, *Friend of Animals* (Ashland, KY: Jesse Stuart Foundation, 1995; original publication, New York: Charles Scribner's Sons, 1942); Steele, Zulma, *Angel in a Top Hat* (New

York: Harper and Brothers, 1942); Turner, James, *Reckoning with the Beast* (Baltimore: Johns Hopkins University Press, 1980).

STEPHEN L. ZAWISTOWSKI

AMPHIBIANS

Many biologists today are concerned by evidence that populations of amphibians around the world are declining and that the welfare of amphibians is seriously affected in their natural habitats by human-caused environmental deterioration. Because the skin of amphibians is not readily resistant to water loss, most species are restricted to streams and ponds or to moist terrestrial and arboreal habitats. The moist skin of amphibians may also make them more vulnerable to injurious ultraviolet rays and chemical pollution than other groups of vertebrates with better protection for the skin. There is general concern that major, global changes in the environment may be specifically injuring amphibian populations throughout the world. For example, ultraviolet (UV) radiation is harmful to humans, and the middle part of the spectrum (UV-B) is particularly dangerous. Recent evidence has shown that the eggs of some species of frogs and toads are very sensitive to UV-B, with high mortality within egg clutches exposed to this radiation. This raises fears that a current reduction in the ozone layer around the earth may subject amphibians to increased levels of UV-B.

There are three groups of amphibians: caecilians, salamanders, and frogs. Caecilians are earthwormlike amphibians that occur in aquatic and terrestrial habitats in Asia, Africa, and America. Little is known about their biology. Therefore, populations may or may not be declining.

About 400 species of salamanders occur in Asia, Europe, North America, and northern South America. Some species are entirely aquatic, living in streams, rivers, or ponds. Other species are semiaquatic or have aquatic larvae with terrestrial adults, while yet others are strictly terrestrial, inhabiting burrows in the soil, or strictly arboreal. The arboreal species, though less well studied, are probably suffering from deforestation in Central and northern South America. Adult males and females of terrestrial species are territorial, defending feeding areas under rocks and logs, and they are aggressive toward some other species of *Plethodon* that appear to be declining. Terrestrial salamanders may not be greatly affected by UV-B or by airborne pollution, due to the buffering influence of the soil.

Streamside salamanders live in habitats that are flushed by flowing water, and thus they too may be relatively protected from airborne pollution, such as acid rain, but not necessarily from UV-B. The salamanders that may be most affected by pollution and UV-B are those that either live in ponds as adults or breed in ponds, having aquatic larvae. If worldwide changes in the environment are occurring, the welfare of pond species might be most at stake.

About 4,000 species of frogs occur throughout North and South America, Europe, Asia, and Australia. They inhabit arboreal, terrestrial, semiaquatic, and aquatic habitats. As with the salamanders, considerable attention has been focused on pond-breeding species in regard to injurious effects of pollution (such as acid rain) and UV-B radiation.

Because of the decline of numerous species of amphibians in nature, scientists who study amphibians in the laboratory have had to reevaluate the ethics of using large numbers of individuals in research or in teaching. For example, a biologist who wishes to conduct an experiment can often estimate just how many frogs or salamanders are needed to obtain significant results; that biologist can then collect or purchase just the minimum number of animals needed to perform the experiment effectively. In the laboratory, animals can often be housed in individual containers, thus reducing mortality due to the spread of infections and contaminants. Another tactic used by laboratory biologists is to cycle the same frogs or salamanders through a series of experiments, rather than obtaining a different set of animals for each individual experiment. This is not always possible when, for instance, surgery is required, but cycling animals among behavioral or ecological experiments is often feasible.

Concern about amphibians takes two basic forms: concern about their welfare in nature and the rights of these animals in the laboratory, given the decline of once-abundant species. More and more species are becoming listed as threatened or endangered, which should help to reduce local human-induced impacts on their populations. Such restrictions will also limit the number and kinds of species that can be used in biological research.

Selected Bibliography. Blaustein, A. R., Amphibians in a Bad Light, *Natural History Magazine* 103(10) (1994): 32–39; Buchanan, B. W., and R. G. Jaeger, Amphibians, in B. E. Rollin and M. L. Kesel (Eds.), *The Experimental Animal in Biomedical Research*, vol. 2 (Boca Raton, FL: CRC Press, 1995), 31–48; Duellman, W. E., and L. Trueb, *Biology of Amphibians* (Baltimore: Johns Hopkins University Press, 1994); Mathis, A., R. G. Jaeger, W. H. Keen, P. K. Ducey, S. C. Walls, and B. W. Buchanan, Aggression and Territoriality by Salamanders and a Comparison with the Territorial Behaviour of Frogs, in H. Heatwole and B. Sullivan (Eds.), *Amphibian Biology*, vol. 2, *Social Behaviour* (Chipping Norton, Australia: Surrey Beatty and Sons, 1995), 633–676; Stebbins, R. C., and N. W. Cohen, *A Natural History of Amphibians* (Princeton: Princeton University Press, 1995); Zug, G. R., *Herpetology: An Introductory Biology of Amphibians and Reptiles* (New York: Academic Press, 1993).

ROBERT G. JAEGER

ANGELL, GEORGE T.

George T. Angell (1823–1909) was president of the Massachusetts Society for the Prevention of Cruelty to Animals. He was the only child of a Baptist pastor who died when the boy was four, and his mother turned to teach-

George T. Angell (1823–1909). Photo courtesy of the Massachusetts Society for the Prevention of Cruelty to Animals.

ing to support herself and her young child. Angell attended Brown University for one year and Dartmouth University for three years. After college, he spent three years teaching while he studied law. He eventually joined the bar in December 1851.

In 1868 George Angell was swept up into a public role protecting animals. On February 22 of that year two horses were ridden to their deaths during a cross-country race in Massachusetts. The *Boston Daily Advertiser* carried a letter from Angell on February 25 decrying the mistreatment of animals and calling for an organized effort for their protection in Massachusetts. Correspondences between Angell, Henry Bergh,* the founder of the American Society for the Prevention of Cruelty to Animals* (ASPCA), and Mrs. William Appleton soon led to the formation of the Massachusetts Society for the Prevention of Cruelty to Animals (MSPCA) with George Angell as president. On May 14 of that year he succeeded in having a law passed that prohibited the cruel treatment of animals in Massachusetts. By June 2 Angell had printed 200,000 copies of the first edition of "Our Dumb Animals." This pamphlet on the proper care of animals is still published as the MSPCA magazine *Animals*.

Throughout his tenure as president of the MSPCA Angell's experience as a teacher was seen in his efforts to promote the importance of "humane education"* in the prevention of cruelty to animals. His belief in humane education was so great that when he visited the Royal Society for the Prevention of Cruelty to Animals (RSPCA)* in England, he felt that the MSPCA had achieved more in one year than the RSPCA had in fifty "because we believed through and through in the power of humane education."

Angell traveled and lectured frequently on the importance of teaching children about kindness to animals. On July 28, 1882, he helped to organize the first American "Band of Mercy." These children's clubs met in schools, helped children learn about animals, and encouraged activities to protect animals. In 1889 Angell organized the American Humane Education Society (AHES) with a special charter granted by the Massachusetts legislature. AHES endured as part of Angell's mission to promote humane education and sponsored the American publication of the classic book *Black Beauty* by Anna Sewell.*

Selected Bibliography. Angell, George T., *Autobiographical Sketches and Personal Recollections* (Boston: Franklin Press; Rand, Avery and Co., 1884); Bank, Julie, and Stephen Zawistowski, The Evolution of Humane Education, *ASPCA Animal Watch*, Fall 1994; McCrea, Roswell C., *The Humane Movement: A Descriptive Survey* (New York: Columbia University Press, 1910); Shultz, William J., *The Humane Movement in the United States, 1910–1922* (New York: Columbia University, 1924); Steele, Zulma, *Angel in a Top Hat* (New York: Harper and Brothers, 1942).

STEPHEN L. ZAWISTOWSKI

ANIMAL BOREDOM

The term "boredom" is used to describe the experience of animals who spend their lives in highly monotonous environments. The animals sleep for prolonged periods of time and can sit in a tense and drooping posture for hours on end. They may also repeat the same pattern of movement over and over (*see* STEREOTYPIES IN ANIMALS), sometimes in ways that damage their own body, or they may damage the bodies of their mates by chewing their tails, ears, or genitals. In discussing such behavior, studies of animal welfare* frequently speak of boredom, suggesting that for a lack of natural "things to do," animals cannot help but fill the time with abnormal patterns of behavior.

Animal Boredom: Pig exhibiting tense and drooping posture of boredom. Photo by Françoise Wemelsfelder.

In the wild, animals face unpredictable and challenging environments. Predators, food shortage, weather, floods, and illness all threaten health and survival and can put the animal under duress. In contrast, animals in captivity tend to live in highly predictable and structured environments where they are challenged infrequently or not at all. To deprive animals of any kind of meaningful activity may leave them bored and continually distressed.*

One problem in studying boredom is its passive nature. Acute emotions such as anger or fear* mostly have clear expressions and are not easily misunderstood. Possibly, animals who impassively sit and stare into space are content rather than bored. The term *boredom* seems to suggest that animals mentally evaluate their passive situation and actively miss a more meaningful life. However, it is very difficult to investigate whether animals can miss what they have never known. Formal models of abnormal behavior (in contrast to informal discussions) therefore prefer to assume that animals experience frustration or distress rather than boredom.

Although the question of mental awareness is important, it is not the only possible approach. Boredom, although admittedly not as easily studied as anger and fear, may be detected from an animal's expression. The question is which signs epitomize an expression of boredom. In human beings, mild boredom results in temporary drowsiness, slight irritation, and the desire to leave a particular situation. But in severe form, boredom borders on depression and is experienced by individuals as a chronic meaninglessness in all that they do. They withdraw from contact with others, are unmotivated, and give a generally despondent and listless impression. The expression of animals in close confinement has similar traits. The hunched, drooping posture of these animals, the way they drowsily half-close their eyes while engaging in repetitive behavior, and their abrupt, aggressive manner of shying back from contact all seem to express withdrawal and dejection, rather than contentment. It seems justified to speak of boredom in this context, even though we do not as yet understand the animal's level of awareness.

Despite the lack of a clear-cut definition, the notion of boredom has inspired many animal keepers and caretakers to provide their animals with more interesting and challenging, enriched environments (*see* ENRICHMENT FOR ANIMALS). When animals are given the chance to organize their own life by seeking food, building nests, finding shelter, and communicating with other animals, their liveliness returns. They appear inquisitive and alert and most likely will not develop abnormally repetitive behaviors. Enclosures and materials that facilitate the animal's natural, species-specific behavior provide the most varied and lasting type of enrichment.

Selected Bibliography. Newberry, R. C., Environmental Enrichment: Increasing the Biological Relevance of Captive Environments, *Applied Animal Behaviour Science* 44(2/4) (1995): 229–243; Van Rooijen, J., Predictability and Boredom, *Applied Animal Behaviour Science* 31(3/4) (1991): 283–287; Wemelsfelder, F., Boredom and Laboratory Animal Welfare, in B. E. Rollin and M. L. Kesel (Eds.), *The Experimental Animal in Biomedical Research*, vol. 1 (Boca Raton, FL: CRC Press, 1990), 243–272; Wemelsfelder, F., The Concept of Animal Boredom and Its Relationship to Stereotyped Behaviour, in A. B. Lawrence and J. Rushen (Eds.), *Stereotypic Animal Behaviour: Fundamentals and Applications to Welfare* (Wallingford: CAB International, 1993), 65–95; Wood-Gush, D. G. M., and K. Vestergaard, Exploratory Behavior and the Welfare of Intensively Kept Animals, *Journal of Agricultural Ethics* 2 (1989): 161–169.

FRANÇOISE WEMELSFELDER

ANIMAL COGNITION

The word "cognition" is derived from the Latin root *cognitio*, which means the ability to learn or know intensively. In modern psychological usage the concept refers to mental faculties whose activities include functions such as conscious awareness, thinking, perception, reasoning, problem solving, complex learning, judgment, and intentional action. The concept of cognition is difficult to pin down because it rests on the fact that processes such as thinking, reasoning, and intention are private events and are not directly observable. Therefore, the existence and action of these processes must be understood from overt behavior. For example, in a classic experiment reported by Wolfgang Köhler in 1925, a small number of chimpanzees* were faced with a behavioral problem where a highly desirable piece of fruit was suspended in view but out of reach from the ceiling of a large room. Around the room were a number of wood crates all capable of accommodating the weight of the animal. The chimpanzees were first observed trying to reach the fruit by jumping and scaling the walls. After a number of failed attempts, some of the chimps oriented toward the suspended fruit and the crates. After a while the animals began to stack the boxes in a position beneath the suspended fruit, creating a series of steps that were then climbed, and access to the fruit was gained.

For Köhler, the sequence provided strong evidence that the chimps had thought about the situation, had gained "insight" about a possible solution strategy, and had then acted out the solution. At the time, this type of cognitive explanation stood in sharp contrast with other "behavioristic" (see BEHAVIORISM) theories of animal learning. These theories argued that it was scientifically improper to talk about hypothetical mental processes. In biology a similar trend existed in which animal behavior was seen as programs that were produced by specific environmental circumstances without cognitive involvement. These mechanistic explanations dominated much of the first half of the 20th century and produced a picture of animals as empty, machinelike entities.

In the 1960s, during what has been called the "cognitive revolution," it began to be appreciated that eliminating any reference to cognitive processes in animals distorted the nature of animal behavior and confused the difficulty in studying these processes with their existence. Since that time, steps have been taken that have begun to reveal important facts about the mental life of animals and have established the domains of cognitive psychology and cognitive ethology as legitimate parts of mainstream science.

Understanding the place of cognition in animals relates to their moral standing* and the ways in which humans might best conduct themselves in relationship to them. In one commonly held point of view, it is believed that as long as an entity is not aware of itself as an individual or able to feel and reflect on its experiences such as pain* and suffering,* what is done to it does not matter ethically. Therefore, an understanding of the cognitive abilities of animals helps to inform the arguments used to justify either including or excluding animals from the protection offered by moral standing.

Selected Bibliography. Bekoff, M., and D. Jamieson (Eds.), *Interpretation and Explanation in the Study of Animal Behavior*, vol. 1, *Interpretation, Intentionality, and Communication*; vol. 2, *Explanation, Evolution, and Adaptation* (Boulder, CO: Westview Press, 1990); Bekoff, M., and D. Jamieson (Eds.), *Readings in Animal Cognition* (Cambridge, MA: MIT Press, 1996); Carruthers, P., *The Animals Issues: Moral Theory in Practice* (New York: Cambridge University Press, 1992); Griffin, D., *Animal Minds* (Chicago: University of Chicago Press, 1992); Griffin D., *Animal Thinking* (Cambridge, MA: Harvard University Press, 1984); Midgley, M., *Animals and Why They Matter* (Athens: University of Georgia Press, 1983).

JOHN P. GLUCK

Intelligence

The word "intelligence" is very difficult to define. Some people use it to mean "smart"; others use it to mean "adaptable." Nonetheless, most people take a commonsense approach to the term. They ask questions such as these: How intelligent are animals? Are they capable of thinking in the same way as humans? For many people, answers to these difficult questions are directly related to issues about animal welfare.* In some basic way, the more intelligent we believe an animal to be, the more likely we are to be concerned about its welfare. The same person who casually swats a mosquito would not dream of killing a dog,* even if the dog were snapping at his heels. There are many reasons for such widespread speciesism,* not the least of which is how we perceive the relative intelligence of the dog and the mosquito.

Some scientists, like Donald Griffin, are not troubled by the lack of an exact definition for words such as intelligence, consciousness, or thought. They believe that we have enough of an intuitive sense of thinking, for example, to look for evidence of it in animals. Griffin's strategy is to look for examples of seemingly "clever" behavior, whether in dogs, cats,* spiders, or wasps, and use these as evidence of animal thinking. Griffin has taken quite literally the suggestion by George Romanes, a 19th-century student of animal intelligence, who viewed observable behavior as the "ambassador of the mind." In its most extreme form, this view holds that all behavior in whatever species reflects a conscious thought process. Not all scientists agree with Griffin (*see* BEHAVIORISM).

In large measure, the scientific study of animal intelligence has been replaced by studies of animal cognition. Animal cognition is concerned with questions about how animals use time, number, space, logic, and memory. Studies of animal cognition typically create experimental situations in which an individual animal must learn to use time, number, space, or logic to solve problems or earn food.

For the past 15 years, researchers have studied how animals "count." Although we cannot say for sure that animal subjects were "counting" in the human sense of the word, we do know that they were very sensitive to the numerical properties of the situations devised. For example, rats and ferrets were trained to eat only *N* pieces of food (either 3, 4, or 5) from a larger array and leave the remainder uneaten. Anyone who has worked with hungry rats or ferrets recognizes how difficult it might be to get an animal to turn its back on a remaining piece of food simply because it exceeds that animal's "target number."

In another experiment, rats were placed in a large enclosure containing 6 movable tunnels. Each rat was trained to take food from a particular tunnel (the 3rd, 4th, or 5th) based on its ordinal position in the array of 6. All rats learned to enter the correct tunnel in the array. Some subjects were retested 12 and 18 months later, and accurate retention was found. Also those rats who were required to enter the 5th tunnel in a row of 6 eventually learned to go to the end of the array and walk back one. It is obviously much easier to "count" backwards from 6 to 5 than it is to count up from 1 to 5. Thus animals may not approach cognitive tasks using strategies observed in human subjects, but they are frequently successful on their own terms.

Most animals have shown considerable evidence of cognitive abilities involving time, space, number, and logic. These results, while impressive in their own right, do not tell us that animals "think" or solve problems like humans. It is important to stress, however, that evidence of the role of thought in higher-order human behavior is also lacking.

Human primates place a high value on what they do well and look for rudimentary evidence of it in other species. We use our own competence to define intelligence. If a rat or dog can do what we do, then we assume that he or she is intelligent. Arguably, this kind of arrogance has no place in our assessments. There may be other forms of intelligence that have little to do with human competence.

Selected Bibliography. Davis, H., Transitive Inference in Rats (*Rattus norvegicus*), *Journal of Comparative Psychology* 106 (1992): 342–349; Davis, H., and R. Pérusse, Numerical Competence in Animals: Definitional Issues, Current Evidence, and a New Research Agenda, *Behavioral and Brain Sciences* 11 (1988): 561–616; Griffin, D. R., *Animal Thinking* (Cambridge, MA: Harvard University Press, 1984); Roitblat, H., T. Bever, and H. Terrace (Eds.), *Animal Cognition* (Hillsdale, NJ: Erlbaum, 1984).
HANK DAVIS

Self-Awareness and Self-Recognition

Under the influence of the ideas of René Descartes,* self-awareness is commonly viewed as a human characteristic not present in animals because they do not have language. Language is reasonably viewed as a means by which people present, maintain, and reflect upon ideas, including ideas about themselves and their relations with others, and thus it is not surprising that animals have been denied self-awareness (and sometimes, any sort of awareness at all). Consequently, ethical theorists such as Immanuel Kant* viewed animals as without self-consciousness and thereby declared them to be inherently unworthy of moral concern. With evolutionary theory came the idea that humans and animals share a common heritage, and the conception of self-consciousness changed from being an all-or-nothing phenomenon to having various meanings, some of which could be had without language.

Turn-of-the-century scientists fascinated by the question of animal psychology produced various theories and measures to come to grips with self-awareness in animals, but were largely unsuccessful. Because of the difficulty of evaluating (or even discerning what would count as) evidence of consciousness, self-awareness, language, or any psychological aspects of animals, scientists in general either assumed that animals had conscious experiences that were unknowable, or subscribed to the belief that psychology should be the study of behavior, such that consciousness and other "psychical" aspects were irrelevant or nonexistent (*see* BEHAVIORISM). However, the questions persisted for some researchers, who attempted to look for humanlike attributes in animals, particularly in chimpanzees* and other great apes who are evolutionarily most closely related to humans.

Language and self-awareness, viewed as the most human of abilities, were sought in great apes (*see* ANIMAL COMMUNICATION; MORAL STANDING OF ANIMALS). Because several researchers had failed to teach apes spoken language, and naturalists found no evidence of linguistic abilities in their natural environment, it was surprising when Beatrix and Allen Gardner provided evidence in the late 1960s that a chimpanzee, Washoe, was able to use and understand aspects of American Sign Language consistent with use by young children. Soon after the discovery of Washoe's abilities, Gordon Gallup found that chimpanzees recognized their image in a mirror as their own; not only did they make faces in the mirror and use it to look at areas of their body that are visually inaccessible without a mirror, but they also wiped away marks they could not feel (but detected in the mirror) that had been placed on their face while they were under anesthesia. Similar experiments have extended self-recognition to at least some members of the other great-ape species as well (and perhaps to some bottlenosed dolphins), but to no other nonhuman species.

The question of how far great apes' self-recognition and language skills can go in the direction of comparable adult human skills is unresolved, but so far their abilities in relation to self-understanding appear to be somewhat limited. Great apes use personal pronouns such as "me" and "you" and use language to describe their current circumstances and their wants and to plan activities, but they do not appear to use language to reflect on their past or present circumstances or to ponder ethical dilemmas. Great apes' (and young children's) abilities to recognize themselves in mirrors and to learn new signs by imitation appear to depend upon their ability to match between their kinesthetic sensations (which tell them of the position and feel of their own body) and their visual experience of themselves (in a mirror) or of another (in imitation), rather than from an extensive psychological understanding of self and other. Great apes can recognize their body as their own and know that they look like another or a mirror image, and they may even be able to produce images of themselves in visual mental representations and use these to plan their (kinesthetically perceived) actions. But it has yet to be shown that great apes can take responsibility for their own actions or be swayed by ethical concerns.

Selected Bibliography. Darwin, C., *The Descent of Man and Selection in Relation to Sex*. (New York: D. Appleton and Company, 1871); Descartes, R., Letter to the Marquis of Newcastle, and Letter to Henry Moore (1649), in R. M. Eaton (Ed.), *Descartes: Selections* (New York: Charles Scribner's Sons, 1927), 355–360; Gardner, R. A., B. T. Gardner, and T. E. Van Cantfort (Eds.), *Teaching Sign Language to Chimpanzees* (Albany: State University of New York Press, 1989); Kant, I., *Lectures on Ethics* (New York: Harper and Row, 1963); Mitchell, R. W., Mental Models of Mirror-Self-Recognition: Two Theories, *New Ideas in Psychology* 11 (1993): 295–325; Mitchell, R. W., N. S. Thompson, and L. H. Miles, (Eds.), *Anthropomorphism, Anecdotes, and Animals* (Albany: State University of New York Press, 1997); Parker, S. T., R. W. Mitchell, and M. L. Boccia (Eds.), *Self-Awareness in Animals and Humans* (New York: Cambridge University Press, 1994); Russon, A., K. Bard, and S. T. Parker (Eds.), *Reaching into Thought* (New York: Cambridge University Press, 1996).

ROBERT W. MITCHELL

Conscious Experience

Animal consciousness is important to many approaches to the ethical treatment of animals. Some of those who study ethics view the prevention of conscious pain* as the highest moral good. Others consider mental capacities such as the ability to form conscious plans, have conscious hopes, or consciously anticipate future harms to be of equal or greater importance. Both supporters and opponents of moral consideration for animals tend to accept the statement that if animals lack consciousness, then they deserve no moral consideration. They disagree about whether animals lack the relevant forms of consciousness.

It is important to realize that in this dispute the term "conscious" is not always used in an ordinary way. Two ordinary uses of consciousness are the difference between wakefulness (consciousness) and sleep (unconsciousness), and the ability of organisms to perceive (and to be conscious or aware of) selected features of their environments. Two remaining technical senses of consciousness are, first, the subjective or personal aspects of conscious experience, and second, self-consciousness. This entry concerns the first notion.

The burden of proof in contemporary science has shifted from those who would deny animal consciousness to those who would accept it. In the 17th century René Descartes* argued that all animal behavior could be explained purely mechanistically without using mental terms such as "consciousness." Current views that agree with Descartes are found in recent arguments by Peter Carruthers that all animal sensations are nonconscious and therefore not worthy of moral consideration. Authors who deny consciousness to animals usually appeal to scientific methodology. The requirement of strict observability made popular by psychological behaviorists (*see* BEHAVIOR-

ISM) such as J. B. Watson and B. F. Skinner further supported the view that the best scientific explanation of an organism's behavior does not involve the attribution of consciousness and that there is therefore no justification for attributing consciousness to animals. However, because consciousness is assumed to be private or personal, it is often taken to be beyond the reach of objective scientific methods.

Questions about animal consciousness are sometimes seen as part of a general problem known as "the problem of other minds"—the problem of how anyone knows about the existence of consciousness besides his or her own. But it is also thought that knowledge of animal minds presents a special problem because one cannot use language to ask animals about their experiences. Descartes and many other philosophers have proved themselves unable to imagine how more sophisticated behavioral experiments could provide knowledge by means other than direct questioning. Advances in cognitive methods originally developed to study cognition in very young children are being applied successfully to nonhuman animals. For instance, by measuring the time spent looking at various displays it is possible to draw conclusions about what the organism knows about what it sees.

The topic of animal consciousness is still taboo for many psychologists, but work between philosophers and those who study behavior is beginning to lay the foundation for treating questions about consciousness in a philosophically sound yet scientific way. The main challenge for those who think that such a strategy is possible is to study and learn more about the relationship between assignments of consciousness and behavioral or neurological evidence.

Selected Bibliography. Allen, C., and M. Bekoff, *Species of Mind: The Philosophy and Biology of Cognitive Ethology* (Cambridge, MA: MIT Press, 1997); Carruthers, P., *The Animals Issues: Moral Theory in Practice* (New York: Cambridge University Press, 1992); Dawkins, M. S., *Through Our Eyes Only: The Search for Animal Consciousness* (San Francisco: W. H. Freeman and Company, 1993); Radner, D., and M. Radner, *Animal Consciousness* (Buffalo, NY: Prometheus Books, 1989); Sorabji, R., *Animal Minds and Human Morals: The Origins of the Western Debate* (Ithaca, NY: Cornell University Press, 1993).

COLIN ALLEN

Consciousness and Thinking

The philosopher David Hume had no doubt that animals were conscious, thinking beings. He wrote: "Next to the ridicule of denying an evident truth is that of taking much pains to defend it; and no truth appears to me more evident, than that beasts are endowed with thought and reason as well as men. The arguments are in this case so obvious, that they never escape the most stupid and ignorant." Although Hume is correct that ordinary common sense finds thinking by animals to be unproblematic, a large number of thinkers have called this view into question. Most famously, the denial of consciousness is associated with René Descartes,* who argued that animals were strictly material bodies, obeying the laws of mechanical physics.

However, it was plain to Charles Darwin,* and to his valued colleague George Romanes, that if anatomical and physiological traits were evolutionarily continuous* (see CONTINUITY) between nonhuman animals and humans, so too were mental ones. This was true not only of intelligence, but also of emotion and feeling, the most morally relevant aspect of thinking, since, as the philosopher Jeremy Bentham claimed, the ability to experience pain,* fear,* anxiety, hunger, thirst, pleasure, and so on is surely what makes a being worthy of moral concern, since what we do to it matters to it. Darwin made his position on animal feeling clear in his book *The Expression of the Emotions in Man and Animals*, and Romanes gathered and critically evaluated stories (anecdotes) about animal thought in his books *Animal Intelligence* and *Mental Evolution in Animals*.

Even though biological science was solidly Darwinian by the end of the 19th century, questions about animal awareness did not vanish and indeed emerged all the more strongly in the early 20th century despite the strength of evolutionary theory in virtue of the rise of positivism. Since mind in animals was not observable, it was argued that it could not be studied scientifically and should not be studied at all.

In a related occurrence, psychology as a science was "losing its mind" with the rise of behaviorism.* In the face of behaviorism, animal consciousness went from scientifically unstudiable to scientifically unreal. The denial of consciousness to animals was given further support by the advent of large amounts of invasive research on animals, which was, as in Descartes's time, much easier to perform if animals were viewed as nonconscious machines who "vocalized" rather than hurt.

The strongest reason for the return of talk about animal mind has been moral (see MORAL STANDING OF ANIMALS). Since the 1960s, society has grown increasingly concerned about animal treatment in the areas of scientific research, agriculture, and toxicity testing, and with that concern has come a social emphasis on issues of animal pain, suffering,* fear, loneliness, boredom (see ANIMAL BOREDOM), and anxiety, which has in turn forced science to reckon with these notions. For example, federal law passed in 1985 compels researchers to control "animal pain and distress." Researchers have thus been led to bring ordinary common sense about animal thought and feeling into science. New approaches in fields like cognitive ethology and studies in primate language (see ANIMAL COMMUNICATION) and animal deception are also leading science back to the Darwinian approach to animal mind and to the use of ordinary common sense.

Selected Bibliography. Bekoff, M., and D. Jamieson (Eds.), *Readings in Animal Cognition* (Cambridge, MA: MIT Press, 1996); Carruthers, Peter, *The Animals Issues: Moral Theory in Practice* (Cambridge: Cambridge University Press, 1992); Chomsky,

Noam, *Cartesian Linguistics* (New York: Harper and Row, 1996); Griffin, Donald, *The Question of Animal Awareness* (New York: Rockefeller University Press, 1976); Rollin, Bernard E., *The Unheeded Cry: Animal Consciousness, Animal Pain, and Science* (Oxford: Oxford University Press, 1989); Rosenfield, Leonora C., *From Beast-Machine to Man-Machine* (New York: Octagon Books, 1968).

BERNARD E. ROLLIN

Recognition of Humans by Animals

There is a growing body of scientific evidence to suggest that animals of many species are capable of telling individual humans apart. These results are impressive because they come from a variety of "lower animals," including invertebrates. There are three reasons why we may care about human recognition among animals. First, such an ability is part of what we call "intelligence" (*see* ANIMAL COGNITION, Intelligence). To the extent that an animal can discriminate among individual humans, he or she may be smarter than we previously believed. For example, one of the reasons many people consider dogs* to be intelligent is the fact that they are capable of telling humans apart and can form deep bonds with their human families. If we can show a similar capability among other animals, that might affect our estimate of other animals' intelligence.

Second, such estimates of intelligence often have a direct bearing on how we treat different species. In short, animal welfare and our estimates of animal intelligence are strongly related (*see* MORAL STANDING OF ANIMALS).

The third reason for wanting to know whether animals can discriminate among humans has to do with research. When animals are used as subjects, many things are done to them. Some are positive. Others may not be. Animals, like humans, try to anticipate such events and to prepare for them in a variety of physical and psychological ways. One of the best predictors for the occurrence of stimuli involving pleasure or pain is the appearance of a particular person. "If Joe comes into my room, I'll suffer pain. On the other hand, if Bill comes in to get me, it's likely that I'll be cuddled or fed."

When such events in an animal's life are reliably associated with a particular person, the stage is set for prediction. If an animal can discriminate Joe from Bill and associate each person with a particular outcome, then a simple form of learning called Pavlovian (named after the Russian physiologist Ivan Pavlov) conditioning will occur. In this case, the conditioning is a little unusual because the predictor or *conditioned stimulus* (CS) is a particular person instead of a bell or a metronome. Psychologist W. Horsley Gantt examined such effects in dogs and used the phrase "Person as CS" to describe them.

If dogs can discriminate between humans, the possibility exists that other species can as well, and such conditioning might have strong and unexpected effects on research. Davis and Balfour (1992) examined research involving a variety of animal species and found that recognition of the scientist or lab technician produced profound behavioral and physiological changes in animal subjects. Furthermore, these effects were frequently overlooked because the scientist failed to allow for the possibility of human recognition.

For example, if someone were studying stress,* he or she would want to know whether blood steroids (glucocorticoids) were elevated. First, the researcher would need some baseline measure of blood values in a nonstressed animal. But it is difficult to find a nonstressed animal if all subjects know what is about to happen to them because they are being handled by a person they have come to associate with pain.

There are data showing that rats, one of the most widely used laboratory animals, can discriminate individual humans. Rats were allowed to explore one of two humans for 10-minute sessions over 14 days. The animals climbed on the handler's body and were talked to and fed treats during these brief exposures. They were then tested on a long table with the familiar person at one end and a stranger at the other. After carefully sniffing the unfamiliar person, all subjects walked to the other end of the table and climbed onto the body of the familiar handler. Five months later all subjects remembered who had handled them initially. In subsequent studies using different rats, subjects were given only five sessions to become familiar with a handler. No food was used. Again, subjects all selected the familiar person during testing. The final test involved exposure to a human for only a single session without food, and the results were the same.

Rats are not the only animals capable of discriminating one human from another; cats,* chickens,* cows, sheep, rabbits, seals, emus, rheas, llamas, pigs,* prairie dogs, chimpanzees,* and domestic dogs all can tell one human from another. As scientific studies continue to replace anecdotes, the evidence for human recognition among animals will become more widely accepted, impacting research design, the assessment of intelligence, and, ultimately, animal welfare.*

Selected Bibliography. Davis, H., and D. Balfour (Eds.), *The Inevitable Bond: Examining Scientist-Animal Interactions* (New York: Cambridge University Press, 1992); Taylor, A., and H. Davis, The Response of Llamas (*Lama glama*) to Familiar and Unfamiliar Humans, *International Journal of Comparative Psychology* 9 (1997): 43–50.

HANK DAVIS AND ALLISON A. TAYLOR

ANIMAL COLLECTORS

An unofficial definition of "animal collector" is an individual who amasses and maintains, over an extended period of time, more animals than he or she can properly care for. Quantity is not necessarily the factor that identifies someone as a collector. In fact, there is no generally accepted number beyond which a responsible custodian of animals automatically becomes reclassified as a collector or addict. Much more significant are the type of care received by the animals being harbored, their physical and psychological conditions, the environment in which they are maintained, and the reaction of their keeper to reasonable attempts to reduce the number of animals.

The term "animal collector" means something different for those working in the humane field than it does for members of the general public. *Collecting* is generally considered a harmless activity, pursued by individuals inclined to accumulate *objects* typically because of a greater-than-average interest in them. Animal *addict*—synonymous with collector when used by many in the humane community—is more likely to alert someone to the fact that reference is being made to an individual whose behavior has serious negative consequences.

Humane officials who investigate situations involving unusual numbers of animals can almost always instantly distinguish the menagerie of a collector from that, for example, of a puppy-mill operator keeping numerous dogs for breeding purposes. Upon entering an animal collector's premises, one commonly observes a number of the following conditions: filthy, overcrowded living quarters for the animals; massive clutter throughout the house or other building where animals are confined; cannibalized carcasses; a lack of ventilation; animals who exhibit unsocialized behavior or depression; inappropriate food, if any is available; and extreme, long-standing health problems, such as a variety of advanced disease, unchecked parasitic infections, untreated injuries, and severe malnutrition, which shows itself most often as emaciation, but occasionally as obesity.

Prosecuting collectors for cruelty to animals is generally a last resort, virtually the only means of separating the collector from his or her victims. What makes legal action almost inevitable is one of the animal addict's most notable traits: a firm refusal to voluntarily part with his or her animals or to see them released from their suffering, no matter how desperately ill or injured they may be. However, even punishment is inadequate in many instances.

There is consensus among those who have confronted animal collectors about the fact that even after one incident has been resolved, repetition is inevitable. For this reason, a lengthy, well-supervised period or probationary arrangement is generally recommended as part of plea bargains or sentencing. During this period and beyond, the services of a social worker, if not a psychiatrist, are usually warranted. Unfortunately, this kind of professional help cannot always be imposed on the collectors against their will.

Since approximately the mid-1980s, humane organizations, health and fire officials, social services agencies, and law-enforcement authorities throughout the United States have shared information and resources concerning cases involving animal collecting. Such networking indicates that the consequences of this phenomenon are both prevalent and profound. The number of animals who suffer at the hands of collectors is impossible to assess.

Selected Bibliography. Cavallo, Janet, When Animal Collectors Go Too Far: Hurting the Ones They Love, *Sunday Recorder* (Amsterdam, NY), September 25, 1994; Lockwood, Randy, The Psychology of Animal Collectors, *American Animal Hospital Association Trends Magazine* 9(6) (1994): 18–21; New Legal Device Can Protect Shelters That Board Animals During Cruelty Cases, *Shelter Sense* (published by the Humane Society of the United States, 1993); New York State Humane Association, Animal Collectors; Adopting Abused Animals: What You Should Know [fact sheets]; Weiss, Lawrence E., Dealing with Collectors, A Cautionary Tale, *CHAIN Letter* (The Collective Humane Action and Information Network), Summer 1995.

SAMANTHA MULLEN

ANIMAL COMMUNICATION

Between Species

Communication between different species has long been observed by humans, who often initiate such dialogue themselves. Twenty-five thousand years ago, during the Late Paleolithic era, the zoological and the sociological were less separate. Human culture recognized nonhuman culture and sought to display a wide array of animals on cave walls, such as those at Lascaux in southern France. The keeping of pets (*see* COMPANION ANIMALS AND PETS) probably coincided with the domestication of both plants and animals.

Early clergyman of countless spiritual traditions all subscribed to a human identification and communication with nonhuman life forms. Saints (*see* RELIGION AND ANIMALS) of the Sinai Peninsula during the so-called Era of Retreat (4th century A.D.) communed with lions, jackals, deer, and even cheetahs.

It took the research of such zoologists as Charles Darwin,* Konrad Lorenz (who lived with a talking crow whom he deemed the smartest bird in the world), Karl von Frisch, Bert Hölldobler, E. O. Wilson, George Schaller, A. Skutch, Jane Goodall, and thousands of others to dignify and reinvest the vast animal and plant kingdoms with their own species-specific communication systems.

Students of the natural world have now produced a large literature that reverses the centuries-old ignorance of Cartesian (*see* DESCARTES, RENÉ) mechanism (thinking of animals as mere machines without feelings, soul, or intelligence). They can confidently attest to the therapeutic benefits of human-nonhuman relationships (most notably between humans and members of the Canidae or dog family) and the existence of animal empathy,* as displayed, for example, between marine mammals and other, nonhuman primates.

Zebras have been observed adopting an orphaned rhinoceros; crocodiles permit, and to a certain degree depend upon, a certain species of birds to clean their teeth; sheep think nothing of being rounded up and organized by sheepdogs. Polar bears have been seen to play with huskies (as opposed to eating them); egrets "hang out" with water buffalo or ride the backs of hippos through African marshes. Every camel herder knows his or her camels; cows and geese, lambs and pigs,* horses and humans, pigeons and sparrows, and prairie dogs and king snakes certainly "speak" to one another. These are not merely effective communications, oriented to the performance of some useful task or avoidance. They also satisfy other (possibly evolutionary) needs, though we still know little about them.

Furthermore, with over eighty million pets in the United States alone, it is clear that a tremendous range of more subtle, emotionally satisfying communication is taking place every day between both domestic and wild members of the animal kingdom, which conforms to a now widely held hypothesis known as *biophilia*: that all life forms tend to focus upon one another, draw pleasure and significance from their relations, and display a distinct preference for the company of others that can claim evidence of affection and affinity, intelligence and candor, and deep feelings and contemplation.

With continued research, the language of elephants and dolphins, of parrots and chimpanzees,* of Babe the pig and his farm friends, and of ants and worms and butterflies and sharks will doubtless be more fully uncovered. Ultimately, as E. O. Wilson has eloquently argued, familiarity will breed tolerance and compassion.

Selected Bibliography. Bruchac, Joseph, Understanding the Great Mystery, in Michael Tobias and Georgianne Cowan (Eds.), *The Soul of Nature* (New York: Plume/Penguin Books, 1996); Glacken, Clarence J., *Traces on the Rhodian Shore: Nature and Culture in Western Thought from Ancient Times to the End of the Eighteenth Century* (Berkeley: University of California Press, 1967); Katcher, Aaron, and Gregory Wilkins, Dialogue with Animals: Its Nature and Culture, in Stephen R. Kellert and Edward O. Wilson (Eds.), *The Biophilia Hypothesis* (Washington, DC: Island Press, 1993); Leroi-Gourhan, André, *Gesture and Speech*, trans. Anna Bostock Berger, introduction by Randall White (Cambridge, MA: MIT Press, 1993; first printed in 1964); Tobias, Michael, *A Vision of Nature: Traces of the Original World* (Kent, OH: Kent State University Press, 1995).

MICHAEL TOBIAS

Language Debates

In recent decades, the question of whether animals are capable of learning language has been intensely debated. Addressing this question requires an understanding of what language is as well as familiarity with leading animal language studies. Many commentators suggest that language is communication that features both (1) content, meaning, or reference and (2) syntax (some set of rules that determines a word's function by its position among other words, while allowing for many new combinations).

In the 1970s, enthusiastic claims were made about the language abilities of chimpanzees* who were trained to communicate in sign language or by manipulating symbols in certain ways. For example, they used sign language or manipulated plastic magnetized symbols that could be moved around on a board (and did not resemble what they indicated). Chimps used symbols in combinations, even original ones, mastered verbs and not just nouns, sometimes referred to things not immediately present, and sometimes apparently took into account a word's position in a string of words (suggesting syntax).

A wave of skeptical interpretations drowned the initially confident claims. One difficulty with these studies was that trainers, when testing animal subjects, often cued them, that is, gave them subtle hints of the correct answer (perhaps unconsciously) with facial expressions or other bodily movements. A second major problem was researchers' overinterpreting test results—reading too much into them. This was due to (1) not determining in advance what results would count as good scientific data, allowing subjective or personal impressions to carry the day, and (2) missing the possibility that subjects used simple rules of thumb without understanding what they were signing.

Suspicion about the results of ape language studies has forced researchers to be more careful. Some recent studies have been impressive. Working with bottlenosed dolphins, Louis Herman and his associates have provided an excellent case for the mastery of both content and syntax in understanding language (*comprehension*, as opposed to language *production*, the focus of most early studies). One dolphin was trained in a language in which words were represented by computer-generated sounds, the other in a language featuring signing with a trainer's hands and arms (eventually replaced by white dots on a screen). Words referred to objects, actions, properties, and relationships; sentences were constructed according to word-order rules allowing for more than 2,000 combinations with different meanings. Thus the same three or four words in different orders had different meanings, requiring syntax to distinguish them; the dolphins showed good comprehension by the accuracy of their responses to different instructions. Cueing was avoided by testing dolphins with computer sounds and abstract television images.

Meanwhile, Sue Savage-Rumbaugh has done leading work with pigmy chimpanzees (a distinct species also known as bonobos; *see* CHIMPANZEES). Kanzi, her star pupil, picked up the use of a keyboard by observation, without direct training, and has also learned to understand spoken multiword English commands. Savage-Rumbaugh has attempted to eliminate cueing during testing sessions with such innovations as communicating to subjects through headphones, with the tester out of sight. In an effort to resemble the language learning of human children, the chimps' education stresses natural exchanges in everyday settings, rather than artificial trials with rewards for correct responses. The subjects are taught not to label things on demand but to ask for things that interest them.

Although chimpanzees have received the most attention in language studies, the other great apes have also been well represented. For example, the gorilla Koko has acquired a vocabulary of over 500 signs, which she combines in strings of up to 6 signs in length. In most of her conversations with humans, Koko has apparently achieved original definitions, abstraction, self-references, and cursing. The orangutan Chantek is reported to have a vocabulary of over 150 signs, to use signs in novel ways, and, occasionally, to sign deceptively.

Selected Bibliography. Bekoff, Marc, and Dale Jamieson (Eds.), *Interpretation and Explanation in the Study of Animal Behavior*, vol. 1 (Boulder, CO: Westview Press, 1990); Cavalieri, Paola, and Peter Singer (Eds.), *The Great Ape Project: Equality beyond Humanity* (London: Fourth Estate, 1993); DeGrazia, David D., *Taking Animals Seriously: Mental Life and Moral Status* (Cambridge: Cambridge University Press, 1996); Rodd, Rosemary, *Biology, Ethics, and Animals* (Oxford: Clarendon Press, 1990); Savage-Rumbaugh, Sue, *Ape Language: From Conditioned Response to Symbol* (New York: Columbia University Press, 1986).

DAVID D. DeGRAZIA

ANIMAL ETHICS COMMITTEES (SWEDEN)

Ethics committees for the review of planned animal experiments became compulsory in Sweden in 1979. Six animal ethics committees (AECs) were set up in the six university regions of Sweden. Each AEC consisted of fifteen to forty-five members, according to how many animal experiments there were in the region. The members represented three categories, equal in size: researchers, technical staff (laboratory assistants and animal technicians), and laymen (societal and animal welfare* laymen, the latter including antivivisectionists [*see* ANTIVIVISECTIONISM] after 1982). The chairman and deputy chairman invariably were researchers.

The mission of the AECs was to review applications of planned experiments, but only those experiments classified as possibly causing pain* or suffering.* The investigators themselves had to classify their experiments in advance. The number of animal experiments reviewed by the AECs included some 50% of the animals used. The AECs made—and still make—decisions, but only "advisory" decisions. Investigators cannot appeal the decisions, but they can send in a new application. An approval cannot be appealed by any party. However, an experiment can be stopped by the authorities if it proves to deviate from what has been approved by the AEC.

After a long and vivid public debate during the 1980s the AECs were changed considerably in 1988. The number of committees was changed to seven (two in the Stockholm region). Every AEC now consists of twelve members representing two categories, the research category and the layman category, equal in size. The chairman and deputy chairman, now judges, are added and do not belong to any of the committee member categories.

In the new AECs all planned animal experiments must be reviewed before they are allowed to start. Also, killing of animals in order to harvest in vitro material now counts as animal experimentation. As from the beginning, there is no openly shown ethical principle used. Some kind of utilitarianism* seems to be the only guide for the decisions. The new animal-protection law says that the AEC shall reject an application if there is another way to solve the scientific problem, or if the problem has no "general interest." However, there is no limit on suffering, which cannot be outweighed by any utility, as is the case in Denmark.

Some consequences of the AECs that can be established are the following: The discussion in the AECs has led to a rise in perception of animal experimentation as a moral problem. Within the laboratories, this has entailed enforced self-policing and an improvement of experimental procedures and care of animals. Outside the laboratories it has generated a discourse regarding animal experimentation as a morally significant issue. At the same time, animal experimentation has been consolidated as an enterprise that is here to stay. The abolitionist protests have abated, and a convergence of opinions has come about. The discussion on animal experimentation has been depoliticized and turned into a more technical discussion concentrating on scientific and animal welfare details. The AECs have led to no pernicious consequences for science and research.

The most significant role that the AECs have played seems to be that of stabilizing the social order. The animal-experimentation issue has turned out to be a question of practical reforms instead of a revolutionary question of either abolition or total acceptance of everything.

Selected Bibliography. Forsman, B., *Research Ethics in Practice: The Animal Ethics Committees in Sweden, 1979–1989*), Studies in Research Ethics no. 4 (Göteborg: Centre for Research Ethics, 1993).

BIRGITTA FORSMAN

ANIMAL EXPERIMENTATION. *See* LABORATORY ANIMAL USE.

ANIMAL INDIVIDUALITY

To be concerned about animal welfare* and animal rights* is to be concerned about individuals. This has ethical implications. For example, it is necessary to decide whether to accept the suffering* of a few individuals if this will provide knowledge that will benefit many other animals. Most people who live, work, and interact with animals develop strong impressions of the animals' individual behavioral characteristics. Scientific studies have also shown, using a variety of measures of behavior, that individual animals have distinctive ways of behaving, or "behavioral styles," which can be likened to the phenomenon of personalities in humans. For example, some individuals are more aggressive or more nervous than others. This means that if we house or treat a group of animals in one particular way, we cannot be sure that they will all respond in the same way. This has several implications.

First, when scientists investigate how a treatment affects the welfare of animals, their findings and conclusions may be specific to the particular individuals they studied and may have limited general applicability. It is also possible that certain individual animals may be selected for study because they have characteristics that favor their inclusion, for example, being docile, even though they may not be typical of the general population. These problems can be overcome by studying a sufficiently large number of individuals and by taking account of behavioral characteristics before the study begins.

Second, the responses of animals may vary such that some individuals readily adapt to a particular situation, while others have to work very hard to cope with it (see ANIMAL WELFARE, Coping). In addition, they may use quite different methods to try to deal with the challenges presented by the situation. This means that if we take the average of the individuals' responses as an indicator of how damaging to animal welfare the situation is, we may end up basing our conclusions on a response that no individual actually showed. The effect of using such conclusions may be that the situation under consideration is deemed to be acceptable in animal welfare terms whereas, in reality, many animals find it difficult to cope with.

A solution to this problem lies in the design of high-welfare systems for animals. If systems can be designed or altered in such a way that they allow animals a certain degree of choice in how they are used, it may be possible to satisfy the requirements and abilities of most individuals. For example, a housing system that provides food at one central location is likely to favor the more aggressive individuals who can control this resource, whereas a system that provides several food sources that are spread out will prevent this problem and will allow timid individuals to feed more freely. Understanding the causes and consequences of individual differences in behavior can lead to effective improvements in animal welfare and remains an important goal of animal welfare science today.

Selected Bibliography. Bekoff, M., Naturalizing and Individualizing Animal Well-Being and Animal Minds: An Ethologist's Naiveté Exposed? in A. Rowan (Ed.), *Wildlife Conservation, Zoos, and Animal Protection: A Strategic Analysis* (North Grafton, MA: Tufts Center for Animals and Public Policy, 1995): 63–115; Benus, R. F., B. Bohus, J. M. Koolhaas, and G. A. van Oortmerssen, Heritable Variation for Aggression as a Reflection of Individual Coping Strategies, *Experientia* 47 (1991): 1008–1019; Jensen, P., Individual Variation in the Behaviour of Pigs—Noise or Functional Coping Strategies? *Applied Animal Behaviour Science* 44 (1995): 245–255; Mendl, M., and R. Harcourt, Individuality in the Domestic Cat, D. C. Turner and P. Bateson (Eds.), *The Domestic Cat: The Biology of Its Behaviour* (Cambridge: Cambridge University Press, 1988), 41–54; Stevenson-Hinde, J., Individual Characteristics: A Statement of the Problem, in R. A. Hinde (Ed.), *Primate Social Relationships: An Integrated Approach* (Oxford: Blackwell Scientific, 1983), 28–30.

MICHAEL MENDL

ANIMAL LIBERATION ETHICS

At the core of animal liberation ethics lies an argument from consistency against the contemporary view of egalitarianism. This view claims that all human beings are equal—whatever their sex, race, or psychological abilities, such as intelligence, skills, and sensitivity. It rejects the view that the members of a particular biological group may be discriminated against because they belong to that group, and it considers ethically offensive the idea that intellectually less endowed individuals, the disabled, small children, or the senile may be routinely taken advantage of by others who are not disabled. Thus neither biological characteristics nor particular psychological properties over and above sentience (*see* SENTIENTISM) are important for equal treatment.

Animal liberation ethics, which became important in the 1970s, was not well received by many people. In response to its challenge, defenders of humanism (*see* ANTHROPOCENTRISM)—the view that human lives and interests should always be given greater weight than nonhuman lives and interests—offered a number of objections. They claimed that humans have special duties toward their closest kin; that, in contrast with race, species differences corresponded with measurable differences; that it is not possible to have rights without the capacity for claiming them; that it is not possible to have rights without the capacity for having duties; or even that nonhuman animals, lacking verbal language (*see* ANIMAL COMMUNICATION, Language Debates), have no conscious interests that may be taken into consideration.

Such objections can be rebutted: the notion of closest kin can be used to justify discrimination against members of the human species as well as members of other species. Also, we grant basic rights to small children, although they certainly cannot claim them or have duties; and in view of the work of Charles Darwin,* the idea of differences in kind rather than in degree between us and all other animals is unlikely (*see* CONTINUITY). Even attempts to draw a line between human infants and nonhuman animals overlook the fact that there are human beings whose mental disabilities cannot be reversed.

All things considered, those who argue against speciesism* believe that there is no argument for discrimination between species that could not be used as an argument for discrimination among humans. They argue that justifications for equality cannot be accepted only up to a point and then be arbitrarily rejected. In highlighting the arbitrariness of the humanist position, animal liberation ethics not only seeks to protect nonhuman beings, but also challenges the direction and basis of much Western moral thinking.

Selected Bibliography. Cavalieri, Paola, and Will Kymlicka, Expanding the Social Contract, *Etica & Animali* 8 (1996): 5–33; Cavalieri, Paola, and Peter Singer (Eds.), *The Great Ape Project: Equality beyond Humanity* (London: Fourth Estate, 1993); Regan, Tom, *The Case for Animal Rights* (Berkeley: University of California Press, 1983); Sapontzis, Steve F., *Morals, Reason, and Animals* (Philadelphia: Temple University Press, 1987); Singer, Peter, *Animal Liberation* (New York: New York Review of Books, 1990).

PAOLA CAVALIERI

ANIMAL MODELS

Biomedical and Behavioral Science

Scientists study animals other than humans to understand (1) animals, (2) humans, or (3) universal processes supposedly true of all animate life. The second goal typically involves the strategy of developing animal models to understand and discover solutions for the treatment of humans.

An ideal in most philosophies of science is to study the object of interest directly, with as little as possible coming between the object and the investigator's immediate observation of it. Model making introduces a screen between the scientist and the actual object of study. This requires a further step in the investigation. The model is used to generate hypotheses that then must be tested in the original.

It is important to note that models as generative or educational devices help us understand through both the similarities and the differences between them and the actual object of study. A filing cabinet limits us to placing information in one location in it, while, with a computer, we can multiply, enter, and store the same information in many "places." The limitation of the filing cabinet allows us to see more clearly this feature peculiar to the computer.

Since its development in the 19th century, the use of animals other than humans as models of human phenomena has been a controversial issue. Contemporary animal rights* activists claim that using animals in this way is wrong both on ethical and scientific grounds. There are shortcomings on both sides in the current debate. With regard to the use of animal models specifically, many scientists and certain professional organizations of scientists claim that an animal model, say, of cancer produces an equivalent of a human condition or disorder "in all respects." Both on logical and empirical grounds, this is not possible. The conditions can be similar only in certain respects. For their part, many animal rights activists fall back on the claim that because a certain model is different in certain respects, then it is of no use in understanding human conditions. Indeed, there have been some good studies showing how the differences between certain models and the respective human conditions have indeed misled scientists. But this does not mean that in some instances, existing similarities, despite differences, might not be informative; differences might even prompt discovery of features of the original not before noticed.

To evaluate the effectiveness of animal models, consider an example in the behavioral sciences. Bulimia is a disorder in which an individual's eating behavior becomes bizarre and his or her body image becomes distorted. In her overconcern about her body image, an adolescent female might eat large quantities of junk food and then vomit (binge-purge behavior). In the animal model of this disorder, a hole is made in the stomach wall; when the animal eats, the food is siphoned off. Through this model of the condition of "eating without calories," scientists attempt to identify and understand various environmental, dietary, and physiological causes of bulimia.

Various forms of evaluation of this animal model were applied with the following results. Through examination of "outcome" studies of current treatments of bulimia, it was found that these treatments are only modestly and temporarily effective. Treatments reduce the frequency of binge-purge behavior but do not eliminate it, and relapse rates to pretreatment behavior a year or two after treatment are high. In any case, examination of the literature involving these treatments showed that they did not derive from animal models of bulimia. Through examination of studies citing this and other animal models of eating disorders, particularly in the literature read by clinicians specializing in the treatment of these disorders, it was revealed that they are infrequently read and, therefore, have little impact on treatment. On the basis of the application of these social scientific and historical methods to this particular animal model, it is clear that the strategy cannot be justified in this case. More work is needed to test the effectiveness of other animal models.

Selected Bibliography. Kaufman, S. (Ed.), *Perspectives on Animal Research* (New York: Medical Research Modernization Committee, 1989–1995); Langley, G. (Ed.), *Animal Experimentation: The Consensus Changes* (New York: Chapman and Hall, 1989); Orlans, F. B., *In the Name of Science: Issues in Responsible Animal Experimentation* (New York: Oxford University Press, 1993); Shapiro, K., *Animal Models of Human Psychology: Critique of Science, Ethics, and Policy* (Seattle: Hogbefe and Huber, 1997).

KENNETH J. SHAPIRO

Real-World Analogies

Models are basic, powerful tools in all areas of technology and science. Research in medicine and psychology commonly uses "models" or model systems. Examples include primate simian immunodeficiency virus (SIV) models for HIV/AIDS, models of Parkinson's disease, animal models of learning and psychopathology, layered network models of brain function, and artificial models for hearing. These very different types of models aid in the discovery of useful principles for addressing real-world problems. Models are necessary in biological and behavioral sciences because living organisms are highly complex.

Despite their wide use and demonstrated usefulness, models—and animal models in particular—are not well understood and hence are often controversial. They are used, for example, to aid understanding of the immune system, of how neurons of the brain work (e.g., to produce learning), of the effects of drugs, and even of how drugs interact under stress—for example, to yield depression, addictions, and lessened immunity to disease. This last phenomenon is psychological and behavioral as well as medical.

A model is a set of several kinds of analogies or similarities between the real-world phenomenon to be understood and the system that is being studied as the model. The key kinds of analogy involved are (*a*) initial analogy and (*b*) causal analogy. In combination, they constitute a model. What is critical is not the degree of material analogy, but the degree to which the model predicts the function of the "real-world" system; this is predictive validity.

Consider a case in psychiatric medicine where there are both similarities and differences. One might note that some set of physiological and behavioral symptoms characterizes patients with a given psychiatric disorder (for example, an inability to cope with challenges and/or loss of memory); one might further note that animals exposed to some drug in a learning experiment exhibit behaviors that are similar to the behavioral symptoms of the patients. A hypothesis that the abnormal behavior of the animal and the abnormal behavior of the patient are similar in important ways would constitute an *initial analogy* in the modeling process. An additional hypothesis might be that the patient's dysfunctional physiological symptom is related to the animal's drug-induced physiological state; this would be a second initial analogy. The degree of *descriptive* similarity between the two sets of behaviors or between the two physiological states would constitute the degree of *material analogy*. If a *causal* relation between the patient's physiology and the patient's behavior is demonstrated that parallels the empirically known relation between the animal's physiology and its behavior, a *causal analogy* can be drawn between these two parallel relations, and we have a formal *model*. It is important to note that an initial analogy alone is not a model. A true model involves both initial analogies and causal analogies. The power of the modeling process is that one can use the known causal relations in one domain as a guide to finding parallel relations in the second domain.

Animal research has in the past validated some theories as well as invalidated other theories of mechanisms and causal architectures for selected human—and animal—physical and mental diseases. Research on animal models continues to contribute substantially to our understanding of viral and bacterial diseases, lifelong development, age-related dementias, chemical-abuse–related dementias such as Korsakoff's syndrome, effects of stress, effects of trauma, and even psychopathology such as phobias and depression. Additionally, animal models have played especially important roles in developing pharmaceutical treatments for some biological and psychological diseases.

Selected Bibliography. Boulton, A. A., G. B. Baker, M. T. Martin-Iverson (Eds.), *Animal Models in Psychiatry* (Clifton, NJ: Humana Press, 1991); Calabrese, E. J., *Principles of Animal Extrapolation* (New York: Wiley, 1983); Fox, M. A., *The Case for Animal Experimentation: An Evolutionary and Ethical Perspective* (Berkeley: University of California Press, 1986); McKinney, W. T., *Models of Mental Disorders: A New Comparative Psychiatry* (New York: Plenum Press, 1988); Mineka, S., and R. Zinbarg, Animal Models of Psychopathology, in C. E. Walker (Ed.), *Clinical Psychology: Historical and Research Foundations* (New York: Plenum, 1991); Understanding Models and Their Use in Science [Special issue], *Journal of Research in Science Teaching* 28(9) (1991).

J. BRUCE OVERMIER

ANIMAL PRESENCE

The importance of animal presence has been best explained by biologist Paul Shepard in his 1978 book *Thinking Animals*. Animals, he says, moved our minds more than anything else in nature as we were evolving toward human beings. Animals fascinated and impressed us, which moved us to think and to speak.

When we lived as foragers with earthbound religions, animals were the first beings, world-shapers, and the teachers and ancestors of people. When we became agriculturalists and looked to the heavens for instruction about the seasons and the elements, we saw animal forms among the stars. Of the forty-eight Ptolemaic constellations, all but a few are organic, and twenty-five are named for animals. Of the twenty-two more that were added in the 17th century, nineteen have animal names. When people built colossal earthworks to appeal to the powers in the heavens, they built them in animal forms. Some in Peru are over a mile long. One in Ohio is in the shape of a giant snake with an egg in its mouth.

In Ice Age caves, the first art shows the human fascination with animal forms. Animals were thought to embody the spirits and powers of nature, and animals have been used to symbolize nature ever since. In ancient Egypt, Hathor, the cow goddess of the sky, was believed to have given birth to the sun. The sky was seen as a giant cow, and her legs were the four corners of the world. Ancient astronomers explained the workings of the universe by reference to the zodiac, which means, literally, "the circle of animals." We can see animals' presence in children's toys, in nursery rhymes, in Aesop's fables, in the medieval bestiaries, and in other moral tales. We can also see the animal presence in language, where they provide the base for some 5,000 expressions—more than any other set of things in nature.

Animals still matter, and in powerful ways that we might want to understand if we want to come to terms with—and live right by—nature. Misothery,* for example, eased exploitation of animals and nature, but it injected nature hating, ruthlessness, and alienation into our worldview. Our older tradition of a greater sense of kinship and of belonging in the world has been cut off, and our feelings for the living world are stunted. Many people feel negative—uneasy at best—about our place in nature.

Misothery makes many people despise much of animals and nature: they despise even the animal and nature that they see within themselves. Fears and hatreds of the "beast" within us can cause us to project the worst of them not only on other animals but also on the Other, that is, people not of our group. Throughout history, we have used animals to symbolize the lust, danger, and deceitfulness we saw in ourselves, but especially that which we imagined in women, Jews, Africans, and various Others.

When animals were seen as spirit powers and as kin, they gave us a vital bond and a sense of belonging to the living world. Animals, then, are much more important than we are prone to think. They are central to our worldview and have been throughout the ages. A better worldview will require humans to have better views of animals. Coming to better terms with nature requires that we come to better terms with animals.

Selected Bibliography. Campbell, Joseph, *The Way of the Animal Powers*, vol. 1 of *Historical Atlas of World Mythology* (New York: Harper and Row, Perennial Library, 1988); Midgley, Mary, *Beast and Man* (Ithaca, NY: Cornell University Press, 1978); Sanday, Peggy Reeves, *Female Power and Male Dominance: On the Origins of Sexual Inequality* (Cambridge: Cambridge University Press, 1981); Shepard, Paul, *Thinking Animals: Animals and the Development of Human Intelligence* (New York: Viking Press, 1978); Thomas, Keith, *Man and the Natural World: A History of the Modern Sensibility* (New York: Pantheon Books, 1983).

JIM MASON

Metamorphosis

Metamorphosis occurs when an individual passes from one state of being into another state of being, as when a caterpillar becomes a butterfly or when, in Ovid's *Metamorphoses*, Pygmalion's ivory statue becomes a living woman (book 10). Here we are concerned with the metamorphosis of humans into animals and animals into humans: a fictional event, strictly speaking.

Ovid's *Metamorphoses*, written around the time of the birth of Christ, is the main source of tales about the transformation from human to animal. Ovid drew upon folktales and the works of other writers to weave stories of metamorphosis into a broad worldview. In *Metamorphoses*, we find the tale of Lycaon, a man who practices cannibalism transformed by Jupiter into a wolf (book 1); and of Actaeon, who sees the goddess Diana naked, is changed by her into a deer, and is torn to pieces by his own hounds (book 3). P. M. C. Forbes Irving argues that in Greek myths, "the transformation into an animal is part of a wider disruption of order" (62): trespass on sacred territory or sexual misconduct, for example. The violation of social law is followed by the offender's "taking to the wilds" (Forbes Irving, 63) in animal form.

Some tales of metamorphosis are etiological; that is, they explain the origins of specific animals or animal features. The tale of Philomela (Ovid, book 6) is an example. Philomela is raped and has her tongue cut out by her brother-in-law Tereus. She and her sister Procne get revenge by killing his and Procne's children and serving them to Tereus for dinner. Discovering what he has eaten, Tereus flies into a rage and pursues the sisters. All three are transformed into birds: Philomela into a swallow, Procne into a nightingale whose mournful song and red feathers signify both her grief and her crime, and Tereus into a hoopoe who appears ever ready for battle and whose typically wide-open beak might symbolize the horror of his cannibalism.

The best-known metamorphosis from human to animal is the werewolf. Originating in preclassical European folklore and popularized in the American film industry, the werewolf is an example of what is involved in the transformation from human to animal in Western culture. In the case of the werewolf, metamorphosis into an animal means the loss of human constraints and regression into pure evil. In the Middle Ages, and even later, the werewolf was seen as the result of the human being's willing submission to Satan, "the Beast." Until the 18th century, "werewolves" were burned at the stake. This practice was in keeping with the medieval belief that humans who were morally degraded took on animal characteristics: the "treachery" of foxes, the "laziness" of the ass. The werewolf served as a warning to Christians to hold onto the rationality and faith that alone elevated humans above animals.

In modern times the person who becomes a werewolf is often pictured as the innocent victim of supernatural forces. Relations between humans and animals have changed so significantly that Ursula Le Guin can give the werewolf tale a twist: in "A Wife's Story" (*Buffalo Gals and Other Animal Presences*), the wife-narrator describes the terrifying vision of her husband metamorphosing from a familiar and sociable wolf into monstrous human form.

In Native North American tradition (*see* NATIVE PEOPLES AND ANIMALS), metamorphoses from human to animal and vice versa are usually more benign. Often tales from the oral tradition show animals becoming "people" when they return to their own world. In the Haida tale "Salmon Boy," as retold by Joseph Bruchac (*Native American Animal Stories*), a boy who has been disrespectful to salmon learns respect when he is transformed into one of them and goes with them to their home. In a Blackfoot tale, "The Piqued Buffalo-Wife" (in *The Storytelling Stone*, edited by Susan Feldman), a human male has sexual relations with a buffalo and must pass through several trials, including death and resurrection, before his buffalo-wife and offspring can be changed permanently into human beings. Boundaries between human and animal are flexible in Native North American tradition. The boundary is flexible in Latin American traditions as well. Modern writers like Julio Cortázar and Carlos Fuentes draw upon myth and legend to make their modern heroes and antiheroes pass through animal phases. Nancy Gray Díaz emphasizes the mutability of the narrative world that permits these writers to take "an extraordinary leap into otherness" (*The Radical Self*, 102).

Metamorphoses from animal to human are rare in modern literature and in Western literature in general, except where the animal was a human being to begin with. Franz Kafka wrote the most famous modern story about metamorphosis, *The Metamorphosis*, which describes the fortunes of Gregor Samsa after he is "transformed in his bed into a gigantic insect." Kafka also wrote "A Report to an Academy" (1917), in which an ape describes to a group of scientists how he "became" human by learning a few simple tricks such as drinking schnapps, smoking cigars, and speaking human language. In John Collier's *His Monkey Wife* (1930), a chimpanzee* receives affirmation of her "humanity" after she has cunningly supplanted a man's fiancée and at last won his love.

In the West, the idea of the great chain of being made it easier to imagine human beings falling through sin into animal form than to imagine animals rising to human level. It has been easier to imagine human consciousness trapped inside an animal body than to disregard the physical shape of the animal so that animals can actually metamorphose into humans. Often, once a human being transformed into an animal has learned a lesson in true humanity, as in Apuleius's *The Golden Ass* (2nd century A.D.), he or she is restored to human shape. In this respect, the metamorphosis can be interpreted as a rite of passage. As modern theorists have concluded, metamorphoses are used in Western literature primarily to explore what it means to be human.

Selected Bibliography. Barkan, Leonard, *The Gods Made Flesh: Metamorphosis and the Pursuit of Paganism* (New Haven: Yale University Press, 1986); Díaz, Nancy Gray, *The Radical Self: Metamorphosis to Animal Form in Modern Latin American Narrative* (Columbia: University of Missouri Press, 1988); Forbes Irving, P.M.C., *Metamorphosis in Greek Myths* (Oxford: Clarendon Press, 1990); Ovid, *Metamorphoses*, translated and with an introduction by Mary M. Innes (Harmondsworth, Middlesex: Penguin Books,

1955); Skulsky, Harold, *Metamorphosis: The Mind in Exile* (Cambridge, MA: Harvard University Press, 1981).

MARIAN SCHOLTMEIJER

ANIMAL RESEARCH. *See* LABORATORY ANIMAL USE.

ANIMAL RIGHTS

Two opposing philosophies have dominated contemporary discussions regarding the moral status of nonhuman animals: (1) animal welfare* (welfarism) and (2) animal rights (the rights view). Animal welfare holds that humans do nothing wrong when they use nonhuman animals in research, raise them to be sold as food, and hunt* or trap* them for sport or profit if the overall benefits of engaging in these activities outweigh the harms these animals endure. Welfarists ask that animals not be caused any unnecessary pain* and that they be treated humanely.

The animal rights view holds that human utilization of nonhuman animals, whether in the laboratory, on the farm, or in the wild, is wrong in principle and should be abolished in practice. Questions about how much pain and death are necessary miss the central point. Because nonhuman animals should not be used in these ways in the first place, any amount of animal pain and death is unnecessary. Moreover, unlike welfarism, the rights view maintains that human benefits are altogether irrelevant for determining how animals should be treated. Whatever humans might gain from such utilization (in the form of money or convenience, gustatory delights, or the advancement of knowledge, for example) are and must be ill gotten.

While welfarism can be viewed as utilitarianism* applied to animals, the rights view bears recognizable Kantian features. Immanuel Kant* was totally hostile toward utilitarianism, not because of what it implies may be done to animals, but because of its implications regarding the treatment of human beings. To the extent that one's utilitarianism is consistent, it must recognize that not only animals may be harmed in the name of benefiting others; the same is no less true of human beings.

Kant abjured this way of thinking. In its place he offered an account of morality that places strict limits on how individuals may be treated in the name of benefiting others. Humans, he maintained, must always be treated as ends in themselves, never merely as means. In particular, it is always wrong, given Kant's position, to harm someone forcefully so that others might reap some benefit, no matter how great the benefit might be.

The rights view takes Kant's position a step further than Kant himself. The rights view maintains that those animals raised to be eaten and used in laboratories, for example, should be treated as ends in themselves, never merely as means. Indeed, like humans, these animals have a basic moral right to be treated with respect, something we fail to do whenever we use our superior physical strength and general know-how to inflict harms on them in pursuit of benefits for humans.

Among the recurring challenges raised against the rights view, perhaps the two most common involve (1) questions about line drawing and (2) the absence of reciprocity. Concerning the latter first, critics ask how it is possible for humans to have the duty to respect the rights of animals when animals do not have a duty to respect our rights. Supporters of the rights view respond by noting that a lack of such reciprocity is hardly unique to the present case; few will deny that we have a duty to respect the rights of young children, for example, even while recognizing that it is absurd to require that they reciprocate by respecting our rights (*see* MORAL AGENCY AND ANIMALS).

Concerning line-drawing issues, the rights view maintains that basic rights are possessed by those animals who bring a unified psychological presence to the world—those animals, in other words, who share with humans a family of cognitive, attitudinal, sensory, and volitional capacities (*see* ANIMAL COGNITION). These animals not only see and hear, not only feel pain and pleasure, they are also able to remember the past, anticipate the future, and act intentionally in order to secure what they want in the present. They have a biography, not merely a biology.

Where one draws the line that separates biographical animals from other animals is bound to be controversial. Few will deny that mammals and birds qualify, since both common sense and our best science speak with one voice on this matter. The rights view can rationally defend the sweeping social changes that recognition of the rights of animals involves—the end of animal model* research and the dissolution of commercial animal agriculture (*see* FACTORY FARMING), to cite just two examples.

Selected Bibliography. Midgley, Mary, *Animals and Why They Matter* (Athens: University of Georgia Press, 1983); Pluhar, Evelyn, *Beyond Prejudice: The Moral Significance of Human and Nonhuman Animals* (Durham, NC: Duke University Press, 1995); Regan, Tom, *The Case for Animal Rights* (Berkeley: University of California Press, 1983); Rollin, Bernard, *Animal Rights and Human Morality*, rev. ed. (Buffalo, NY: Prometheus Books, 1992); Singer, Peter, *Animal Liberation* (New York: New York Review of Books, 1990).

TOM REGAN

Distinguishing Animal Rights from Animal Welfare

The notion of "animal welfare" dates back far before "animal rights." In fact, "rights" in their modern sense did not enter common usage until the 1700s. It was most notably through the publication of *Animal Liberation* by Australian philosopher Peter Singer in 1975 that the animal liberation movement as we know it coalesced. There were several reasons for the new radical view, all of which directly influenced the content of Singer's important book: (1) using the liberation movements on behalf of blacks and women as models, the animal liberation movement rejected "speciesism"* (arbitrary discrimination on the basis of species membership) as well as racism, sexism, homophobia, and ableism; (2) advances in evolutionary biology blurred species boundaries between humans and other animals; (3) rebellions occurred within human organizations (e.g., the Royal Society for the Prevention of Cruelty to Animals'* support of hunting*—many of its wealthy patrons were fox hunters—led to the formation of the Hunt Saboteurs Association in 1963); and (4) modern animal cruelties were documented in Ruth Harrison's 1964 book *Animal Machines*, which exposed factory farming,* and in Richard Ryder's 1975 *Victims of Science*, which revealed horrors in the laboratory.

Technically, "animal rights" can refer to any list of rights for animals, although currently, the term is widely understood to refer to the idea of abolishing all use or exploitation of animals, a view reflected in Tom Regan's *The Case for Animal Rights*. "Animal welfare" is generally understood as advocating "humane use" of animals, at minimum upholding animal well-being by prohibiting "unnecessary cruelty" (a common legal phrase). In spite of this general meaning, there remains a whole spectrum of alternative views as to what "animal welfare" is: (1) *animal exploiters' "animal welfare,"* which amounts to the reassurance by those who use animals as commercial or recreational resources that they care for animals well; (2) *commonsense animal welfare*, which is the average person's vague concern to avoid cruelty and perhaps to be kind to animals; (3) *humane animal welfare*, which is more principled, deep, and disciplined than commonsense animal welfare in opposing cruelty to animals, but does not reject most animal-exploitive industries and practices (fur and hunting are occasional exceptions, along with the worst farming or laboratory abuses); (4) *animal liberationist animal welfare*, championed by Peter Singer, which would minimize suffering while accepting, for example, some types of vivisection; (5) *new welfarism (see* ANIMAL RIGHTS, Animal Rights and New Welfarism); and (6) *animal welfare/animal rights views*, which do not clearly distinguish the two. Richard Ryder subscribes to both ideas, although he is a complete abolitionist regarding animal use. Both animal welfare and animal rights, he says, "denote a concern for the suffering of others," and he evidently does not see the value of using the term to distinguish abolitionists from nonabolitionists who are still humanitarians.

Selected Bibliography. Carson, Gerald, *Men, Beasts, and Gods: A History of Cruelty and Kindness to Animals* (New York: Charles Scribner's Sons, 1972); Finsen, Lawrence, and Susan Finsen, *The Animal Rights Movement in America: From Compassion to Respect* (New York: Twayne, 1994); Jasper, James M., and Dorothy Nelkin, *The Animal Rights Crusade: The Growth of a Moral Protest* (New York: Free Press,

1992); Ryder, Richard D., *Animal Revolution: Changing Attitudes towards Speciesism* (Oxford: Basil Blackwell, 1989).

DAVID SZTYBEL

Animal Rights and New Welfarism

Although the theory of animal rights is basically different from that of animal welfare,* there is a significant chasm between the theory of animal rights and the social phenomenon that we call the "animal rights movement." Despite its apparent acceptance of the rights position, the modern animal-protection movement has failed to translate the theory of animal rights into a practical and theoretically consistent strategy for social change. The language of rights is, for the most part, used to describe virtually any measure that is thought to lessen animal suffering. So, for example, a proposal to provide a bit more cage space to animals used in experiments is regarded as promoting animal rights even though such a measure represents a classic example of welfarist reform.

It would be simplistic, however, to say that the modern animal rights movement* is no different from its classical welfarist predecessor. The modern animal "rights" movement has clearly rejected the philosophical doctrine of animal rights in favor of a version of animal welfare that accepts animal rights as an ideal state of affairs that can be achieved only through continued adherence to animal welfare measures. This hybrid position—that the long-term goal is animal rights but the short-term goal is animal welfare—is called the "new welfarism" and its advocates the "new welfarists." The new welfarists believe, for example, that there is some causal connection between cleaner cages today and empty cages tomorrow. As a result, the animal "rights" movement, despite its use of rights language and its long-term goal of abolishing institutionalized animal exploitation, continues to pursue an ideological and practical agenda that cannot be distinguished from measures endorsed by those who accept at least some forms of animal exploitation.

Selected Bibliography. Francione, Gary L., Animal Rights and Animal Welfare, *Rutgers Law Review* 48 (1996): 397–469; Francione, Gary L., *Animals, Property, and the Law* (Philadelphia: Temple University Press, 1995); Francione, Gary L., *Rain without Thunder: The Ideology of the Animal Rights Movement* (Philadelphia: Temple University Press, 1996); Pluhar, Evelyn, *Beyond Prejudice: The Moral Significance of Nonhuman Animals* (Durham, NC: Duke University Press, 1995); Regan, T., *The Case for Animal Rights* (Berkeley: University of California Press, 1983); Rollin, Bernard, The Legal and Moral Bases of Animal Rights, in Harlan B. Miller and William H. Williams (Eds.), *Ethics and Animals* (Clifton, NJ: Humana Press, 1983); Singer, Peter, *Animal Liberation* (New York: New York Review of Books, 1990).

GARY L. FRANCIONE

Biological Perspectives

Some philosophers, theologians, anthropologists, and authorities in other fields have suggested that various characteristics make the human animal a unique species. Because of our uniqueness, they argue, we are justified in "exploiting" other animals, which do not possess these capacities, for our own purposes. Other philosophers have argued that the human animal is really not unique, and that differences between humans and other animals are quantitative rather than qualitative. According to this argument, the human animal cannot claim uniqueness as the basis for justifying the "exploitation" of other species. From the biological perspective, however, these arguments about human superiority or lack thereof are largely irrelevant.

The human animal is obliged to do what other animals must do to ensure their survival as individuals and as a species: They must struggle to survive against the forces of natural selection that operate relentlessly to drive all species into extinction. These forces are very effective: Well over 99.9% of all the plant and animal species that have ever existed on this planet have become extinct. Extinction is the rule of life, not the exception.

All species are distinguished by certain features that give them an adaptive advantage in the struggle for survival. For humans, our main adaptive advantage is our big brain, with its inherent intellectual capacities. An important, and possibly unique, feature of our intellect is our insatiable curiosity. We are driven by a need to learn all that we can about the universe around us—both the living and the nonliving. This need is the reason that we study animals and plants, bacteria and fungi, and parasites and viruses, as well as rocks, the weather, the atom, the solar system, and beyond. The knowledge that we gain from such studies is used to assure our survival as a species.

Four forces of natural selection threaten our existence as a species and that of other species as well: microbes; parasites (for example, those that cause malaria and schistosomiasis); insects (which act as vectors for diseases and can destroy food crops); and natural calamities (for example, the asteroid strike that is thought to have caused the extinction of the dinosaurs). It has been suggested that we ourselves now pose a new threat to our own existence because of our development of thermonuclear weapons, overpopulation, and pollution, with consequent environmental destruction. The only way that we can ensure our survival in the face of these threats is to constantly acquire new knowledge about the ever-changing nature of these threats. The new knowledge can then be used to protect us from new (and old) dangers. Using other species for study is one way to acquire this new knowledge.

Another argument used by advocates for animal equality holds that there are no morally "relevant" differences between human and nonhuman animals. Holding nonhuman animals in lesser regard than humans therefore constitutes an unjustifiable form of discrimination called speciesism.* To bolster this argument, speciesism is analogized with racism and sexism.

Counterarguments from a biological perspective can be made. One can question whether or not discrimination on the basis of species is immoral by observing the behavior of other species. Virtually all animal species studied by humans show preference to their own kind in various ways. They prefer to associate with and to mate with their own kind, and predatory species (including plants) rarely prey upon their conspecifics (members of their own species) (*see* PREDATION).

Equating discrimination on the basis of species membership with sexism or racism can be questioned. It is argued that preferential treatment of other humans on the basis of gender, race, or ethnicity is morally unjustifiable because the people who are being discriminated against are of the same species with the same capacities as those who are doing the discriminating. Furthermore, arguments are made that only humans can know that they are suffering from unjust discrimination or are being subjected to prejudicial treatment, and that to treat fellow human beings as subhuman is a crime against humanity. To treat nonhuman animals as nonhuman animals may be considered a logical consequence of our biology.

Selected Bibliography. Nicoll, C. S., and S. M. Russell, Animal Rights, Animal Research, and Human Obligations, *Molecular and Cellular Neurosciences* 3 (1992): 271–277; Nicoll, C. S., and S. M. Russell, Editorial: Analysis of Animal Rights Literature Reveals the Underlying Motives of the Movement: Ammunition for a Counter Offensive by Scientists, *Endocrinology* 127 (1990): 985–989; Nicoll, C. S., and S. M. Russell, Mozart, Alexander the Great, and the Animal Rights/Animal Liberation Philosophy, *Federation of American Societies for Experimental Biology Journal* 5) (1991): 2888–2892; Nicoll, C. S., and S. M. Russell, The Unnatural Nature of the Animal Rights/Liberation Philosophy, *Proceedings of the Society for Experimental Biology and Medicine* 205 (1994): 269–273; Russell, S. M., and C. S. Nicoll, A Dissection of the Chapter "Tools for Research" in Peter Singer's *Animal Liberation*, *Proceedings of the Society for Experimental Biology and Medicine* 211 (1996): 109–138.

CHARLES S. NICOLL AND SHARON M. RUSSELL

Profiles of Animal Rights Advocates

Studies have shown that regardless of gender, those who adopt the traditional feminine sex role (more caring and sensitive to the concerns of others, in contrast to the more masculine domination and nondifferentiation as defined by the Bem Sex Role Inventory) are most likely to support animal rights ideals. Not surprisingly, animal rights advocates are often vegetarians (*see* VEGETARIANISM). They are often concerned about domination by one individual or group over others. Generally liberal, both religiously and politically, supporters of the animal rights movement* are more likely to be ecologically concerned and to have a more negative view of the military than those who oppose this movement. As a group, animal rights advocates tend to be more empathic and are likely to rely more on their feelings and intuitions (to be classified as feeling and intuitive types on the Myers-Briggs

Type Inventory and as sensitive and imaginative on the 16 Personality Factor Inventory, psychological tests for assessing personality type) than those who support other positions in this debate.

Animal rights advocates are more likely than those who support the animal welfare* position to be more rigid in their thinking, very idealistic, and less likely to explore multiple sides of issues. However, those who support animal rights also tend to be more skeptical in their evaluations of science and the scientific method. More focused on how research applies to the real world than on reliability and consistency, animal rights advocates tend to adopt a more global perspective than their opponents.

Selected Bibliography. Adams, C. J., *The Sexual Politics of Meat: A Feminist-Vegetarian Critical Theory* (New York: Continuum Press, 1990); Broida J., L. Tingley, R. Kimball, and J. Miele, Personality Differences between Pro and Anti Vivisectionists, *Society and Animals* 1 (1993): 129–144; Collard, A., and J. Contrucci, *Rape of the Wild* (Bloomington: Indiana University Press, 1988); Galvin, S. L., and H. A. Herzog, Jr., Ethical Ideology, Animal Rights Activism, and Attitudes toward the Treatment of Animals, *Ethics and Behavior* 2 (1992): 141–149; Galvin, S. L., and H. A. Herzog, Jr., The Ethical Judgment of Animal Research, *Ethics and Behavior* 2 (1992): 263–286.

JOHN P. BROIDA

Ecofeminists' Perspectives

Ecofeminism, or ecological feminism, represents the position that there are important connections between the oppression of women and the domination of nature. Within the ecofeminist literature, these connections are described as being historical (causal), experiential (empirical), symbolic (literary and religious), theoretical (conceptual, epistemological, and ethical), political, and/or practical. While not all ecofeminists agree about the kinds of connections that can be drawn between the oppression of women and the domination of nature, all agree that any feminist theory or environmental ethic that fails to recognize some connection is incomplete or inadequate.

Some feminist theorists object to the connection ecofeminists make between the domination of women and the domination of nature because it appears to move women closer to nature and animals. This, critics say, is wrong because to be "animal-like" is to make a negative statement in Western culture, and because closeness to nature implies distance from culture. Yet being conceptually "close to animals" is a problem only if animals are seen as less than human.

A specific concern in ecofeminism about nonhuman animals has developed only recently as ecofeminists work to include animals in wider environmental discussions. Many of these women have begun to develop theories and practices that link ecofeminism to animal defense. Part of this work involves highlighting parallels between the specific ways that women and animals are oppressed. For example, the practice of killing furbearing animals for their skins is justified through sexist rhetoric, while menopausal women are encouraged to use the drug Premarin, which is produced through large-scale exploitation of pregnant horses.

An ecofeminist animal defense theory draws on traditional animal defense theories, such as the rights (*see* ANIMAL RIGHTS) approach of Tom Regan and the utilitarian (*see* UTILITARIANISM) approach of Peter Singer, and emphasizes the importance of animal suffering.* However, ecofeminist analyses go farther than these theories in that ecofeminists are concerned about broader questions of animal oppression and the relationship between this type of oppression and the oppression of women, people of color, and the natural world.

Selected Bibliography. Adams, Carol J., *Neither Man nor Beast: Feminism and the Defense of Animals* (New York: Continuum, 1994); Adams, Carol J., *The Sexual Politics of Meat: A Feminist-Vegetarian Critical Theory* (New York: Continuum, 1990); Adams, Carol J., and J. Donovan (Eds.), *Animals and Women* (Durham, NC: Duke University Press, 1995); Birke, Lynda, *Feminism, Animals, and Science: The Naming of the Shrew* (Philadelphia: Open University Press, 1994); Gaard, Greta (Ed.), *Ecofeminism: Women, Animals, Nature* (Philadelphia: Temple University Press, 1993).

LORI GRUEN AND LYNDA BIRKE

Moderation

The idea of moderation suggests a middle ground between extreme positions. If we are going to consider what moderation with regard to human treatment of animals means, we need first to define the extremes. Those who argue the extreme position for animal rights claim that animals' lives are to be valued equally with human lives, and thus all forms of animal use should be ended. This would mean an end to the eating of animal flesh; the use of animals for research, entertainment, or their skins or fur; and the commercial sale of animals as pets (*see* COMPANION ANIMALS AND PETS). Those who argue the extreme position against animal rights claim that animal lives have no value independent of their usefulness to humans, and therefore any forms of animal use that are beneficial to humans are acceptable, and that humans have no obligations whatsoever to animals. When it comes to animal rights, most people reject the extremes and argue for a position of moderation, suggesting that animals do have moral status (*see* MORAL STANDING OF ANIMALS) and that we do have some obligations to them, but that our obligations to animals are fewer and less strong than our obligations to human beings. What exactly these obligations are is answered in many different ways, and thus there is a range of moderate positions, rather than a single view that can be called "moderate."

One of the advantages of moderation with regard to animal rights is that its goals appear to be more achievable than those of more extreme positions.

Abolishing all uses of animals in research or ending the eating and hunting* of animals would involve dismantling powerful industries and changing the everyday habits of billions of people in fundamental ways. Regulating the uses of animals in such industries, on the other hand, seems a more achievable goal. Clearly, moderation is closely associated with the animal welfare* movement and with the idea of humane treatment and avoidance of cruelty. Regulation of the use of animals generally involves the Three Rs of reduction, refinement, and replacement (*see* ALTERNATIVES TO ANIMAL EXPERIMENTS). Some who advocate taking a moderate approach see this as a means of achieving the goals of animal rights in a gradual, step-by-step manner. But some philosophers, such as Gary Francione and Tom Regan, argue that such a moderate approach is hopeless as a means of achieving the ends of animal rights, since welfare approaches do nothing to dismantle the system that treats animals as the property of humans.

In spite of the difficulties in finding a moderate position, scientists and philosophers continue to grapple with the complexities of the ethical use of animals. Biologists and ethologists provide information about the nature of animals that is crucial in the ongoing ethical discussions. How intrusive should they be? Should ethologists studying carnivores, for example, create or interfere with predatory situations (*see* PREDATION)? Should wild animals be brought into captivity? What, if anything, counts as respectful and ethical use of animals? These are the sorts of complex and difficult questions that those taking the moderate position must attempt to answer.

Selected Bibliography. Daniels, Thomas J., and Marc Bekoff, Domestication, Exploitation, and Rights, in Marc Bekoff and Dale Jamieson (Eds.), *Interpretation and Explanation in the Study of Animal Behavior* (Boulder, CO: Westview Press, 1990), 345–377; Dresser, R., Standards for Animal Research: Looking at the Middle, *Journal of Medicine and Philosophy* 13 (1988): 123–143; Finsen, Susan, On Moderation, in Marc Bekoff and Dale Jamieson (Eds.), *Interpretation and Explanation in the Study of Animal Behavior*, vol. 2 (Boulder, CO: Westview Press, 1990), 394–419; Francione, G., *Rain without Thunder* (Philadelphia: Temple University Press, 1996); Regan, Tom, *The Case for Animal Rights* (Berkeley: University of California Press, 1983); Rowan, Andrew, *Of Mice, Models, and Men: A Critical Evaluation of Animal Research* (Albany: State University of New York Press, 1984).

SUSAN FINSEN

ANIMAL RIGHTS ADVOCATES. *See* ANIMAL RIGHTS.

ANIMAL RIGHTS MOVEMENT[1]

The first animal rights* movement began well over 100 years ago in England. The early movement was primarily antivivisectionist (*see* ANTIVIVI-

[1]Adapted from *Animals' Agenda*, July/August 1996.

SECTIONISM) and inspired protests, legislative reforms, antivivisectionist hospitals, and a broad base of support. Earlier humane leaders and antivivisectionists worked together, but by 1910 humane leaders withdrew from criticizing institutional cruelties such as vivisection. Although humane societies and some antivivisection societies from the early era survive to this day, it was not until the 1960s that the modern animal rights movement re-emerged.

The first organization to speak for animal rights in the modern sense was the British Hunt Saboteurs Association, formed in 1963. In philosophy and tactics, this organization represented a radical shift from the welfarism of the humane organizations. The Hunt Sabs used confrontational tactics, disrupting hunts and confronting hunters. At around that time a number of environmental organizations (e.g., Greenpeace) emerged, and Cleveland Amory founded the Fund for Animals and launched campaigns against hunting* and trapping.*

In 1970 Richard Ryder coined the term "speciesism,"* and by 1972 the Animal Liberation Front was operating in Britain. By the end of the Vietnam War the animal rights movement began to take hold in the United States. Those questioning the war and the justice of a system oppressive to women and minorities were intellectually disposed to extend their challenge of the status quo, and it only required that someone should point out the connections as they revealed the severe exploitation of animals hidden in factory farms and laboratories. In the 1970s, two philosophers, Peter Singer and Tom Regan, provided foundations for this challenge. Singer's book *Animal Liberation* and Regan's articles propelled animal issues into serious discussion within academic circles, and many in the movement date their awakening to reading these philosophers.

Early campaigns focused upon experimentation, targeting well-documented cases of laboratory animal suffering with protest and legislation. In 1975 Henry Spira and United Action for Animals investigated Museum of Natural History–sponsored research involving blinding, deafening, and mutilating the sex organs of cats.* Demonstrations highlighting the research's apparent futility and high cost and the animals' suffering* eventually convinced Mayor Ed Koch and 120 members of Congress to question it, and the National Institutes of Health halted its funding.

In the first few years of the 1980s important national organizations originated, including People for the Ethical Treatment of Animals (PETA), Transpecies Unlimited, Farm Animal Reform Movement (FARM), Feminists for Animal Rights, Mobilization for Animals, and In Defense of Animals. Vegetarian leaders, such as Alex Hershaft, joined the movement, and farm-animal suffering and vegetarianism* joined experimentation as central issues. Meanwhile, the Animal Liberation Front (ALF) first appeared in the United States with a 1977 raid releasing two dolphins from a Hawaii research lab.

The 1980s was a decade of protests and high media visibility. The public learned about animal research through investigative work by PETA, which succeeded in gaining the first conviction of a researcher for cruelty in U.S. history (*see* SILVER SPRING MONKEYS). This and other cases (such as the University of Pennsylvania Head Injury Lab) helped mobilize a national movement. Massive annual protests on World Day for Laboratory Animals drew thousands to targeted laboratories. Protests, civil disobedience, and ALF raids brought the movement into mainstream awareness. Animal rights became a familiar mass-media topic. The movement drew in diverse groups, including feminists, gays, environmentalists, Buddhists, celebrities, and artists. Organizations—some multi-issue (e.g., PETA and Fund for Animals), others more specialized (e.g., FARM and Last Chance for Animals)—sprang up across the country.

In addition to taking the message to the streets, organizations such as the Culture and Animals Foundation, the Ark Trust, the Rutgers Animal Rights Law Center, and the Animal Legal Defense Fund successfully brought animal rights into movie houses, museums, courts, and classrooms. Although not drawing the media coverage of protests and civil disobedience, this activity produced lasting accomplishments as well as mechanisms for continued progress in bringing an end to animal exploitation.

The 1990s began with as many as 75,000 turning out at a March for the Animals in Washington, D.C. The animal rights movement was becoming a social force to be reckoned with. More than the march, other movement accomplishments attest to its endurance and promise. Cruelty-free cosmetics are now readily available, and many large companies have given up animal testing. Fur sales have dropped drastically. Vegetarianism and veganism are on the increase, and the meat and dairy industry's Basic Food Groups propaganda has been replaced with the "New Four Food Groups for Optimal Nutrition" promoted by Physicians' Committee for Responsible Medicine, emphasizing grains, vegetables, fruits, and legumes as healthiest. Animal rights has become a focus of academic discussion, with much scholarship devoted to the topic.

The decade of the 1990s has seen a shift within the movement and a change in political climate. The media coverage of the 1980s has dwindled, as have numbers gathered each year for events such as World Day for Laboratory Animals. Demonstrations do not earn the press they garnered in the 1980s. The ALF is rarely found in the headlines with sensational raids. The movement is changing as it matures. Organizations such as Earthsave and Farm Sanctuary take a different approach, using the media to expose animal abuse but supplanting protest with education, while other organizations take yet different approaches. Many activists now look for new avenues to welcome adherents, such as sanctuaries and informational fairs.

As the movement matures, both its tactics and the details of its message have changed somewhat, though the core idea of liberating animals from human oppression remains unchanged. A greater emphasis on the interconnectedness of different forms of exploitation is increasingly found in the movement's message. Proponents of animal rights have always maintained that speciesism is analogous to racism and sexism. Some urge that a linkage stronger than mere analogy is crucial to the future success of the movement.

Selected Bibliography. Carson, Gerald, *Men, Beasts, and Gods: A History of Cruelty and Kindness to Animals* (New York: Scribner's, 1972); Finsen, Lawrence, and Susan Finsen, *The Animal Rights Movement in America: From Compassion to Respect* (New York: Twayne, 1994); Lansbury, Coral, *The Old Brown Dog: Women, Workers, and Vivisection in Edwardian England* (Madison: University of Wisconsin Press, 1985); Ryder, Richard, *Animal Revolution: Changing Attitudes towards Speciesism* (Oxford: Basil Blackwell, 1989); Turner, James, *Reckoning with the Beast: Animals, Pain, and Humanity in the Victorian Mind* (Baltimore: Johns Hopkins University Press, 1980).

SUSAN FINSEN AND LAWRENCE FINSEN

Sociology of the Animal Rights Movement

Behavioral scientists have used several approaches to understanding the sociology and psychology of those who oppose the use of animals. Some researchers have distributed surveys; other investigators have collected data based on extended interviews with animal activists. All of these studies show that animal activists are a diverse group with varying philosophies and approaches toward the treatment of animals but often share some common characteristics.

Virtually all recent research indicates more involvement by females than by males. (This was also true of the Victorian antivivisection [*see* ANTIVIVISECTIONISM] movement.) The reasons for the predominance of women among rank-and-file activists are unclear. Animal activists are also much more likely than the average American to be Caucasian; 95% of *Animals' Agenda* readers were reported to be white, as were 93% of attendees of the 1990 March for the Animals.

The majority of activists have middle- and upper-socioeconomic-class backgrounds. As a group, they come from households with higher-than-average median incomes. The survey of *Animals' Agenda* readers revealed that almost 40% lived in households with an income of over $50,000, as compared with 5% of the general public. They are generally better educated than the average American. Over 80% of activists have attended college, and about a fourth have graduate degrees. Many activists hold professional positions. Relatively few activists live in rural areas or towns with a population of less than 10,000. The overwhelming majority share their homes with companion animals.*

Data from several studies of activists attending the 1990 March for the Animals indicate that more often than not, activists identify with the moderate to left side of the political spectrum. A majority of activists indicate that they also support the goals of some other social movements. Among these are the environmental, women's, and gay rights movements. The majority of animal activists do not appear to support the antiabortion movement. Most animal activists are not religious in a conventional sense; several studies have reported that the majority of activists are not affiliated with mainstream organized religions, and a substantial proportion report being atheists or agnostics.

Public attitudes toward the animal rights movement are mixed. Several surveys have reported that a majority of Americans have generally positive attitudes toward the animal rights movement. For example, a 1994 public opinion poll reported that most respondents had either a very favorable (23%) or a mostly favorable (42%) view of the animal rights movement. On the other hand, only 7% of a 1990 survey said that they agreed with both the agenda of the animal rights movement and its strategies. Eighty-nine percent of the respondents felt that activists were well meaning, but either disagreed with the movement's positions on issues or on strategies for accomplishing specific goals.

Selected Bibliography. Herzog, H. A., Jr., "The Movement Is My Life": The Psychology of Animal Rights Activism, *Journal of Social Issues* 49 (1993): 103–119; Jamison, W., and W. Lunch, Rights of Animals, Perceptions of Science, and Political Activism: Profile of Animal Rights Activists, *Science, Technology, and Human Values* 17 (1992): 438–458; Jasper, J. M., and D. Nelkin, *The Animal Rights Crusade: The Growth of a Moral Protest* (New York: Free Press, 1992); Plous, S., An Attitude Survey of Animal Rights Activists, *Psychological Science* 2 (1991): 194–196; Richards, R. T., and R. S. Krannich, The Ideology of the Animal Rights Movement and Activists' Attitudes towards Wildlife, *Transactions of the North American Wildlife and Natural Resources Conference*, 1991, 363–371; Sperling, S., *Animal Liberators: Research and Morality* (Berkeley: University of California Press, 1988).

HAROLD A. HERZOG, JR.

ANIMAL SACRIFICE. *See* LABORATORY ANIMAL USE; RELIGION AND ANIMALS.

ANIMAL SELF-AWARENESS AND SELF-RECOGNITION. *See* ANIMAL COGNITION.

ANIMAL SHELTERS. *See* SHELTERS.

ANIMAL SYMBOLISM

The use of animal symbolism is as old as human consciousness. Throughout history, the animals that people observed and interacted with have been used as powerful vehicles for the expression of ideas and concepts. Although the spread of urbanization in modern times has sharply limited the extent of many people's contact with and knowledge about animals, their use as a frame of reference in thought and speech remains common. However, when a particular animal is used symbolically, the image of the animal that is projected may differ from its actual biological counterpart. Indeed, for a large share of the population in the industrialized world, relationships with animals as they are symbolically perceived have to a great extent replaced interactions with actual animals. Common beliefs about particular species, rather than personal experience, often determine attitudes toward animals.

The power of the symbolism assigned to animals has strong and important implications for the welfare of animals and even for their very survival. Metaphoric interpretation of an animal's form or behavior frequently results in the creature being classified in human terms such as "good" or "evil," with associated effects upon the preservation or eventual extinction of the particular species. Symbolism attributed to a species can act to increase positive affiliation, resulting in the animal's survival, or it can cause alienation of that animal from the human sphere, with consequent persecution and/or destruction. People concerned with animal welfare* and animal rights* not only try to understand the biological characteristics of various animals, but also appreciate the key role that symbolism plays in influencing people's views of animals and the consequent treatment of animals in society.

Selected Bibliography. Lawrence, Elizabeth Atwood, The Sacred Bee, the Filthy Pig, and the Bat out of Hell: Animal Symbolism as Cognitive Biophilia, in Stephen R. Kellert and E. O. Wilson (Eds.), *The Biophilia Hypothesis* (Washington, DC: Island Press, 1993); Lopez, Barry, *Of Wolves and Men* (New York: Scribner's, 1978); Turner, Victor, *Dramas, Fields, and Metaphors* (Ithaca, NY: Cornell University Press, 1975); Willis, Roy, *Man and Beast* (New York: Basic Books, 1974); Willis, Roy (Ed.), *Signifying Animals: Human Meaning in the Natural World* (New York: Routledge, 1994).

ELIZABETH ATWOOD LAWRENCE

ANIMAL THEOLOGY. *See* RELIGION AND ANIMALS.

ANIMAL WELFARE

Dictionaries define "welfare" and "well-being" by using phrases such as "the state of being or doing well" and "a good or satisfactory condition of existence" (*see* WELL-BEING OF ANIMALS). These phrases tell us that the "welfare" or "well-being" of animals has to do with their quality of life. To be more precise about the meaning of welfare and well-being requires that we go beyond the issue of how the terms are used and address the value issue of what we consider important for animals to have a good quality of life.

Three main approaches to this question have emerged. Some people emphasize how animals *feel* (*see* FEELINGS OF ANIMALS). According to this view, the affective states of animals ("feelings" or "emotions") are the key elements in quality of life. Thus a high level of welfare requires that animals experience comfort, contentment, and the normal pleasures of life, as well as being reasonably free from prolonged or intense pain,* fear,* hunger, and other unpleasant states. A second approach emphasizes the *biological functioning* of the animal. According to this view, animals should be thriving, capable of normal growth and reproduction, and reasonably free from disease, injury, malnutrition, and abnormalities of behavior and physiology. A third approach emphasizes *natural living* and considers that animals should be kept in reasonably natural environments and be allowed to develop and use their natural adaptations and capabilities.

These three approaches, although formulating the issue in different ways, often agree in practice. However, there are some real differences between the three views of welfare. For example, a pig farmer using criteria based on biological functioning might conclude that the welfare of a group of confined sows is high because the animals are well fed, reproducing efficiently, and free from disease and injury. Critics using other criteria might conclude that the welfare of the same animals is at risk because they are unable to lead natural lives, or because they show signs of frustration and discomfort (*see* PIGS).

Scientific knowledge about animals can often help in assessments of animal welfare. However, knowledge alone cannot turn judgments about the quality of life into purely factual matters. Science cannot, for instance, prove whether freedom of movement is better or more important than freedom from disease.

This inevitable involvement of values in the assessment of animal welfare does not mean that we cannot do objective scientific research in assessing the welfare of animals. For example, housing calves in individual stalls has many effects on their degree of movement, disease transmission, levels of "stress"* hormones, and so on, and these can be studied as objectively as other scientific variables. But how we use the measures to draw conclusions about the animals' welfare, and even which ones we choose to study, involve value judgments about what we think is important for the animals.

There are confusing semantic differences concerning the use of "welfare" and "well-being." Scientists and others commonly write about a certain "level of" welfare and thus use the term as a kind of scale, running from high to low. Thus one might speak of "poor welfare." This usage will sound strange to those who think of "welfare" as referring only to the good end of the scale. However, we do not have a distinctive term for the scale, and using "welfare" (or "well-being") in this dual sense fills the need. A comforting precedent is the word "health," which means both (1) freedom from illness and injury and (2) the general condition of an organism with reference to its degree of freedom from illness and injury.

Confusion also arises because people have tried to distinguish between "welfare" and "well-being" in various ways. One approach uses "well-being" for the state of the animal and "welfare" for the broader social and ethical issues; thus one might say that the well-being of animals is at the heart of animal welfare controversies. A second approach uses "welfare" to refer to the long-term good of the animal and "well-being" for its short-term state, especially how the animal feels; hence a painful vaccination may enhance an animal's welfare but reduce its feelings of well-being. A third approach, often followed in Europe, uses "welfare" exclusively because it is the traditional term in ethical and scientific writing, in most legislation, and in the names of animal welfare organizations. A fourth approach, often followed in the United States, uses "well-being" instead of "welfare" because welfare sounds like a political hot potato. Finally, many people treat the two terms as synonymous, following the lead of many dictionaries. Treating "welfare" and "well-being" as synonyms is probably the simplest and conforms best to everyday usage of the terms, but that will not stop scholars and scientists from continuing to use more specialized meanings.

Selected Bibliography. Dawkins, M. S., *Animal Suffering* (London: Chapman and Hall, 1980); Duncan, I. J. H., and M. S. Dawkins, The Problem of Assessing "Well-Being" and "Suffering" in Farm Animals, in D. Smidt (Ed.), *Indicators Relevant to Farm Animal Welfare* (The Hague: Martinus Nijhoff, 1983), 13–24; Fraser, A. F., and D. M. Broom, *Farm Animal Behaviour and Welfare*, 3rd ed. (London: Baillière Tindall, 1990); Fraser, D., Science, Values, and Animal Welfare: Exploring the "Inextricable Connection," *Animal Welfare* 4 (1995): 103–117; Rollin, B. E., *Farm Animal Welfare* (Ames: Iowa State University Press, 1995).

DAVID FRASER

Assessment of Animal Welfare

The role of animal welfare science is to provide information about the biology of animals—their perceptual and mental abilities, their needs and preferences, their responses to how they are treated—that should help people to make decisions about animal welfare issues in an informed way and perhaps lead to a consensus of opinion. The question arises as to whether animal welfare science can go one step further than this to provide information that shows, beyond reasonable doubt, that welfare in one situation is better than that in another.

Animal welfare scientists measure the animal's behavior, physiology, and physical state in order to get an idea of how animals respond to the ways in which they are treated. Putting all these data together to provide a single measure of the animal's welfare remains a technically challenging and still-unsolved task. Solutions to this problem are a major goal of animal welfare science today.

Similar problems exist for scientific attempts to specify absolute cutoff points at which welfare becomes unacceptable. An additional problem here is in identifying conditions where welfare is agreed to be good and acceptable that can act as standards against which other conditions can be compared. An obvious suggestion is to take the animal in its natural environment as the baseline condition. However, for many domestic species, it is difficult to identify what a "natural" environment actually is, and in most environments that we might call "natural," animal welfare is far from perfect. Animals living in the wild are often under threat from starvation, temperature variations, injury, and predation,* and in many cases, it would seem inappropriate to use measures of their behavior or physiology in the wild as benchmarks for defining acceptable welfare in animals under our care.

An alternative approach has been to ask the animal how it values different features of its environment. Scientists have developed ways of measuring how hard animals will work to get access to resources such as food, shelter, or companions. They have shown that animals will continue to maintain access to the same amount of certain resources even if they have to work very hard for them. In the same way, the extent to which animals work to avoid things can also provide valuable information about how aversive or damaging these are. However, the problem still exists of deciding at exactly what level of work a resource becomes important enough for it to be considered an essential feature of the animals' captive environment.

The scientific assessment of animal welfare has much to offer in terms of informing us about how animals perceive their environments and what they find stressful (*see* STRESS). This information can be used to argue that the welfare of animals kept in one way is better or worse than that of animals kept in a different way. Agreement may be more easily achieved in some situations than others.

Selected Bibliography. Broom, D. M., Animal Welfare: Concepts and Measurements, *Journal of Animal Science* 69 (1991): 4167–4175; Fraser, D., Science, Values, and Animal Welfare: Exploring the "Inextricable Connection," *Animal Welfare* 4 (1995): 103–117; Mason, G., and M. Mendl, Why Is There No Simple Way of Measuring Animal Welfare? *Animal Welfare* 2 (1993): 301–319; Mendl, M., Some Problems with the Concept of a Cut-off Point for Determining When an Animal's Welfare Is at Risk, *Applied Animal Behaviour Science* 31 (1991): 139–146; Rushen, J., and A. M. B. de Pasillé, The Scientific Assessment of the Impact of Housing on Animal Welfare, *Canadian Journal of Animal Science* 72 (1992): 721–743.

MICHAEL MENDL

Coping

To cope is to have control of mental and bodily stability. This means that all of the various control systems are functioning effectively. The term "coping" refers to the process of controlling the environmental effects. Sometimes this is achieved very easily because the environmental impact is slight in relation to the adaptive ability of the individual. In such cases there is only a minor effect on the welfare of the animal. If, on the other hand, coping is very difficult, then the welfare is considerably poorer. In some circumstances the environmental effects on the individual may be such that there is only a brief period when coping is not possible, but prolonged failure to be in control of mental and bodily stability leads to reduced life span and reproduction. The individual is then said to be stressed (see STRESS), and this is a further situation where welfare is poor.

Every individual human or other animal is exposed to impacts of the environment that require action. Some of these environmental effects are physical, such as changes in temperature or painful blows, while others are mental, like frightening threats or the loss of a social companion. For most effects, the animal has a system that, when activated, tends to reduce any damage that might result from the environmental impact.

The scientific use of the word "coping" that is described here reflects the popular use of the word but is more precise and refers to the full range of environmental impacts on the individual. It is often combinations of difficulties that make coping difficult. This is true for all species of animals. The methods of coping that are used may help with several problems at once. For example, many emergency responses require more energy than normal to allow the animal to utilize skeletal muscle more efficiently, make the heart pump faster, and reduce response time. Such general physiological methods of trying to cope are usually combined with one or more of a variety of physiological responses that are specific to the effect that the environment is having upon the animal. Hence if it is too cold, the animal may raise its hair, shiver, and reduce blood supply to peripheral parts of the body, but in extreme circumstances, adrenal responses are involved as well.

Coping methods may be behavioral and mental as well as physiological. If normal responses are not effective, other changes may be brought about that affect the mental state of the individual. For example, a pig subjected to repeated unavoidable contact with a frightening conspecific or human may show a severely reduced range of behavior and abnormal lack of responsiveness. Close confinement of pigs,* with consequent reduction in ability to show various regulatory responses, often results in the animals showing high levels of stereotypies* such as bar biting or sham chewing. Such abnormal behavior is likely to be an attempt at coping but may continue, despite being ineffectual, as a behavior pathology.

Some behavioral coping methods may be closely associated with physiological changes and biochemical actions in the brain such as those of the naturally occurring opioids β-endorphin and the enkephalins. Another important coping system that has links with other systems is the immune system, since T-lymphocyte activity is modified by both adrenal hormones and opioids. Hence the efficiency of the body's fight against disease may be

changed by environmental effects on the individual that are quite uncon-
nected with the pathogens involved.

Failure to cope ultimately results in death, but many changes occur before
this extreme is reached, and some of these are detectable. Injury and disease
can be recognized, as can extreme modifications of behavior. However, it
may be difficult to recognize depression in some individuals who are not
coping with their environment. Responses to problems involving reduced
activity and failure to act appropriately will eventually be reflected in obvious
signs but are less conspicuous to an observer than active responses. We now
know that both active and passive coping responses may be used in a given
situation. Some individuals tend to use mainly active or mainly passive responses, but others use both at different times.

Selected Bibliography. Broom, D. M., and K. G. Johnson, *Stress and Animal Wel-
fare* (London: Chapman and Hall, 1993); Fraser, A. F., and D. M. Broom, *Farm
Animal Behaviour and Welfare*, 3rd ed. (London: Baillière Tindall; New York: Saun-
ders, 1990); Lazarus, R. S., and S. Folkman, *Stress, Appraisal, and Coping* (New York:
Springer, 1984); Monat, A., and R. S. Lazarus (Eds.), *Stress and Coping*, 3rd ed. (New
York: Columbia University Press, 1991).

DONALD M. BROOM

Freedom

Freedom means the possibility to determine actions and to make responses. An animal's welfare is affected by the extent to which the individual
has freedom. Those freedoms that are given to an individual by others, for
example, those given to a farm animal by its human keepers, tend to result
in better welfare. The idea of providing animals with freedom carries with
it a suggestion of moral obligation toward the animals. It assumes that the
provider ought to give the animals certain opportunities and resources.

The idea of specifying the freedoms that should be given to animals was
put forward in the Brambell Committee Report that was presented to the
Government of the United Kingdom in 1965. These freedoms have been
incorporated into the United Kingdom Ministry of Agriculture, Fisheries,
and Food Welfare Codes supplied to farmers and others for many years.
The version of these detailed by the Farm Animal Welfare Council in 1992
is listed here:

1. Freedom from hunger and thirst by ready access to fresh water and a diet to
   maintain full health and vigor

2. Freedom from discomfort by providing an appropriate environment including
   shelter and a comfortable resting area

3. Freedom from pain, injury, or disease by prevention or rapid diagnosis and treatment

4. Freedom to express normal behavior by providing sufficient space, proper facilities, and company of the animal's own kind

5. Freedom from fear and distress by ensuring conditions and treatment that avoid mental suffering

These freedoms are described as being ideals that anyone with responsibility for animals should aim to provide, and it is further explained that animal welfare will be better if those who have care of livestock practice the following:

1. Caring and responsible planning and management

2. Skilled, knowledgeable, and conscientious stockmanship

3. Appropriate environmental design

4. Considerate handling and transportation

5. Humane slaughter (see TRANSPORTATION AND SLAUGHTER)

These lists identify the principal requirements of animals in relation to significant environmental factors to which they have to adapt, and the obligations of people toward the animals.

Selected Bibliography. Broom, D. M., Needs, Freedoms, and the Assessment of Welfare, *Applied Animal Behaviour Science* 19 (1988): 384–386; Fraser, A. F., and D. M. Broom, *Farm Animal Behaviour and Welfare*, 3rd ed. (London: Baillière Tindall; New York: Saunders, 1990); Webster, J., *Animal Welfare: A Cool Eye towards Eden* (Oxford: Blackwell, 1995).

DONALD M. BROOM

ANIMAL WELFARE ACT. *See* LABORATORY ANIMAL WELFARE ACT.

ANIMAL WELFARE INFORMATION CENTER (AWIC)

In 1985, an amendment to the Laboratory Animal Welfare Act* was included in the Farm Bill and signed into law. This amendment, the Improved Standards for Laboratory Animals Act (Public Law 99–189), asks researchers who do biomedical research using animals to try to reduce pain* and distress* that animals experience in the laboratory (see LABORATORY ANIMAL USE). To help researchers determine if alternative methods are available, the amendment established the Animal Welfare Information Center (AWIC) in 1986. The U.S. Congress wrote into the law that AWIC's main missions are to provide information that can be used for (1) training researchers who use animals about more humane animal care and use and (2) improving methods of animal experimentation that can reduce or replace animal use or minimize pain or distress to the animals.

As part of the National Agricultural Library, AWIC has a third mission. It provides information about animals to anyone who requests it. Information requests cover a wide range of topics, including care, use, and natural history; animal transportation*; legislation; and animal-protection philosophy.

AWIC produces many publications such as lists of books, articles, reports, and videotapes; information resource guides; fact sheets; and a quarterly newsletter. The bibliographies contain literature citations on topics ranging from housing, husbandry, and welfare of particular animals to ethical and moral issues concerning animal use in research, teaching (*see* EDUCATION AND THE USE OF ANIMALS), and product testing. The information resource guides contain lists of organizations, product suppliers, and publications. The *Animal Welfare Information Center Newsletter* is published quarterly with articles about animal care and use, legislation, and funding for research.

Not only does AWIC perform literature searches for researchers, but it also teaches researchers, institution administrators, veterinarians, and librarians how to perform them at their own facilities. AWIC also offers a workshop, "Meeting the Information Requirements of the Animal Welfare Act." The workshop is designed to answer questions participants may have about the Laboratory Animal Welfare Act, the information requirements, and how to perform an alternatives literature search.

AWIC continues to exchange materials and information with organizations and government agencies within the United States and abroad. As new technology becomes available, AWIC will improve the quality of its products and reach more people than ever. All of AWIC's publications (except for its CD-ROM), workshops, consultations, and less extensive literature searches are free of charge. The customer is always AWIC's priority because by providing useful information to the customer, AWIC plays a role in improving the way animals are housed, handled, and cared for by people.

Selected Bibliography. Allen, T., Meeting the Information Requirements of the Animal Welfare Act: A Workshop, *Animal Welfare Information Center Newsletter* 5(3) (1994): 6; Kreger, M., and T. Allen, Electronic Information for Animal Care and Use, *Lab Animal* 22(10) (1993): 53–54; Larson, J., The Animal Welfare Information Center of the National Agricultural Library, in B. T. Bennett, M. J. Brown, and J. C. Schofield (Eds.), *Essentials for Animal Research: A Primer for Research Personnel*, 2nd ed. (Beltsville, MD: National Agricultural Library, 1994), 67–72; U.S. Code of Federal Regulations (1995), Title 9, Part 1, Subchapter A, Animal Welfare.

MICHAEL D. KREGER

ANIMAL-ASSISTED THERAPY

An area of human-animal interaction receiving much attention is animal-assisted therapy (AAT), and there is an ever-increasing trend to permit, and even encourage, animal contact with people in a variety of institutional settings and for those whose mobility is limited by age or disability. To date, there are few reports of negative effects of AAT programs on animals. Good programs provide appropriate animal selection, care protocols, and emergency handling procedures.

Historically, the first AAT programs were in hospital settings. Small companion animals* are the most common, but farm animals are also employed in some nonurban settings. Resident animals, often employee-owned mascots, pose some of the ethical problems that face all owned animals. They must be well maintained with appropriate food, water, shelter, social interaction, and veterinary care. Also, unlike the usual pet, institutionalized animals may be on "duty" much of the day, and there is the potential for fatigue. It has been suspected that overt abuse may occur in programs in mental hospitals and prisons, though there is little documentation as to the extent of the problem.

Visitation programs have become more common, especially in nursing-home settings. Such programs often use animals from local humane societies, believing that such programs provide a positive public image of their organization, which improves public support and public donations. Nevertheless, the use of shelter* animals may be in conflict with the societies' stated missions by occupying vehicles and personnel that would otherwise be used to investigate cruelty toward animals* and capture loose animals. The animals used in the visits may otherwise be adopted, and often there is little known about the behavior or past health history of the animals used. There is now a trend to use animals known to the handlers, such as the volunteer's own animals.

There is growing evidence that animals play a positive role for elderly persons living alone, and there are programs to help aged people adopt animals from local humane societies. Unfortunately, older people often have trouble finding housing that accepts animals and planning for animals.

Horseback-riding programs are different from other AAT programs in that they require the client to visit the horse's facility, not the other way around. Consequently, most concerns address appropriate husbandry at the stable or barn. There is also a concern for overwork, but all programs appear sensitive to the problem, and therapeutic riding is only a small part of the horse's riding experience. There are established organizations for support with therapy protocols, client safety, and insurance.

Nondomesticated animals have been used both in their captive settings and placed with people in their own homes. Dolphin swimming programs receive positive media coverage, but the ethics of keeping dolphins in captivity is a concern. Dolphins do show stress in captivity and often do not show the increase in life span seen with other captive wildlife.

Service simians (monkeys) are being used for people with special needs, usually paraplegia. Ethical concerns include the potential of zoonotic diseases from primates, the use of shock packs for training, and the need to remove the canine teeth to reduce bite injuries to the human users or their visitors. The animals appear not to remain in service for more than a few years.

The most important ethical consideration regarding the animals used in therapy is no different from the concern we have for all animal use: are the animals treated with the respect they deserve? It has been long documented that stroking an animal lowers one's blood pressure, presumably an indication of reduced stress. Dogs* and horses being petted demonstrate a similar response, presumably for the same reason.

One of the most important humane considerations for companion animals is that they are part of a social group. This is why one requirement of the new Laboratory Animal Welfare Act* for research dogs is that the dogs have access to exercise and socialization, and it appears that the animals used in AAT receive frequent and rewarding human social contact. After a therapeutic session has ended, all involved, the recipient of the service, the therapist, and the animal must have benefited from the experience.

Selected Bibliography. Beck, A. M., The Therapeutic Use of Animals, *Veterinary Clinics of North America: Small Animal Practice* 15(2) (1985): 365–375; Beck, A. M., and A. H. Katcher, *Between Pets and People: The Importance of Animal Companionship*, rev. ed. (West Lafayette, IN: Purdue University Press, 1996); Beck, A. M., and N. M. Meyers, Health Enhancement and Companion Animal Ownership, *Annual Review of Public Health* 17 (1996): 247–257; Fredrickson, M., *Handbook for Animal-assisted Activities and Animal-assisted Therapy* (Renton, WA: Delta Society, 1992); Iannuzzi, D., and A. N. Rowan, Ethical Issues in Animal-assisted Therapy Programs, *Anthrozoös* 4(3) (1991): 154–163.

ALAN M. BECK

ANIMAL / HUMAN RECOGNITION. *See* ANIMAL COGNITION.

ANIMALITY

Animality as a concept is derived from the human/animal division in Western thought. Because humankind has been interested in explaining and reinforcing its preeminence, animality has been defined by what it lacks. The search for the special defining trait of humanity has involved underestimating other animals. Animality is an abstraction, or concept, loosely based upon observations of actual animals and sometimes relevant to human beings.

Since the middle of the 19th century, when Charles Darwin* asserted the idea of continuity* between humans and other animals, researchers have attempted to rebuild the idea of animality. It has been necessary to establish that animals experience emotions beyond basic ones: love, grief, resentment, hope, and the like. Darwin himself began this work in *The Expression of the Emotions in Man and Animals*. Jeffrey Masson and Susan McCarthy have gathered together much of the anecdotal evidence for animal emotion in *When Elephants Weep*.

Evidence of consciousness (*see* ANIMAL COGNITION, Consciousness and Thinking) in animals has also been debated. The idea of instinct has been a significant obstacle to determining consciousness in animals to the satisfaction of scientific authorities. Even when animals appear to exhibit consciousness, the argument can still be made that they are acting on instinct. Daisie and Michael Radner cover the history of this debate in *Animal Consciousness*.

As emotional complexity and consciousness in animals are becoming established, language use is taken to be the main factor separating humans from other animals (*see* ANIMAL COMMUNICATION). Certain animals, notably chimpanzees,* have been trained to respond to and employ human language in the form of signs. Critics of these experiments say that defenders are simply seeing the "Clever Hans phenomenon." Clever Hans was a horse who seemed to be able to count and add up numbers, signaling his answers by stamping a hoof. Observers noted that Hans was in fact responding to barely visible cues from his trainer instead of understanding words and responding appropriately. Even higher-order uses of language, such as appropriate manipulations of words, have been set down as accident or the Clever Hans phenomenon.

Argument to the effect that animals do employ language has resulted in attempts to redefine the meaning of language. Thomas Sebeok states that "animals demonstrably employ symbols," citing as one example the waggle dance of bees by means of which bees inform other members of the hive where they might find a food source. Michael Bright concludes his 1984 survey of animal language by stating that "[w]hen an animal can make the jump into talking about something that isn't there we would say it has language" (*Animal Language*, 231). Arguably, some neotropical birds pass this test by uttering alarm calls when no predator is around to distract fellow birds from a flying insect.

Some chimpanzees pass one of the tests designed to demonstrate the existence of a self-concept. If a red dot is placed on their forehead while they are anesthetized and then, once awake, they are stood in front of a mirror, they touch the red dot on their own foreheads, suggesting that the image in the mirror is an image of themselves—they appear to engage in self-recognition (*see* ANIMAL COGNITION, Self-Awareness and Self-Recognition). Mimicry in animals challenges the idea that animal identity is fixed. A healthy bird can pretend to be an injured bird to protect her young.

Evidence of moral values, social structures, and even culture among animals is mounting. Konrad Lorenz (in his book *On Aggression*) noted certain rituals among animals (geese in particular) that convey aggression but prevent outright injury. Frans de Waal has studied gestures and social devices designed to forestall hostilities among primates (see *Peacemaking among Primates*). Jane Goodall observed tool use among chimpanzees, notably the use of a peeled stick to draw ants out of their nests—a practice studied and imitated by young chimpanzees. Much of this evidence is covered in the book *Chimpanzee Cultures*.

The whole idea of animality is being actively debated, and it will continue to be important to humankind to think seriously about the meaning of humanity. Available information shows that humans can no longer be quite as bold in their claims about the nature of animals as they have been in the past.

Selected Bibliography. Bright, Michael, *Animal Language* (London: British Broadcasting Corporation, 1984); Darwin, Charles, *The Expression of the Emotions in Man and Animals* (Chicago: University of Chicago Press, 1965); Griffin, Donald R., *Animal Minds* (Chicago: University of Chicago Press, 1992); Masson, Jeffrey, and Susan McCarthy, *When Elephants Weep: The Emotional Lives of Animals* (New York: Delacorte Press, 1995); Mitchell, Robert W., and Nicholas S. Thompson (Eds.), *Deception: Perspectives on Humans and Nonhuman Deceit* (Albany: State University of New York, Press, 1986); Radner, Daisie, and Michael Radner, *Animal Consciousness* (Buffalo: Prometheus Books, 1989); Sebeok, Thomas A., Zoosemiotics: At the Intersection of Nature and Culture, in Thomas A. Sebeok (Ed.), *The Tell-Tale Sign: A Survey of Semiotics* (Netherlands: Peter De Ridder Press, 1975); Wrangham, Richard W., F. de Waal, and W. C. McGrew (Eds.), *Chimpanzee Cultures*, with a Foreword by Jane Goodall (Cambridge, MA: Harvard University Press, 1994).

MARIAN SCHOLTMEIJER

ANTHROPOCENTRISM

The term "anthropocentrism" refers to the traditional orientation of Western thought about and attitudes toward humans' relationship to nature. Anthropocentrism is also characterized as homocentrism, human chauvinism, speciesism,* and human-centered ethics. The underlying assumption is that humans are at the center of things: either apart from nature as a different order of being altogether or at the top of a hypothetical species hierarchy or ladder. More specifically, anthropocentrism means in its crude expression that human interests, needs, and desires are all that matter, and that if any life form can be said to possess intrinsic value, only *Homo sapiens* can. We may identify three general varieties of anthropocentrism.

1. *Dominionism.** Rooted in the Old Testament and in ancient Greek philosophy, dominionism is the position that nature and individual things in nature exist only in order to serve the needs and interests of humans. Dom-

inionism is also referred to in the literature as "strong anthropocentrism" and is commonly associated with such ideas as mastery of nature and nature's possessing merely instrumental (or use) value, and with the collective pride of species self-glorification. Dominionists think of nature as a boundless storehouse of resources. The frontier mentality and entrepreneurism are representative modes of dominionism.

2. *Stewardship*. A milder form of anthropocentrism may also be traced to the Judeo-Christian (*see* RELIGION AND ANIMALS) tradition and is found in others as well, for example Islam, which holds that humans are nature's caretakers, the vice-regents of Allah, for whose glory all acts are performed. It is present too in the thought systems of Indigenous Peoples. Often labeled "weak anthropocentrism" in the literature, the stewardship view is manifested in such ideas as husbandry, wise management, and the conservation and preservation of nature. Within weak anthropocentrism, however, while the human species and individual human beings still matter most, other species matter and possess value as well. Concern may be expressed within this framework for biodiversity and sustainable development.

3. *Evolutionary Perspectivism*. It is natural for each species, according to the outlook of evolutionary perspectivism, to act as if its survival, flourishing, and reproduction are the highest goods. Clashes are inevitable since there could not be an ecosphere as we know it without conflict and competition. Some infer from this that whatever humans choose to do in nature is simply a reflection of their own species-specific behavioral repertoire, the same as would be the case for any other animal. Others suggest that nature's well-being is an important overall consideration in the scheme of things, but that it in fact coincides with humans' enlightened self-interest, so that there need be no ultimate opposition between humans and nature. That is, when humans pursue their "proper end," they will then act in the best interests of nature as a whole.

Many philosophers and social critics perceive anthropocentrism as a belief that, if it ever had a purpose, has now outlived its usefulness and become outmoded. Others maintain that anthropocentrism is in some sense inescapable. Just as spiders, if they could evaluate the world around them conceptually and articulate the result in language, would be arachnicentric (spiders are arachnids), so would wolves (genus *Canis* and species *lupus*) be lupucentric and cows (bovids) bovicentric. How, then, could humans be other than homocentric? But while we may, and perhaps must, accept that human values and experience determine the standpoint from which we project outward, it does not necessarily follow that overcoming our anthropocentrism is impossible. The human outlook is an essential reference point, but this does not mean that all values must be human centered. We cannot conclude that it is impossible to empathize (*see* EMPATHY FOR ANIMALS) and connect with nonhuman nature just because we happen to belong to the species

Homo sapiens any more than we can conclude that it is impossible to empathize and connect with other human beings simply because we all happen to be individual and separate subjects of consciousness. How far we can and should try to project outward is something that cannot be decided in advance.

Selected Bibliography. Attfield, R., *Environmental Philosophy: Principles and Prospects* (Brookfield, VT: Avebury, 1994); Knudtson, P., and D. Suzuki, *Wisdom of the Elders* (Toronto: Stoddart, 1992); Norton, B., Environmental Ethics and Weak Anthropocentrism, *Environmental Ethics* 6 (1984): 131–148; Passmore, J., *Man's Responsibility for Nature: Ecological Problems and Western Traditions*, 2nd ed. (London: Gerald Duckworth, 1980); Taylor, P. W., *Respect for Nature: A Theory of Environmental Ethics* (Princeton: Princeton University Press, 1986).

MICHAEL ALLEN FOX

Humanism

By thinking of ourselves as human, we identify with all other "human" beings: we take their point of view and think of the world as it appears to "humans." That world is one structured by our desires, our memories, and our symbolic associations. When we are forced to remember that the weather, the animals, and the stars are utterly indifferent to us, we console ourselves with dreams of controlling, domesticating, and colonizing everything (and so concealing from ourselves what will still be true: that the universe has its own rules, not ours).

"Civilized morality" progressively outlaws violence as a mode of human interaction: we should not humiliate, torture, rape, or even kill another human being, because no human being could enjoy being victimized like that. "Civilized morality," in fact, is the creation of a "human" point of view that counts for more with its believers than any other bias or desire. "Humanism," as the form of civilized morality, requires us to believe that "human beings" and "humanity" as such must matter more to any "civilized being" than any other kind or quality.

Humanism rests in the conviction that it is human or humane values that should be pursued, and that nothing "in nature" should be accepted as a limitation on "the human spirit." Those who seek to reinvent such limits are likely to be thought "inhuman" or "reactionary" because they thereby imply that there are other forms of life and being that demand or should be given more respect than "ours." We should not think, so humanists contend, that "beastly behavior" is as much a thing to be revered as "humane behavior."

Selected Bibliography. Barrow, John, and Frank Tipler, *The Anthropic Cosmological Principle* (Oxford: Oxford University Press, 1986); Crook, John, *The Evolution of Human Consciousness* (Oxford: Clarendon Press, 1980); Gould, Stephen Jay, *Wonderful*

Life (London: Hutchinson, 1991); Trigg, Roger, *The Shaping of Man* (Oxford: Blackwell, 1982), Wills, Christopher, *The Runaway Brain* (New York: Basic Books, 1993).

STEPHEN R. L. CLARK

Animals as Subjects-of-a-Group-Life

The notion of animal rights* carries the risk of redefining animals in human, even Western human, terms. Animals live in social and physical domains that may differ significantly from circumstances that humans would appreciate. As a result, animals are harmed, not just as individuals (for example, by the infliction of pain*), but also in their ecological and social relations with other animals and other nature.

Canadian naturalist and environmental philosopher John Livingston opposes the notion of animal rights, although he respects the animal rights movement.* His view of the animal's self contrasts with Tom Regan's notion of animal subjectivity.* According to Regan, animals are "subjects-of-a-life," by which he means that life matters to animals—that certain forms of life are better or worse for them as individuals. This is what entitles them to rights.

John Livingston does not dispute that things matter to animals. What he would argue is that the notion of "subject-of-a-life" is too limited, that it fails to take note of the animals' otherness. He argues that the wild animal— even the solitary wild animal—is unthinkable as just an individual. He proposes that animals have several integrated forms of self. Only one of these selves is an individual self. A flock of birds is not a sum of individuals, nor does a flock consist of individual birds reduced to being passive parts of a big group machine. Livingston believes that there exists a form of group awareness shared by the whole flock. In a way the individual *is* the group. The everyday consciousness of wild beings is participatory rather than self-centered.

Moreover, Livingston believes that animals also have a sense of biocommunity, an awareness of other species, as well as an awareness of the biosphere, the planet. Thus they are not just subjects-of-an-individual-life, but also subjects-of-a-group-life, subjects-of-a-community-life, and subjects-of-a-planet-life, and they have needs accordingly. To define them as humanlike individuals is to do them a disservice, to downplay their otherness. Instead of lifting animals up to our level, it actually reduces them to humanness. For Livingston, giving animals rights plays directly into the hands of the anthropocentric colonizers of the world. It dedicates the entire planet to the primacy of the (Western) individual and embodies the final conquest of nature. It cuts off numerous animals from their own kind and their natural surroundings.

Selected Bibliography. Benton, Ted, *Natural Relations: Ecology, Animal Rights, and Social Justice* (London and New York: Verso, 1993); Livingston, John A., *Rogue Pri-*

mate: An Exploration of Human Domestication (Toronto: Key Porter, 1994); Noske, Barbara, *Beyond Boundaries: Humans and Animals* (Montreal: Black Rose, 1997); Regan, Tom, *The Case for Animal Rights* (London: Routledge, 1983).

                                                                                    BARBARA NOSKE

ANTHROPOMORPHISM

Anthropomorphism, in its most general sense, refers to thinking in human terms about an object that is not human. That human qualities are applied to something nonhuman explains why the terms "anthropomorphism" and "anthropomorphic" are almost always used in a negative sense. In the field of animal studies, anthropomorphism is a term of criticism applied to those who assign human qualities to nonhuman animals. Animal lovers, animal liberationists, and even cognitive ethologists who assign mental states of a human variety to nonhuman animals are commonly accused of anthropomorphic thinking. In whatever field it has occurred, anthropomorphic thinking has been regarded as naïve, primitive, and mistaken by many philosophers and many scientists. It is also associated with immaturity and emotionality.

Clearly, humans are unique. But also, equally clearly, humans have many properties in common with other sorts of beings and objects (e.g., having a shape, having anatomical parts). If we assign to nonhumans those qualities that only humans have, we are mistaken. But if we assign to nonhumans human properties that those nonhumans also have, we hav not made a mistake.

The charge of anthropomorphism concerning animals most regularly comes up in connection with discussions of animal mentality, especially consciousness and thinking (*see* ANIMAL COGNITION). The complaint that this is anthropomorphic thinking, and a mistake, then rests on the claim that animals do not have mental characteristics. Many people assume that animals are like humans in having intentions, emotions, and reason. But according to antianthropomorphic thinkers, this is merely a naïve assumption or an emotional projection onto animals of our own characteristics, no more justified than the desire to think of physical processes as goal directed just because when we act in the world, we are goal directed.

Anthropomorphism concerning animals is not a simple mistake of thinking that a nonhuman is a human. To understand what mistake it is, it is useful to first note that a certain specific form of inference, which could be called "the anthropomorphic inference," is at the basis of anthropomorphism concerning animals. Ascribing mental predicates (or terms for mental states) to animals does not occur by accident. It is guided by the observer's perception of the situation. We assign mental predicates to an animal on the basis of the situation and behavior of the animal. We explain an animal's behavior by doing this, and we do so as we would explain our own behavior if we (humans) were in similar circumstances and/or behaved in similar ways.

We want to learn if there are appropriate ways to describe nonhuman animals. For example, can we claim that the octopus is "curious" or the dog* is "angry"? All such descriptions may be rejected as too anthropomorphic by those who wish to avoid anthropomorphic thinking. It is especially difficult to establish, in an empirical way, that such descriptions are justified. But if we reject these types of descriptions, we need to consider how far we are to go in using a purified or nonanthropomorphic language to describe animals. Just using words describing bodily movements leaves out much context—the situations in which animals are behaving. Even if some animals have mental states, it is important to remember that humans can misinterpret their behavior in many ways. In the end, just as it is difficult for anthropomorphic thinkers to show that these types of descriptions are justified, it is difficult for antianthropomorphic thinkers to suggest a good replacement for this sort of language.

Anthropomorphism, under the assumption that animals do have mental states, is still a concern. Supporting the decision to engage in anthropomorphism by further knowledge of the animals involved and the context in which they behave is called "critical anthropomorphism" (*see* ANTHROPOMORPHISM, Critical Anthropomorphism) by some people.

Selected Bibliography. Asquith, P., Why Anthropomorphism Is Not Metaphor: Crossing Concepts and Cultures in Animal Behavior Studies, in R. W., Mitchell, N. S., Thompson, and H. L. Miles (Eds.), *Anthropomorphism, Anecdotes, and Animals: The Emperor's New Clothes* (New York: State University of New York Press, 1997), 22–34; Bilgrami, A., Other Minds, in J. Dancy and E. Sosa (Eds.), *A Companion to Epistemology* (Oxford: Blackwell, 1993); Fisher, J. A., Disambiguating Anthropomorphism, in P. P. G. Bateson and P. H. Klopfer (Eds.), *Perspectives in Ethology* 9 (1991): 49–85; Kennedy, J. S., *The New Anthropomorphism* (Cambridge: Cambridge University Press 1992); Mitchell, R. W., N. S. Thompson, and H. L. Miles (Eds.), *Anthropomorphism, Anecdotes, and Animals: The Emperor's New Clothes* (New York: State University of New York Press, 1997); Morton, D. B., G. M. Burghardt, and J. A. Smith, Critical Anthropomorphism, Animal Suffering, and the Ecological Context, in S. Donnelly and K. Nolan (Eds.), Animals, Science, and Ethics, *Hastings Center Report* 20 (1990): 13–19.

JOHN ANDREW FISHER

Critical Anthropomorphism

Anthropomorphism can be useful in studying and interpreting animal behavior if it is applied critically. This means anchoring anthropomorphic statements and inferences in our knowledge of species's natural history, perceptual and learning capabilities, physiology, nervous system, and previous individual history. That is, if we ask what we would do in the animal's position, or how we would feel if we were treated like the animal, we must apply all the information we know about the animal as well as our own experience. For example, given what we know about dogs,* it would be safe to infer that a kicked dog writhing and squealing is feeling pain.* We would not be safe in concluding that the dog is feeling pain in exactly the same way we do, however. But given what little we know about earthworms, it would not be safe to conclude that an earthworm on a fishing hook is feeling pain in any way comparable to our pain when we are stuck. We could, though, conclude that the experience is an aversive or painful one to the worm since it avoids or tries to remove itself from such situations.

Critical anthropomorphism helps us to pose and formulate questions and hypotheses about animal behavior. Although we can never experience directly what another animal, including another human being, thinks or feels, we can make predictions as to what the animal or person would do using anthropomorphic methods. Insofar as we ground these predictions on real similarities across individuals, they may be very accurate and replicable. Enough research may even allow us to claim that the subjective mechanisms are comparable as well as the behavioral responses. Many of the greatest comparative psychologists and ethologists have acknowledged their use of anthropomorphic insights in formulating ideas and generating experiments in animal behavior. However, this is rarely stated in scientific reports, especially in this century.

Critical anthropomorphism seems to be a necessary practice. Certain behavior patterns such as watching for possible predators, greeting, aggression, fear,* indecision, and dominance can only be recognized once we know the normal behavioral repertoire. Thus courtship and fighting have been confused and mislabeled in species. Mating behavior, which involves neck biting in many mammals, may be anthropomorphically mislabeled aggression or fighting. Dominance wrestling in rattlesnakes was considered mating because observers did not know the sexes of the participants. The entwining of the snakes certainly appeared sexual anthropomorphically, and the snakes never bit or tried to injure each other as seriously fighting animals should try to do. Now we know that rattlesnakes are not immune to their own venom, and biting would quickly kill both antagonists. The wrestling allows the stronger male to obtain access to female snakes without either animal being killed.

Selected Bibliography. Burghardt, G. M., Animal Awareness: Current Perceptions and Historical Perspective, *American Psychologist* 40 (1985): 905–919; Hart, L. (Ed.), *Responsible Conduct of Research in Animal Behavior* (Oxford: Oxford University Press, 1998); Lockwood, R., Anthropomorphism Is Not a Four Letter Word, in M. W. Fox and L. D. Mickley (Eds.), *Advances in Animal Welfare Science* (Washington, DC: Humane Society of America, 1985), 185–199; Mitchell, R. W., N. S. Thompson, and H. L. Miles (Eds.), *Anthropomorphism, Anecdotes, and Animals: The*

Emperor's New Clothes (Albany: State University of New York Press, 1996); Ristau, C. (Ed.), *Cognitive Ethology: The Minds of Other Animals* (Hillsdale, NJ: Erlbaum, 1991).

GORDON M. BURGHARDT

ANTIVIVISECTIONISM

Antivivisectionism is the generally accepted label for opposition to the use of animals in scientific research. "Vivisection" literally means cutting into or cutting up live organisms. Historically, this is an accurate description of the way in which experiments upon unanesthetized animals were carried out. Antivivisectionism became a very strong movement during the 19th century in Victorian England. Relatively little of today's scientific research using animals is of this highly invasive sort. "Vivisection" has tended over time to take on a wider meaning and now refers to all experimental procedures that result in the injury or death of animals. "Antivivisection" has correspondingly evolved in meaning.

Antivivisectionists tend to be abolitionists (those who demand the total end of animal experimentation, whether accomplished immediately or gradually), but they may also have more limited and practical goals, such as the ending of certain kinds of experiments deemed morally unacceptable (e.g., cosmetics testing on rabbits' eyes by the Draize test, burn experiments on animals, or pain* experiments performed without anesthesia or analgesia). In contrast, *animal welfarists*, though they oppose cruelty, generally accept the use of animals in research but campaign for their more humane treatment and for reduction, refinement, and replacement (the Three Rs; *see* ALTERNATIVES TO ANIMAL EXPERIMENTS) in regard to overall animal usage.

Animal experimentation has been opposed by antivivisectionists on very many grounds: (*a*) inapplicability or limited applicability of data to humans owing to cross-species differences; (*b*) methodological unsoundness (being unscientific); (*c*) dangerously misleading and harmful results; (*d*) wastefulness, inefficiency, and expense; (*e*) triviality; (*f*) redundancy; (*g*) motivation by mere curiosity; (*h*) cruelty; (*i*) availability of alternatives; and (*j*) desensitization of researchers and their coworkers. Scientists who are animal users regularly argue that great advances in medicine and human (and animal) health would not have occurred without animal experimentation. However, antivivisectionists claim that most of the important breakthroughs (e.g., increased longevity, control of infectious diseases) would have occurred, or even did occur, without animal experimentation, and furthermore that animal experimentation has in many instances retarded progress. However, some antivivisectionists acknowledge that medical science has benefited from animal experimentation, but still argue that the future need not resemble the past.

In recent decades much more attention has been paid to the ethics of animal experimentation. Virtually every scientist using live animals in research today is subject to some form of ethical regulation and scrutiny, whether the system in place is mandatory or voluntary and based on peer review or nonspecialist review, and granting agencies and professional organizations and journals generally assign standards that must be adhered to for activities under their control. At the same time, many professional philosophers and others have focused on the issues surrounding animals' moral status (*see* MORAL STANDING OF ANIMALS), with important meaning for the ethics of animal research. Animal rights* and animal liberation* theories draw very strict limits to what is morally permissible by way of animal experimentation and not infrequently forbid it altogether. Several radical action groups, a few of which practice guerrilla tactics (e.g., secret raids on laboratories to free animals), have secured a prominent place in the public protest arena. All of these influences have generated considerable controversy, with constructive debate and change being the result.

Two philosophical issues in this larger debate are cost-benefit analysis and the central ethical dilemma. Generally, attempts to justify animal experimentation from an ethical standpoint weigh the costs to animals (in terms of harm, suffering, and death) against the benefits to humans of the research in question. But in the ethics of research using live human subjects, two conditions must be met: (*a*) subjects must give their voluntary, informed consent; and (*b*) costs and benefits must be calculated with reference to the individual subjects concerned. The point may be made that it is never ethically acceptable (because of justice considerations) to make some worse off in order by that same act to make others better off when no benefits make up for the losses to those worse off. The central ethical dilemma is that the more we learn from the biological and behavioral sciences, the more similarities we see between humans and other animal species, and hence the greater is our motivation for continuing to do animal research in order to understand ourselves better, but by the same token closer similarity creates a greater onus of human moral responsibility toward nonhumans. It is very difficult to argue, on the one hand, that animals are very like us, and on the other, to deny that they should be treated very much as we would wish to be treated.

However these issues are to be sorted out socially, certain things are clear. Knowledge is not an end in itself. If it were, horrible research in the name of science carried out on animals or humans could be morally justified. Therefore, the burden of moral responsibility and justification always lies with animal (as with human) experimenters.

Selected Bibliography. Fox, Michael Allen, Animal Experimentation: A Philosopher's Changing Views, *Between the Species* 3 (1987): 55–60; Francione, Gary L., and Anna Charlton, *Vivisection and Dissection in the Classroom: A Guide to Conscientious Objection* (Jenkintown, PA: American Anti-Vivisection Society, 1992); Orlans, F. Barbara, *In the Name of Science: Issues in Responsible Animal Experimentation* (New York: Oxford University Press, 1993); Ruesch, Hans (Ed.), *1000 Doctors (and Many More) against Vivisection* (Massagno, Switzerland: CIVIS, 1989); Sperling, Susan, *Animal Liberators: Research and Morality* (Berkeley: University of California Press, 1988); Turner, James, *Reckoning with the Beast: Animals, Pain, and Humanity in the Victorian Mind* (Baltimore: Johns Hopkins University Press, 1980).

MICHAEL ALLEN FOX

ARGUMENT FROM MARGINAL CASES. *See* MARGINAL CASES.

ASPCA. *See* AMERICAN SOCIETY FOR THE PREVENTION OF CRUELTY TO ANIMALS.

ASSOCIATION OF VETERINARIANS FOR ANIMAL RIGHTS (AVAR)

The Association of Veterinarians for Animal Rights (AVAR) was founded in 1981 by Nedim C. Buyukmihci and Neil C. Wolff. The term "rights," as opposed to "welfare," was chosen for the title of the organization because it exemplified the different philosophy of this approach. Although veterinarians* are already involved in animal welfare, this is clearly inadequate to protect the nonhuman animals' interests.

In veterinary medicine, the standard of caring for nonhuman animals is usually based on what is deemed "adequate veterinary care." Nonhuman animals are treated as the property of the "owners." Although there usually is a sincere attempt to relieve suffering* and improve the quality of life for these animals, there are no meaningful limits to what may be done with them. When one examines the issues without prejudice and with humility, there do not appear to be any morally relevant differences between human and other animals that justify denying other animals similar rights, consideration, or respect, based upon their interests or upon whether what we propose to do matters to the individual (*see* SPECIESISM).

Selected Bibliography. Buyukmihci, Nedim C., Consistency in Treatment and Moral Concern, *Journal of the American Veterinary Medical Association* 206(4) (1995): 477–480; Mason, Jim, and Peter Singer, *Animal Factories*, 2nd ed. (New York: Harmony Books, 1990); Pluhar, Evelyn B., When Is It Morally Acceptable to Kill Animals? *Journal of Agricultural Ethics* 1(3) (1988): 211–224; Regan, Tom, *The Case for Animal Rights* (Berkeley: University of California Press, 1983); Singer, Peter, *Animal Liberation* (New York: New York Review of Books, 1990).

NEDIM C. BUYUKMIHCI

ATTITUDES TOWARD ANIMALS

Pre-Christian Attitudes

Attitudes toward animals among past, preliterate societies can only be determined indirectly from the traces of cultural practices, art, and artifacts that have survived in the archaeological record. Direct comparisons with recent or current cultures are only appropriate where obvious similarities in animal-related attitudes, beliefs, and values seem to exist.

For example, artistic representations of wild mammals—mammoths, bison, wild horses, and cattle—are the most prominent feature of the famous Paleolithic cave and rock paintings of Europe, which range in age from 12,000 to 30,000 years B.P. (before present). Many theories have been put forward to explain the significance of all of this animal-oriented artwork, but probably the most plausible account comes from recent studies of the Bushman rock art of southern Africa. According to Bushman informants, these rock paintings are the work of shamans, and they picture the content of dreams or visions experienced during shamanic trance states. Animal figures predominate because animals are thought to be the living, material embodiment of these powerful spirit beings.

The idea that animals are fully conscious (see ANIMAL COGNITION) beings who possess spiritual power is widespread among hunting* and gathering societies. Not surprisingly, it also appears to engender considerable anxiety and guilt about killing animals for food. Most of these cultures engage in complex rituals and taboos designed either to relieve the guilt arising from hunting or to honor the spirits of deceased animals. Failure to treat animals with appropriate ritual respect is thought to invite spiritual anger in the form of accidents, ill health, or loss of success in future hunting. Most likely, prehistoric hunters shared similar beliefs about animals.

The advent of agriculture and animal husbandry roughly 12,000 years ago (see DOMESTICATION) produced a dramatic shift in the balance of power between humans and the animals they depended on for food. From being essentially independent coequals or superiors, animals became slaves or subordinates, entirely dependent on humans for care and protection. This shift in power relations was reflected in religious belief systems that became increasingly hierarchical throughout the ancient world. The original shamanic animal spirits were progressively elevated to the status of zoomorphic (animal-looking) gods with increasingly awesome powers. Wholesale animal sacrifice was widely practiced during this period, supposedly as a means of currying favor with these deities and promoting success in agriculture and other endeavors. In reality, only the blood or small portions of the carcass were reserved for the gods. The rest of the meat was usually sold or redistributed to the populace.

In many, if not all, of these ancient civilizations, it appears that the consumption of unsacrificed meat was largely taboo, so the priesthood tended to exercise relatively exclusive control over meat production, slaughter, and distribution. At least some of this division of labor seems to have reflected continuing moral concerns about the practice of killing animals for food. Surviving accounts of sacrificial rituals, for example, indicate that ideally the sacrificial animal was supposed to approach the altar willingly without coercion, and that it was often encouraged to nod its head as if assenting to its own slaughter. Following the sacrifice, the priests who performed the act sometimes whispered apologies in the animal's ear, and it was not uncommon for the sacrificial knife to be "punished" by being destroyed. More direct evidence of ambivalence regarding the ethics of animal sacrifice can also be discerned in early literature. For example, the oldest sections of the Rig Veda, the most ancient religious text from India, are primarily descriptions of how, when, and where to perform animal sacrifices. Later sections thought to date from about 2800 B.P. categorically reject sacrifice and advocate the practice of *ahimsa* (noninjury) toward all living things, an idea that subsequently became integral to the philosophies of three major contemporary Indian religions: Buddhism, Jainism, and the yogic branches of Hinduism (*see* RELIGION AND ANIMALS).

Evidence of similar concerns is also apparent in classical Greek literature from about 2500 B.P. The early Pythagorean and Orphic schools of Greek philosophy believed in the Eastern concept of reincarnation—the idea that the soul or spirit is eternally reborn after death in different bodies, including those of animals. According to some accounts, Pythagoras and his followers were not only opposed to animal sacrifice for this reason, but also advocated a vegetarian diet. Opposition to animal sacrifice and vegetarian advocacy continued to recur as themes in classical philosophical literature until the third century A.D. However, their influence was counterbalanced and eventually overwhelmed by Aristotle's (384–322 B.C.E.) hierarchical and purposeful view of nature as an ascending scale of living beings, each created to serve as food or labor for those higher up the scale. According to this view, rational humans had a natural or God-given right to use less rational and therefore "lower" organisms for food or other purposes. Aristotle's unusually human-centered worldview has continued to dominate Judeo-Christian and Islamic thought and philosophy ever since (*see* RELIGION AND ANIMALS).

Selected Bibliography. Jacobsen, K. A., The Institutionalization of the Ethics of "Non-injury" toward All "Beings" in Ancient India, *Environmental Ethics* 16 (1994): 287–301; Lewis-Williams, D., and T. Dowson, *Images of Power: Understanding Bushman Rock Art* (Johannesburg: Southern Book Publishers, 1989); Manning, A., and J. A. Serpell (Eds.), *Animals and Human Society: Changing Perspectives* (London: Routledge, 1994); Serpell, J. A., *In the Company of Animals*, 2nd ed. (Cambridge: Cambridge University Press, 1996); Sorabji, R., *Animal Minds and Human Morals: The Origins of the Western Debate* (Ithaca, NY: Cornell University Press, 1993).

JAMES A. SERPELL

Changing Attitudes throughout History

Human attitudes toward animals are tied to questions of human identity. What we think about animals depends upon how we define ourselves. This is as true today as it was in the early Christian centuries. When early Christian church fathers explored the issue of people's relationship with animals, they departed from the classical position and claimed that humans are very different from animals because humans have souls and animals do not. The characteristic that church fathers determined most defined humanity in contrast with animals was what they called "reason." This meant intelligence and the ability for abstract, logical thought. They believed that reason was the property of the soul, and that reason more than anything else separated humans from animals.

In discussing people's relationship to animals, medieval thinkers quoted the biblical verse that gave Adam and Eve "dominion over the fish of the sea and over the birds of the air and over every living thing that moves upon the earth" (Genesis 1:28). However, they believed that people had dominion not simply because the Bible said so, but because people possessed intelligence* and reason that set them apart from animals in profound and definite ways.

This attitude can be seen in the most popular artistic portrayal of animals in the early Middle Ages. The image most often used was an illustration of the biblical moment in which Adam named the animals. The illustration included here is from a medieval manuscript and shows this biblical scene that medieval people believed defined the relationship between humans and the animal world. Medieval thinkers believed that Adam had the right to "name" the animals because his intellect allowed him to understand the nature and purpose of each animal and give the animal a name that was appropriate to its purpose. Medieval thinkers took a functional approach to animals, categorizing them by how they were useful to humans.

In the Middle Ages, people believed that the main functions of animals were as follows: They were to work, to be food, and to supply skins (or wool) for use by humans. In addition, some animals (like horses or hawks) could provide status for their owners. This functional approach to animals pervaded much of medieval people's thinking and actions. For example, Thomas Aquinas said that there would be no animals in heaven because people would not need to work, eat, or wear clothes. Furthermore, here on earth there was no need to preserve animals that were seen as "useless." Wolves fell in this category since they did not work for humans, were useless

Changing Attitudes toward Animals throughout History: "Adam Naming the Animals." *Source:* By permission of the British Library, Royal 12 F XIII f34v.

as food, and had no value independent of their value to humans. Therefore, people believed that it was a good thing to hunt them to extinction.

The idea that humans were qualitatively different from animals did not last. By the late Middle Ages (the 12th century and later) the literature began to show signs of the beginnings of a blurring of the lines between humans and animals. By the late Middle Ages and the Renaissance, people began to see an animal side of human nature and more similarities between humans and animals than had previously been allowed.

There is not a smooth development in the history of attitudes toward animals from distance to closeness. People in each period of time defined animals in large part according to how they wanted to see themselves. In the 18th century (the Age of Reason) people prided themselves again on having reason and intellect that set them apart from animals. During this time many people believed that animals could feel no pain* since they did not have human intellect. Therefore, there was no such thing as cruelty* to animals.

In the 19th century, however, things changed again. People began to define humans as creatures of feeling and passions, rather than just intellect. This they shared with animals, and thus animals might be treated with care for feelings. When people began to see themselves in their animals, they increasingly began to have a different relationship with them. Animals became the source of and outlet for affection as people emphasized their relationship with their pets.

The line between humans and animals was finally eliminated in the 19th century with the work of Charles Darwin.* From then on, creation has been seen as a large continuum that joins humans with all the animals from the simplest protozoa to the complex great ape. The chasm of difference that was described in the early Christian centuries has gone. Now, for example, some theologians argue that there is a place for animals in an afterlife that was once promised only to humans.

Selected Bibliography. Ham, Jennifer, and M. Senior, *Animal Acts: Configuring the Human in Western History* (New York: Routledge, 1997); Linzey, Andrew, *Christianity and the Rights of Animals* (New York: Crossroad, 1987); Ritvo, Harriet, *The Animal Estate: The English and Other Creatures in the Victorian Age* (Cambridge, MA: Harvard University Press, 1987); Salisbury, Joyce E., *The Beast Within: Animals in the Middle Ages* (New York: Routledge, 1994); Thomas, Keith, *Man and the Natural World: A History of the Modern Sensibility* (New York: Pantheon Books, 1983).

JOYCE E. SALISBURY

Attitudes among Children

Young children certainly show a great deal of interest in animals. But just because animals are interesting to youngsters does not mean that they are automatically loved or respected. Children below 4 or 5 years of age are quite capable of displaying overtly callous and cruel behavior, with little apparent concern for the suffering they might be inflicting. Such behavior is generally discouraged in modern Western society.

As children's natural empathic (*see* EMPATHY) and nurturant tendencies emerge in later childhood, more caring, emotional attitudes toward many kinds of animals appear to develop. This more positive, caring view of animals, however, does not apply similarly to all animals. Adult society's feelings and beliefs about which animals are "nice" and which are "nasty" appear to be readily transmitted to children. For example, in a recent British survey it was found that wolves and rats were two of the species most likely to be disliked by children, while two very similar animals, dogs* and rabbits, were voted as being two of the most liked species. Personal experience of the animals concerned cannot account for these differences: although most of the children had encountered dogs and rabbits, few, if any, had come face-to-face with a wolf (now extinct in Britain) or even a rat.

Pet keeping is particularly common among middle-childhood children (around 8 to 12 years) (*see* COMPANION ANIMALS AND PETS). This is probably the age at which children's emotional interest in animals is at its highest and when, especially for girls, big-eyed, cuddly, furry animals are particularly attractive. After this age, in the teenage years, interest in moral issues surrounding animals and their use by humans becomes more prominent. This is the time when young people are most likely to take "stands" on animal issues (and, indeed, other issues such as political ones) by, for example, adopting vegetarian or vegan diets (*see* VEGETARIANISM) or becoming involved in environmental or animal rights* campaigns. As interests outside the home take prominence in teenagers' lives, interest and involvement in pet keeping often wane a little. But childhood experience of pets nevertheless appears to retain an influence. In an recent study it was found that university students who had grown up with pets, especially cats* and dogs, were more likely to have greater concerns about the welfare of animals than those who had not grown up with such pets. They were also more likely to show "ethical food avoidance" practices such as vegetarianism, and they were more likely to belong to environmental and animal welfare* charities or organizations.

Selected Bibliography. Kellert, S., Attitudes toward Animals: Age-related Development among Children, *Journal of Environmental Education* 16(3) (1985): 29–39; Morris, P. H., V. Reddy, and R. C. Bunting, The Survival of the Cutest: Who's Responsible for the Evolution of the Teddy Bear? *Animal Behaviour* 50 (1995): 1697–1700; Paterson, D., Assessing Children's Attitudes towards Animals, in D. Paterson and M. Palmer (Eds.), *The Status of Animals: Ethics, Education, and Welfare* (Wallingford, UK: CAB International, 1989), 58–63; Paul, E. S., and J. A. Serpell, Childhood Pet Keeping and Humane Attitudes in Young Adulthood, *Animal Welfare* 2 (1993): 321–337; Paul, E. S., and J. Serpell, Why Children Keep Pets: The Influence of Child and Family Characteristics, *Anthrozoös* 5(4) (1992): 231–244.

ELIZABETH PAUL

Attitudes among Students

Since the publication of Peter Singer's *Animal Liberation* in 1975, print and electronic news media, movies and television sit-coms, and textbooks and popular books increasingly have concerned themselves with issues relating to the treatment of animals other than humans. As a result, students have been exposed to and have formed opinions about issues ranging from hunting* and trapping* to the use of animals in research, product testing, and the classroom. The diversity of their views is indicated by a study that distinguished 10 different attitudes toward animals found in the American public, for example, ecologistic, humanistic, moralistic, dominionistic (*see* DOMINIONISM), aesthetic, utilitarian (*see* UTILITARIANISM), and negativistic. While there is a considerable diversity of attitudes, individuals hold *hard* attitudes. This means that at an early age individuals form strong views toward animals and that these particular views are enduring.

Numerous studies have established that the gender of a person is the most powerful predictor of his or her general attitude toward animals. For example, one investigator found that in 10 to 15 countries studied, with a trend in the same direction in the remaining 5 countries, women significantly more than men opposed animal research. The reasons for this "gender gap" are not fully understood but involve differences in parental views of girls and boys, such as the importance given in the socialization of girls to developing caring and nurturing relationships.

Age is also an important variable, younger people being more concerned with animal welfare.* Although there is a clear link to age, the relation of attitudes toward animals and amount of education, specifically science education, is unclear. One study found no significant relation between degree of scientific knowledge and attitude, while a second found that more scientifically knowledgeable young adults were less likely to oppose animal research.

Attitudes toward animals are also related to political positions. Liberalism as compared to conservatism is associated with more proanimal views. As compared to a group of college students, animal rights* activists attending a large national protest are more likely to believe that moral behavior will really produce positive results. Further, those who take up the cause of animals are also more likely to be concerned about discrimination against certain classes of people. Support for animal rights is associated with more tolerance of human diversity, specifically, acceptance of rights for women, homosexuals, and ethnic minorities. Concerns for the welfare of human and nonhuman animals are typically held by the same individual. One final variable is personality type. People who rely more on intuition and feeling and are more focused on relationships are more likely than thinking types to oppose animal research.

In terms of actual positions on the issues, there is, as indicated, a diversity of views. Taking attitudes toward animal research as an example, evidence as to the general level of opposition to the use of animals in research is mixed. Although a number of studies found that on average, individuals support a middle position, an extensive study of individuals in 15 countries, discussed earlier, found a high level of opposition.

Finally, in terms of the impact of these attitudes, there is some evidence of a decline in the use of animals in biomedical and psychological research. One group of investigators suggests that "decline in work with animals stems largely from changing student attitudes" and that these attitudes "are in tune with current widely shared concerns for the natural environment and animal welfare."

Selected Bibliography. Driscoll, J., Attitudes toward Animal Use, *Anthrozoös* 5 (1992): 32–39; Galvin, S., and H. Herzog, Ethical Ideology, Animal Rights Activism, and Attitudes toward the Treatment of Animals, *Ethics and Behavior* 2 (1992): 141–149; Kellert, S., Perceptions of Animals in America, in R. Hoage (Ed.), *Perceptions of Animals in American Culture* (Washington, DC: Smithsonian Institution, 1989), 5–24; Pifer, L., K. Shimizu, and R. Pifer, Public Attitudes toward Animal Research: Some International Comparisons, *Society and Animals* 2 (2) (1994): 95–113.

KENNETH J. SHAPIRO

AUTONOMY OF ANIMALS

The original meaning of *autonomy* as applied to ancient Greek city-states is *self-rule*. More recently, the term has been applied to individuals, actions, and desires. To answer the question "Are any animals autonomous beings who are capable of performing autonomous actions?" requires not only carefully studying animals, but also determining what sorts of actions qualify as autonomous.

Autonomous actions must at least be intentional actions. Every intentional action involves a desire and a belief that help to explain why the action was performed. Tom Regan argues that beings capable of intentional action are capable of one kind of autonomy—what he calls "preference autonomy" (*preference* being another word for *desire*). On this analysis, assuming that a dog can (1) desire a bone and (2) believe, as she trots into the backyard, that she can find a bone there, then the dog is capable of acting autonomously.

But one can be capable of acting autonomously but fail to do so for any of several reasons. For example, physical constraints such as locked doors can prevent a dog from going into the backyard. Force can prevent intentional actions from being autonomous. If you intentionally give money to someone, but only because he threatened you with a gun, your action is coerced, not free or autonomous. Moreover, sometimes we act intentionally, and even freely, but without sufficient understanding of what we are doing for our action to be autonomous. If a hospital patient intentionally and freely signs a form that states agreement to participate in psychiatric research, but the patient believes that the form simply entitles her to therapy following hospitalization, the patient has not autonomously agreed to participate in research.

Autonomous action clearly involves more than simply intentional action. One analysis, favored by Tom Beauchamp, is that actions are autonomous if they are performed (1) intentionally, (2) with understanding, and (3) without controlling influences (e.g., force) that determine the action. But certain other writers, such as Gerald Dworkin and David DeGrazia, would argue that these conditions are not sufficient for autonomous action. Apparently, on the present analysis, a bird feeding her young would, under normal circumstances, count as acting autonomously (assuming that birds can act intentionally).

Because autonomous beings are beings capable of acting autonomously, one's answer to the question "Are any animals autonomous beings?" will depend, in part, upon one's view of autonomous action. Those with relatively undemanding requirements are likely to conclude that many animals are autonomous. The view that anyone capable of intentional action is autonomous implies that all animals capable of having the appropriate sorts of desires and beliefs qualify. Which animals have such desires and beliefs is an extremely complex question, involving difficult conceptual issues in the philosophy of mind and various kinds of scientific evidence regarding animals. Tom Regan somewhat cautiously argues that normal mammals beyond the age of one year are capable of intentional action. David DeGrazia contends that most or all vertebrates and perhaps some invertebrates can act intentionally.

On a multitier account, animals are autonomous beings only if they can critically evaluate the preferences that move them to act and sometimes modify them on the basis of higher-order preferences and values. This is a high standard, requiring considerable capacity for abstraction and an advanced form of self-awareness (see ANIMAL COGNITION). Perhaps such abstraction and self-awareness require language. There is a strong case that some apes have achieved language comprehension and production and that some dolphins have achieved language comprehension. The most suggestive evidence from the language studies of the possibility of animal autonomy may be evidence that apes apologized for such actions as biting a trainer and going to the bathroom indoors (see ANIMAL COMMUNICATION). Typically, apologies express regret for one's actions, but one might also regret the motivations that moved one to act. At present it seems unclear, from the multitier view, (1) whether autonomy might be possible for the languageless and (2) whether any animals are, in fact, autonomous beings.

Selected Bibliography. Beauchamp, Tom L., The Moral Standing of Animals in Medical Research, *Law, Medicine, and Health Care* 20(1–2) (1992): 7–16; Christman, John (Ed.), *The Inner Citadel: Essays on Individual Autonomy* (New York: Oxford University Press, 1989); DeGrazia, David D., *Taking Animals Seriously: Mental Life and Moral Status* (Cambridge: Cambridge University Press, 1996); Dworkin, Gerald, *The Theory and Practice of Autonomy* (Cambridge: Cambridge University Press, 1988); Regan, Tom, *The Case for Animal Rights* (Berkeley: University of California Press, 1983).

DAVID D. DEGRAZIA

AVAR. *See* ASSOCIATION OF VETERINARIANS FOR ANIMAL RIGHTS.

B

BAITING. *See* BEAR BAITING.

BEAK TRIMMING. *See* CHICKENS.

BEAR BAITING

Baiting is the controversial practice of using food and decaying animal carcasses to attract nonhuman animals so that sport hunters may shoot them at close range. Baiting is a highly effective hunting* method and is commonly practiced on bears. Bear baiting is currently permitted on millions of acres of national forests in 10 states: Alaska, Idaho, Maine, Michigan, Minnesota, New Hampshire, Utah, Washington, Wisconsin, and Wyoming. Eight of these states also permit the use of hounds of hunt bears. During a pursuit season, hunters are allowed to use bait to attract a bear that is then chased and treed by trained dogs.

Bait stations are generally composed of two parts, a tree stand and one or more piles of bait. The tree stand provides a place for the hunter to hide and wait. Baiting sites are designed to give off a strong odor so that they act as an attractant to bears. Typically, the bait consists of meat scraps, dough-nuts and other sweet foods, and rotting fruits and vegetables. In some in-stances, hunters use "walk-in" baits such as horses, sheep, or cows that are walked into the forest and then are shot and left tied to a tree.

Bear baiting is known to be an extremely effective method of luring and shooting black bears. Baiting usually is done in the spring and fall, before and after hibernation, when bears must consume large amounts of food as a way to increase body weight. Hunters who use baits claim that the baiting method allows them to distinguish species and sex of bears and helps to avoid shooting female bears with cubs. However, a study done by the Colorado Division of Wildlife found that a number of female bears killed over baits were lactating, indicating that the bears had recently given birth to cubs.

Black bears currently live in 32 states. Very little information exists on the health of black bear populations in states where bear baiting is allowed. Demand for the species as a game animal is high and increasing. For all of North America, the annual number of bears killed increased from 25,000 in 1972 to 41,000 in 1989. Illegal poaching of bears is also considered to be a problem in the United States. Colorado wildlife managers estimate that poaching and "predator-control" efforts result in 400 to 600 unreported bear kills each year in that state. The Asian belief in the healing powers of the bear has led to international trade in bear body parts in which one gall bladder can sell for upwards of $3,500. Dealers also pay $250 each for bear paws, which are considered a Chinese food delicacy.

Bear baiting is known to affect adversely a number of endangered species* that feed on carrion. These include grizzly bears, northern bald eagles, and gray wolves; a gray wolf was killed over a black bear bait in 1994 in Maine, and in 1982, four grizzly bears were killed over baits in the Shoshone National Forest in Wyoming. It has been observed that even if a grizzly that comes to black bear bait is not killed directly, it can become conditioned to people smells and activities, resulting in habituation. Habituation disrupts normal wild animal behavior and may lead to animal/people conflicts, the majority of which are resolved to the detriment of the wild animal.

There is increasing public scrutiny of the ethics of sport hunting. Public opinion polls consistently find the public opposed to baiting. Bear baiting does not fit with notions of fair chase (*see* HUNTING, Fair Chase) that are supported by the majority of the public. This includes a substantial number of hunters themselves. A study of Colorado hunters showed that 90% objected to bait hunting because it gives the hunter an unfair advantage. Bear baiting continues, however, largely because of the strength of the professional hunter and outfitter-guide associations.

Selected Bibliography. Colorado Division of Wildlife, Black Bear Management Plan, 1990; Glitzenstein, Eric, and John Fritschie, The Forest Service's Bait and Switch: A Case Study on Bear Baiting and the Service's Struggle to Adopt a Reasoned Policy on a Controversial Hunting Practice within the National Forests, *Animal Law* 1(1) (1995): 45–77; Jonkel, Charles, The Colorado Black Bear Amendment, Ursid Research Center, White Paper, 1992; Medlock, Aaron, Use of Bait in Hunting on National Forest Lands: A Report Opposing the Proposed Policy of the Forest Service and Supporting a Ban on Bear Baiting, unpublished paper on file with Animal Law,

Lewis and Clark College of Law, 1994; University of Wyoming Survey Research Center, Public Attitude Survey on Black Bear Management in Wyoming, 2, 1992.

LEILA STANFIELD

BEASTLINESS

The term "beast" has generally had negative connotations. By strict dictionary definition, beasts are simply land animals, as distinct from humans, reptiles,* fish,* birds, and insects. "Beast" is a neutral term in such expressions as "not fit for man or beast" and "bless the beasts and children." "Beast" acquired negative connotations from its association with the Antichrist, most familiarly invoked in the "mark of the beast," 666 (Revelation 13:18). Human beings who were violent or lewd were said to be behaving like beasts.

As Michel Foucault observes in *Madness and Civilization*, in the 18th century the insane were likened to beasts, chained and caged and put on display. Eighteenth-century political theory tended to picture people as beasts needing a firm ruling hand. Perhaps taking his cue from Niccolò Machiavelli (see Clarke and Linzey, 12–14), Friedrich Nietzsche asserted the idea of the "blond beast," a masterful figure who scorned notions of decency to ruthlessly conquer lesser beings (*The Genealogy of Morals*, 1887). In these instances, the beast is associated with lawlessness, for good or ill.

Previously signifying humankind's "descent" into animal ways, "bestiality"* now refers primarily to humans having sexual relations with nonhuman animals. "Beast" has undergone another transformation in the 20th century that has rendered the term passé as a descriptor for violent people. These days, violent people are said to be behaving like animals, partly because "beast" has acquired lightly comic and largely British overtones, as in "Oh, Freddie, you *are* a beast" or "this beastly weather."

Selected Bibliography. Adams, Carol J., *Neither Man nor Beast: Feminism and the Defense of Animals* (New York: Continuum, 1994); Clarke, Paul A. B., and Andrew Linzey (Eds.), *Political Theory and Animal Rights* (London: Pluto Press, 1990); Dekkers, Midas, *Dearest Pet: On Bestiality*, trans. Paul Vincent (London: Verso, 1994); Foucault, Michel, *Madness and Civilization: A History of Insanity in the Age of Reason*, trans. Richard Howard (New York: New American Library, 1965); Krafft-Ebing, Richard von, *Psychopathia Sexualis* (New York: Scarborough Books, 1978).

MARIAN SCHOLTMEIJER

BEHAVIORISM

Behaviorism is the theory and practice of psychological research that considers behavioral responses to external stimuli as the only justifiable area of psychological study. This view ignores the mind and mental states, especially consciousness (*see* ANIMAL COGNITION, Conscious Experience) because they are considered to be inaccessible to scientific study or to be only by-products of brain function. The organism itself is treated as a "black box."

Behaviorism dominated American psychology for some 50 years (from the 1920s through the 1970s) and spread to many other countries. Behaviorism sent a strong message to the scientific community that considering any mental states of animals, such as consciousness and feelings, is unscientific and therefore inappropriate. This message was accepted by a number of bio-medical and other researchers practicing animal research because it allowed them to take the view that animals were not conscious or did not experience pain* or suffering.* The influence of behaviorism has decreased because of growing interest in human and nonhuman animal cognition (thinking and consciousness), which led to the development of the field of cognitive ethology, the study of animal minds.

Selected Bibliography. Boakes, R., *From Darwin to Behaviorism: Psychology and the Minds of Animals* (Cambridge: Cambridge University Press, 1984); Griffin, D., *Animal Minds* (Chicago: University of Chicago Press, 1992); Rollin, B., *The Unheeded Cry: Animal Consciousness, Animal Pain, and Science* (New York: Oxford University Press, 1989).

ANDRZEJ ELZANOWSKI

BERGH, HENRY

Born to an aristocratic shipbuilding family in New York, Henry Bergh (1813–1888) helped to change the way Americans thought about animals. As a youngster he fought to stop boys who brought cats* and dogs* to the river in sacks to drown them. During his service in Russia, Bergh witnessed a peasant beating his cart horse, and he jumped from his own carriage and stopped him. Bergh underwent a transformation that would change the rest of his life and change life for animals in the United States. Upon leaving his post in Russia, Bergh stopped in London to visit the earl of Harrowby, the president of the Royal Society for the Prevention of Cruelty to Animals (RSPCA),* to learn the organization and functions of that society. When he returned to New York in June 1865, he began the business of organizing a similar society in America. Bergh garnered the moral and financial support of many of his influential friends, and on April 10, 1866, he secured a charter from the state of New York for the formation of the American Society for the Prevention of Cruelty to Animals (ASPCA),* the first animal-protection organization in the United States. April 19 saw the passage of a new law that prohibited cruelty to animals, and the ASPCA was given the power to enforce that law.

One of the first cases that Bergh brought to court was that of a ship captain and his crew. They had transported sea turtles by punching holes

Henry Bergh (1813–1888). Photo courtesy of the American Society for the Prevention of Cruelty to Animals.

through their fins and tying them on their backs on the deck of the ship. The judge threw the case out of court on the grounds that turtles were not animals and therefore were not covered under the new law. As Bergh's activities to protect animals increased, so did the opposition against his work. The butchers who shipped animals to market with their legs tied and stacked on top of one another, the "sportsmen" who enjoyed watching and betting on dogfights, the transport companies that overworked the horses that pulled the city's trolleys, and even P. T. Barnum all came under the scrutiny of the ASPCA's president. His activities were such that they soon earned him the nickname "the Great Meddler." Bergh persevered against these opponents and even came to earn the grudging respect of some.

Henry Bergh's reputation for standing against cruelty was so great that in 1874, Etta Wheeler, a social worker, brought a most extraordinary case to his attention. In a celebrated trial Bergh was able to win the release of Mary Ellen* McCormack from her abusive foster parents. He soon after founded the Society for the Prevention of Cruelty to Children with his attorney, Elbridge Gerry, in December 1874.

By 1873, 25 states and territories had used the ASPCA as a model for the start of similar societies. During a great snowstorm on March 12, 1888, Bergh died early in the morning. Bergh's legacy includes the many hundreds of societies in America that work to protect animals and children.

Selected Bibliography. Franz, William C., Bergh's War: The First Crusade for Animal Rights, *Elks Magazine*, October 1980; Loeper, John J., *Crusade for Kindness: Henry Bergh and the ASPCA* (New York: Atheneum, 1991); Pace, Mildred Mastin, *Friend of Animals: The Story of Henry Bergh* (Ashland, KY: Jesse Stuart Foundation, 1995; original publication, New York: Charles Scribner's Sons, 1942); Steele, Zulma, *Angel in a Top Hat* (New York: Harper and Brothers, 1942); Turner, James, *Reckoning with the Beast* (Baltimore: Johns Hopkins University Press, 1980).

STEPHEN L. ZAWISTOWSKI

BERNARD, CLAUDE

Claude Bernard (1813–1878) is called the father of modern biomedicine. More than a century after his death, his ideas still influence the theory and practice of biomedicine. Bernard claimed that genuine biomedical sciences must be conducted in the laboratory and not in hospitals. That is, he viewed the biomedical sciences as sciences on the same footing as chemistry or physics. He also thought that laboratory experiments on animals were directly relevant to human biomedicine. In principle, no other method (save immoral and illegal human experimentation) could yield the same results.

Bernard's beliefs about the limitations of clinical medicine and the importance of laboratory investigation were framed by larger 19th-century methodological debates. Bernard was an early supporter of hypothesis testing, and nonhuman animals were the "matter" of physiological investigation. Bernard also believed that physiology should aim for laws as rigorous as those found in physics. He thought that the fundamental properties of "vital units" were the same for all species. Livers may come in different sizes and shapes, but they all respond to stimuli in basically the same way. Put differently, species differences are quantitative, differences in degree, not qualitative, differences in kind. Once we make suitable appropriate adjustments for quantitative differences (e.g., body weight or dose), we can apply experimental findings from one species to another: we may presume same effect from same cause, even when the test subjects belong to different species.

Bernard's assumption that species differences were ultimately explained using universal laws is tied, in important ways, to his rejection of the theory of evolution (*see* DARWIN, CHARLES). He rejected evolution because it did not, in his day, have consequences that could be tested in controlled laboratory experiments. However, the biological sciences are now held together and unified by evolutionary theory, which is supported by both laboratory and field research. It is now recognized that Bernard's conception of species differences is too simplistic for scientific purposes.

Selected Bibliography. American Medical Association (AMA), *The Use of Animals in Biomedical Research: The Challenge and Response*, rev. ed. (Chicago: American Medical Association, 1992); Bernard, C., *An Introduction to the Study of Experimental Medicine* (1865; Paris: Henry Schuman, 1949); Burggren, W. W., and W. E. Bemis, Studying Physiological Evolution: Paradigms and Pitfalls, in M. H. Nitecki (Ed.), *Evolutionary Innovations* (Chicago: University of Chicago Press, 1990), 198–228; Elliot, P., Vivisection and the Emergence of Experimental Medicine in Nineteenth Century France, in N. Rupke (Ed.), *Vivisection in Historical Perspective* (New York: Croom Helm, 1987), 48–77; LaFollette, H., and N. Shanks, *Brute Science: Dilemmas of Animal Experimentation* (London: Routledge, 1996); Schiller, J., Claude Bernard and Vivisection, *Journal of the History of Medicine and Allied Sciences* 22 (1967): 246–260.

HUGH LaFOLLETTE AND NIALL SHANKS

BESTIALITY

Though the term "bestiality" originally referred to a broad notion of earthy and often distasteful otherness, its meaning is nowadays confined to sexual relations between humans and nonhuman animals. Bestiality is also described as "zoöphilia," "zoöerasty," "sodomy," and "buggery." It can occur in a wide variety of social contexts, including adolescent sexual exploration, typically by young males in rural areas; eroticism, a rare event where animals are the preferred sexual partner of humans; cruelty, especially by young males or in cases of partner abuse; and commercial exploitation, as in pornographic films or in live shows of women copulating with animals in bars or sex clubs. In some societies, such as in New England from the Puritan 1600s until the mid-19th century, bestiality was regarded with such alarm that even the very mention of it was condemned. It is thus also referred to as "that unmentionable vice" or "a sin too fearful to be named" or "among Christians a crime not to be named."

The earliest and most influential censures of bestiality are the Mosaic commandments contained in Deuteronomy, Exodus, and Leviticus. Deuteronomy, for example, declared, "Cursed be he that lieth with any manner of beast" (27:21), while Exodus ruled that "[w]hosoever lieth with a beast shall surely be put to death" (22:19). Besides mandating death for humans, Leviticus dictated that the offending animal must also be put to death. Though it is difficult to know the precise intentions of those who originally condemned bestiality, historically there have been three chief beliefs about why it is so wrongful a behavior: (1) that it is a rupture of the natural, God-given order of the universe; (2) that it violates the "procreative intent" required of all sexual relations between Christians; and (3) that it produces monstrous offspring that are the work of the Devil.

Sociological information about the occurrence of bestiality is quite unreliable, especially given its private nature and the social stigma still attached to it. Very little solid information about its prevalence exists. Tentatively, it appears that bestiality is practiced mostly by young males in rural areas and that its prevalence depends on such factors as the level of official and popular tolerance, opportunity, proximity to animals, and the availability of alternative sexual outlets. Some sexologists have claimed, with the use of interviews and questionnaires, that 8% of the male population has some sexual experience with animals but that a minimum of 40% to 50% of all young rural males experience some form of sexual contact with animals, as do 5.1% of American females. But because of the poor sampling techniques of such studies, these figures are likely to be overestimates.

Sexual relations involving humans and animals have always been condemned and investigated—or, in the interests of "tolerance," ignored—exclusively from an anthropocentric (see ANTHROPOCENTRISM) perspective. Even the modern animal rights movement* has been silent on the issue of bestiality. But sexual relations with humans often cause animals to suffer great pain and even death, especially in the case of smaller creatures like rabbits and hens. Moreover, as it is impossible for us to know whether animals can ever assent to sexual relations with humans, it is probably best to treat all such cases as forced sex. Sexual relations involving humans and animals are thus more appropriately termed "interspecies sexual assault."

Selected Bibliography. Adams, Carol J., Bestiality: The Unmentioned Abuse, *Animals' Agenda* 15(6) (1995): 29–31; Dekkers, Midas, *Dearest Pet: On Bestiality*, trans. Paul Vincent (London: Verso, 1994); Kinsey, Alfred C., Wardell B. Pomeroy, Clyde E. Martin, and Paul H. Gebhard, *Sexual Behavior in the Human Female* (Philadelphia: W. B. Saunders, 1953); Krafft-Ebing, Richard von, *Psychopathia Sexualis*, trans. Franklin S. Klaf (New York: Stein and Day, 1886); Liliequist, Jonas, Peasants against Nature: Crossing the Boundaries between Man and Animal in Seventeenth- and Eighteenth-Century Sweden, *Journal of the History of Sexuality* 1(3) (1991): 393–423.

PIERS BEIRNE

History of Attitudes

Bestiality refers first to people acting like animals, in a bestial way. However, its second meaning, sexual contact between humans and nonhuman animals, is the most frequent current use of the word. Attitudes about bestiality have changed over time, and these attitudes are revealing of people's general perception of animals.

The early Christian medieval world inherited both texts and traditions that described human/animal intercourse. In the classical Greco-Roman texts, gods in the form of animals had intercourse with humans, and tales drawn from folklore also preserved anecdotes of such sexual contact. Pagan Germanic tradition also preserved tales of bestiality, whether between human and animal or between humans, one of whom took the shape of an animal.

The Christian tradition did not accept bestial intercourse, but there was a change over time in the perception of the severity of the sin. During the earliest prohibitions, bestiality was regarded as no more serious than masturbation. By the 13th century, however, Thomas Aquinas ranked bestiality as the worst of the sexual sins, and the law codes recommended harsh penalties for the practice.

There seem to be two primary reasons for this change. The first is that by the late Middle Ages churchmen became more concerned with the presence of demons interacting with humans. As part of this preoccupation, tales of bestiality increasingly referred to intercourse with demons, the succubi and incubi that seemed ubiquitous. The increased concern with bestial intercourse seems also to reflect a growing uncertainty about the separation of humans and animals. Preoccupation with and legislation against bestial intercourse expressed an attempt to secure the separation of species when it seemed endangered.

As church laws were taken over in the late Middle Ages by kings who wanted to exert more authority over their kingdoms, what had once been identified as sinful then became identified as illegal. It is in this form that laws against bestiality persisted into the modern world.

Selected Bibliography. Aelian, *On the Characteristics of Animals* (Cambridge, MA: Harvard University Press, 1959); Brundage, James, *Law, Sex, and Christian Society in Medieval Europe* (Chicago: University of Chicago Press, 1987); Dekkers, M., *Dearest Pet* (London: Verso, 1994); Payer, Pierre, *Sex and the Penitentials* (Toronto: University of Toronto Press, 1984); Salisbury, J. E., *The Beast Within* (New York: Routledge, 1994).

JOYCE E. SALISBURY

BIO-CARTESIANISM

Bio-Cartesianism is the idea that the mind is a nonphysical object, separate from the physical brain. In this view, the human brain, although purely physical like the remainder of the body, nevertheless evolved in ways that neither reflected nor caused evolutionary changes elsewhere in the organism. Bio-Cartesianism lies at the heart of the tension between the scientific and moral justifications of animal research. Experimenters defend their practices by claiming that cognitive differences between humans and nonhuman animals are the differences that morally justify their practice (*see* ANIMAL COGNITION). Yet they claim that there are similarities elsewhere in the organism that scientifically justify generalizations from animals to humans.

The moral dilemma is this: if the cognitive abilities of humans and animals are sufficiently different to morally justify experimentation, then these differences will both reflect and promote other biological differences that compromise straightforward generalizations of findings in animals to humans. On the other hand, if underlying biological mechanisms are sufficiently similar to justify reasonably direct scientific inferences from animals to humans, then the higher-order traits of the test subjects are likely sufficiently similar to human traits to make research morally troublesome.

Selected Bibliography. Edey, M. A., and D. Johanson, *Blueprints: Solving the Mystery of Evolution* (New York: Penguin, 1989); LaFollette, H., and N. Shanks, *Brute Science: Dilemmas of Animal Experimentation* (London: Routledge, 1996); Mayr, E., How Biology Differs from the Physical Sciences, in D. Depew and B. Weber (Eds.), *Evolution at a Crossroads: The New Biology and the New Philosophy of Science* (Cambridge, MA: MIT Press, 1986); Rachels, J., *Created from Animals* (Oxford: Oxford University Press, 1990); Schiller, J., Claude Bernard and Vivisection, *Journal of the History of Medicine and Allied Sciences* 22 (1967): 246–260; Sober, E., *Philosophy of Biology* (Boulder, CO: Westview Press, 1993).

HUGH LaFOLLETTE AND NIALL SHANKS

BROOME, ARTHUR

Arthur Broome (1780–1837) was an Anglican priest and founder of the Royal Society for the Prevention of Cruelty to Animals (RSPCA),* the first national animal welfare* society in the world. He called together the first meeting in June 1824 that led to the foundation of the society. Broome's work was immensely sacrificial. He gave up his London living (in Bromley-by-Bow) to work full-time (unpaid) for the society as its first secretary. He paid for the first inspectors to police Smithfield meat market in London out of his own pocket, thus inaugurating the tradition of anticruelty inspectors known to this day. He ended up in prison, paying for the society's debts. Broome was indebted to the thought of Humphry Primatt,* whose historical work *The Duty of Mercy and the Sin of Cruelty to Brute Animals* (1776) he revised for its second edition. The first Prospectus of the Society, penned by Broome himself, makes clear its origin in the Primatt-like doctrine of Christian benevolence: "Our country is distinguished by the number and variety of its benevolent institutions . . . all breathing the pure spirit of Christian charity. . . . But shall we stop here? Is the moral circle perfect so long as any power of doing good remains? Or can the infliction of cruelty on any being which the Almighty has endued with feelings of pain and pleasure consist with genuine and true benevolence?" (1824, 197). Indeed, the First Minute Book recorded (though not through unmixed motives) that "the proceedings of this Society are entirely based on the Christian Faith and on Christian Principles." Broome, together with other luminaries such as William Wilberforce, Lord Shaftesbury,* and Richard Martin,* changed the conscience of a nation, and consequently and indirectly the conscience of other nations by the establishment of sister SPCAs throughout the world. Broome was sadly forgotten by the society and eventually died in obscurity— in the words of historians, "unwept, unhonoured and unsung" (Fairholme and Pain, 64).

Selected Bibliography. Broome, Arthur, First Prospectus of the SPCA, June 25, 1824, *RSPCA Records* 2 (1823–1826): 196–198; Fairholme, E. G., and Wellesley Pain, *A Century of Work for Animals: The History of the RSPCA, 1824–1924* (London: John Murray, 1924); Linzey, Andrew, *Animal Theology* (London: SCM Press; Urbana: University of Illinois Press, 1995); Moss, Arthur, *Valiant Crusade: The History of the R.S.P.C.A.* (London: Cassell, 1961); Primatt, Humphry, *The Duty of Mercy and the Sin of Cruelty to Brute Animals*, 2nd rev. ed. by Arthur Broome (Edinburgh: T. Constable, 1832); Turner, James, *Reckoning with the Beast: Animals, Pain, and Humanity in the Victorian Mind* (Baltimore: Johns Hopkins University Press, 1980).

ANDREW LINZEY

BROPHY, BRIGID

Brigid Brophy (1929–1995) was a British author and social critic who pioneered the modern tradition of animal rights. Her *Sunday Times* article "The Rights of Animals," published in 1965, heralded a new ethical sensitivity to animals. Brophy was a dedicated vegetarian (*see* VEGETARIANISM), antivivisectionist (*see* ANTIVIVISECTIONISM), and an unsparing opponent of all blood sports. Her speeches, reviews, and articles articulated an uncompromising view of animal rights: "Those rights are inalienable and irreducible. You can't do arithmetic that trades six of one sort of rights for two of another. If it were justifiable to sacrifice one laboratory animal for the good of humans, then it would be justifiable to sacrifice one laboratory human for the good of a hundred humans" ("Brigid Brophy and Vivisection," 135). Her first novel, *Hackenfeller's Ape* (1953), which won first prize at the Cheltenham Literary Festival, is the story of a distinguished scientist who risks his academic career to save an ape from a rocket experiment.

Selected Bibliography. Brophy, Brigid, Brigid Brophy and Vivisection [text of a speech to the Annual Public Meeting of the National Anti-Vivisection Society, May 30, 1970], *Animals' Defender* 14(4) (July/August 1970): 133–138; Brophy, Brigid, The Darwinist's Dilemma, in David Paterson and Richard D. Ryder (Eds.), *Animals' Rights: A Symposium* (London: Centaur Press, 1979), 63–72; Brophy, Brigid, The Ethical Argument against the Use of Animals in Biomedical Research, in *The Rational Use of Living Systems in Biomedical Research* (Potters Bar: Universities Federation for Animal Welfare, 1972), 51–57; Brophy, Brigid, *Hackenfeller's Ape* (London: Rupert Hart-Davis, 1953; London: Penguin Books, 1969); Brophy, Brigid, In Pursuit of a Fantasy, in Stanley Godlovitch, Roslind Godlovitch, and John Harris (Eds.), *Animals, Men, and Morals: An Enquiry into the Maltreatment of Non-Humans* (London: Gollancz,

1980), 124–145; Brophy, Brigid, The Rights of Animals, *The Sunday Times*, October 10, 1965, reprinted in *Don't Never Forget: Collected Views and Reviews* (London: Jonathan Cape, 1966), 5–21.

ANDREW LINZEY

BUDDHISM. *See* RELIGION AND ANIMALS.

BULLS. *See* RODEOS.

BUSHMEAT

Bushmeat is a general term for wild animals caught and killed in their home forests. A few human groups in central and western Africa have a tradition of occasionally eating gorillas and chimpanzees,* but the last decade has seen an increase in the number of killings. The great apes are now hunted for profit with modern weapons, and the victims' bodies are sold as part of a commercial bushmeat trade. The killing of great apes is illegal in every country where it takes place, but prosecutions are almost unknown.

The upsurge in the number of deaths—now thousands every year—is a result of increased logging activities, until recently mostly by European firms. These companies build new roads into previously inaccessible forests and allow hunters to travel on company vehicles to remote areas where gorillas, chimpanzees, and other large animals can still be found. The hunters kill all but the smallest animals and transport the meat to logging camps and onward to distant urban markets. As the logging expands, so does the death rate.

Although the effects of the bushmeat trade have been known for years, until very recently, attempts to interest the relevant organizations and the world press have come to nothing. There are now signs of increased interest and action. Most conservationists now argue that commercial bushmeat hunting will clear the forests of wildlife long before the trees are felled.

Selected Bibliography. Ammann, K., Saving the Great Apes, in Karl Ammann, *Gorillas*, Insight Topics (Hong Kong: Apa Publications, 1997), 151–161; Fa, J. E., J. Javier, J. P. Delval, and J. Castroviejo, Impact of Market Hunting on Mammal Species in Equatorial Guinea, *Conservation Biology* 9 (October 1995): 1107–1115; Kano, T., and R. Asato, Hunting Pressure on Chimpanzees and Gorillas in the Motaba River Area, Northeastern Congo, *African Study Monographs* 15(3) (November 1994): 143–162; McRae, M., and K. Ammann, Road Kill in Cameroon, *Natural History Magazine* 106(1) (February 1997): 36–47, 74–75; Rose, A. L., The African Great Ape Bushmeat Crisis, *Pan African News* 3(2) (November 1996): 1–6.

MICHAEL GARNER, ANTHONY ROSE, AND PAUL WALDAU

C

CAGING. *See* CHICKENS.

CARROLL, LEWIS (CHARLES L. DODGSON)

Lewis Carroll (Charles L. Dodgson; 1832–1898) was a don (meaning "fellow") of Christ Church, Oxford, who achieved fame through his Alice books. Carroll was also a major figure in the antivivisection (*see* ANTIVIVISECTIONISM) controversy at Oxford. His campaign against experimentation on animals led to the publication of his savage satire on vivisection (1875). He was a forerunner of the view that animal experiments would lead inexorably to experimentation on human subjects (1875, 14–16). Carroll opposed cruelty* to animals on theological grounds, maintaining that vivisection was the result of "secular" education that neglected Christian virtues. He was also adamantly opposed to hunting* and shooting animals for sport (see Cohen, 397).

Selected Bibliography. Carroll, Lewis, *Some Popular Fallacies about Vivisection* (Oxford: Printed for Private Circulation, June 1875); Carroll, Lewis, Vivisection as a Sign of the Times [letter], *Pall Mall Gazette*, February 12, 1875; Cohen, Morton N., *Lewis Carroll: A Biography* (London: Macmillan, 1995).

ANDREW LINZEY

CARSON, RACHEL

Rachel Carson (1907–1964) was a naturalist whose work *Silent Spring* (1962) exposed the public to the biocidal impact of chemical insecticides.

Carson became an advisor to the Animal Welfare Institute in the late 1950s. The success of her most important book was a bellwether of rising ecological consciousness in the postwar era and a catalyst of political and organizational change in the realm of environmental protection. Although the language and rhetoric of her most celebrated work is anthropocentric (*see* ANTHRO-POCENTRISM), Carson's personal philosophy incorporated a deeper moral perspective. Carson contributed a foreword to Ruth Harrison's early exposé of factory farming,* *Animal Machines* (1964), in which she sternly criticized the objectification* and reduction of nonhuman animals for human purposes. In addition, she was a strong supporter of federal guidelines to regulate the use of animals in laboratories (*see* LABORATORY ANIMAL USE). Carson dedicated *Silent Spring* to Albert Schweitzer,* whose reverence-for-life philosophy had deeply influenced her own life and work. Fittingly, she was the recipient of the Animal Welfare Institute's Albert Schweitzer Medal in 1963.

Selected Bibliography. Brooks, Paul, *The House of Life: Rachel Carson at Work* (Boston: Houghton Mifflin, 1972); Carson, Rachel, *Silent Spring* (Boston: Houghton Mifflin, 1962); Norwood, Vera, *Made from This Earth: American Women and Nature* (Chapel Hill: University of North Carolina Press, 1993).

BERNARD UNTI

CATS

The domestic cat is the most popular companion animal* in the United States today, with nearly 60 million of them living in American households. Many people are concerned about the welfare of cats in our society. Two important issues in the welfare of cats are their use in biomedical research and the growing number of homeless cats.

In 1881, zoology professor St. George Mivart published a textbook called *The Cat: An Introduction to the Study of Backboned Animals, Especially Mammals.* He described the cat as "a convenient and readily accessible object for reference" in studying mammals, including humans. Since the publication of Mivart's book, cats have been used primarily to learn about the specific functions of nerve cells and about how the brain processes visual information. Research with cats has contributed to advances in treating various disorders of the eye, including "lazy eye," glaucoma, and cataracts, as well as recovery from damage to the brain and spinal cord from injuries and strokes. Cats also have been used to study particular medical problems they have in common with humans, such as hearing disorders, diabetes, and acquired immune deficiency syndrome (AIDS). Research in these areas is contributing to both feline and human health.

Relatively few cats are used for the purpose of research for human health

Feral cat colony caretaker John Jones has befriended many of his wards. Photo courtesy of Paul Glassner/San Francisco SPCA.

(compared to other nonhuman animals). In 1995 fewer than 30,000 cats were used for research purposes in the United States, representing only 2% of all animals used in research that year (not including rats and mice*). Furthermore, the institutions conducting research with cats in the United States, Great Britain, and many other countries must comply with the strict regulations for animal care and use specified by their respective animal welfare* laws.

An even greater issue in cat welfare today is pet overpopulation, particularly the problem of free-roaming, unowned, feral cats (see FERAL ANIMALS). Millions of cats are living in city streets and parks without close human contact. Several factors may account for the existence of so many homeless cats. Many people believe that cats can survive easily on their own and choose to abandon their pets when it is inconvenient to keep them. Also, pet cats with access to the outdoors sometimes stray from home. If these animals are not identified with a tag, microchip, or tattoo and do not return home on their own, they may become permanently lost and resort to life on the streets. In addition, if pet cats are allowed outdoors without having been spayed or neutered, they can mate with stray cats whose litters may be born outside, further contributing to the homeless cat population.

The question of what to do about these free-roaming or feral cats is being hotly debated among the humane community, wildlife agencies, and cat advocacy groups. Two primary management philosophies exist. Many groups support TTVAR (trap, test, vaccinate, alter [spay/neuter], release) as long as there are people willing to feed and provide veterinary care for outdoor cat colonies. The arguments in favor of this method are that, as domestic animals, these cats deserve our assistance and, even if a colony is removed, other cats will move into the area.

Others, however, believe that it is better to trap and humanely kill these animals. The advocates of this policy are concerned that even with help from human caretakers, these animals suffer and die a miserable death. In addition, questions remain concerning the spread of disease, both within the cat population and to humans, and the impact of these animals on wildlife populations, especially birds and small mammals.

Selected Bibliography. AVMA Animal Welfare Forum, Veterinary Perspectives on the Use of Animals in Research, *Journal of the American Veterinary Medical Association* 206(4) (1995): 457–482; Berkeley, E. P., *Maverick Cats: Encounters with Feral Cats* (New York: Walker, 1982); Clifton, M. (Ed.), Seeking the Truth about Feral Cats and the People Who Help Them, *Animal People*, November 1992; Fitzgerald, B. M., Diet of Domestic Cats and Their Impact on Prey Populations, in D. C. Turner and P. Bateson (Eds.), *The Domestic Cat: The Biology of Its Behaviour* (New York: Cambridge University Press, 1988), 123–144; Mivart, S. G., *The Cat: An Introduction to the Study of Backboned Animals, Especially Mammals* (London: John Murray, 1881); U.S. Department of Agriculture, *Animal Welfare Enforcement: Fiscal year 1995*, APHIS Publication No. 41–35–042 (Washington, D.C.: U.S. Department of Agriculture, 1996).

R. LEE ZASLOFF

CENTER FOR ALTERNATIVES TO ANIMAL TESTING. *See* LABORATORY ANIMAL USE.

CHICKENS

Until relatively recently, most chickens were raised outdoors in small, free-ranging flocks. The primary product from these flocks was eggs. Poultry meat was scarce and expensive. But poultry meat and eggs are now the most abundant and least expensive animal food products, due largely to the development in the last 40 years of a highly intensified, large-scale poultry-production industry.

The poultry industry is the largest (in terms of animal numbers) and most highly automated of all of the animal-production industries. In the United States alone, nearly 8 billion poultry, mainly chickens and turkeys but also waterfowl, game birds, ostriches, and emus, are raised each year. Chickens have undergone intense genetic selection, and two distinct types of chickens are now used, one for egg production and a faster-growing bird (a broiler) for meat production. Chickens and turkeys are produced by an increasingly smaller number of companies that oversee all phases of production, from hatching to slaughter. Turkeys, broilers, and breeder flocks are typically housed in large groups on the floor in enclosed or semienclosed buildings, while almost all chicken hens used for egg production are housed in "battery" cages.

Cage production systems provide hens with protection from soil-borne diseases and predation.* However, caging is also the poultry-production practice that has been most widely criticized. Typical battery cages are barren and lack the features that the hen requires to perform behaviors like dustbathing, perching, and nesting. Space allowances have also been criticized. In the United States, hens are typically provided with about 48 square inches of space per bird, whereas in the European Community the legislated minimum space requirement is 72 square inches per hen. However, the hen needs at least that much space to turn around comfortably and more to groom or perform other comfort behaviors. Caged hens also develop osteoporosis because of a lack of exercise combined with a calcium deficiency associated with their high rate of egg laying. In consequence, up to 24% of hens sustain bone breakage when they are removed from their cages to be transported to the processing plant.

Several alternatives to conventional cages are being investigated, varying from more intensive systems like modified battery cages containing perches, dustbaths, and nestboxes to more extensive systems like aviaries (similar to battery cages, but tiered so that the hens can occupy several levels) and free-range production systems. Problems in more extensive systems include higher egg costs, reduced egg quality, increased feather pecking and cannibalism, and, in indoor systems, poorer air quality leading to respiratory illness in both hens and farm workers.

Beak trimming, which is routinely used to reduce injuries and mortality associated with feather pecking and cannibalism in both cage-housed and more extensively housed hens, also poses a welfare problem. Approximately one-half of the beak is removed using either a hot cauterizing blade or a precision trimmer. The latter makes a small hole in the beak, causing the tip to fall off several days later. Although the pain* associated with beak trimming was once thought to be minor and of short duration, it is now known that hens that have their beaks trimmed using a hot blade experience both acute and chronic pain.

Another controversial practice is induced molting, which is used to extend the period of egg production in a flock. Birds in the wild normally molt their feathers periodically, but in commercial poultry the molt is induced artificially so that all hens will molt simultaneously and return to egg production quickly. The most common procedure used to induce molt in the United States is to withdraw feed for several days to several weeks.

Welfare issues in broiler and turkey production are also now receiving

Typical housing for broiler-breeders. Approximately 5,000 birds are kept in a floor pen of this type. Photo courtesy of Joseph Mauldin.

Laying hens in a free-range production system in England. Free-range systems like this one, in which 1,200 hens are housed, are becoming more common in Europe. Photo courtesy of Arnold Elson.

Laying hens housed in so-called battery cages. More than 90 percent of laying hens in the United States are housed in cages similar to these throughout the production period.

attention. These include health problems, like leg weakness and other skeletal deformities in broilers and turkeys and cardiovascular and metabolic problems in broilers, that appear to result primarily from genetic selection for rapid growth rate. Considerable research is being conducted on these problems, and improvements might be possible by moderating growth during certain periods or selecting for increased skeletal strength.

Last, there has been increasing interest in improving poultry transport and slaughter methods (*see* TRANSPORTATION AND SLAUGHTER). Birds being sent to slaughter are hand captured, hung upside down in groups by the legs while being carried to the transport crates, crated, and then transported by road over varying distances to the processing plant. At the processing plant, the birds are uncrated, hung upside down on a shackle line, stunned using an electrical current, and then killed by a mechanical knife. Rough handling and poor transport conditions can cause stress,* bruising, bone breakage, and mortality. Mechanical catchers have been invented that cause less stress than human handling, although problems have been encountered with the maneuverability of these machines. Improved closed-transport vehicles that allow closer control over temperature and humidity during hot or cold weather have also been developed, although most poultry in the United States are still transported in open vehicles. Because electrical stunning is not always effective in producing unconsciousness, carbon dioxide is being studied as an alternative. Similarly, while unhealthy or surplus chicks used to be killed at the hatchery by suffocation, this practice has largely been abandoned in favor of more humane methods.

Selected Bibliography. Appleby, M. C., and B. O. Hughes, Welfare of Laying Hens in Cages and Alternative Systems: Environmental, Physical, and Behavioral Aspects, *World's Poultry Science Journal* 47 (1991): 110–128; Appleby, M. C., B. O. Hughes, and H. A. Elson, *Poultry Production Systems: Behaviour, Management, and Welfare* (Wallingford, Oxon: CAB International, 1992); Mench, J. A., The Welfare of Poultry in Modern Production Systems, *Poultry Science Reviews* 4 (1991): 107–128; Nicol, C., and C. Saville-Weeks, Poultry Handling and Transport, in T. Grandin (Ed.), *Livestock Handling and Transport* (Wallingford, Oxon: CAB International, 1993); Savory, C. J., and B. O. Hughes (Eds.), *Proceedings of the Fourth European Symposium on Poultry Welfare, Edinburgh, September 18th–21st, 1993* (Potters Bar, Herts: Universities Federation for Animal Welfare, 1993).

JOY A. MENCH

CHIMPANZEES

The chimpanzee is the species biochemically and genetically closest to humans. Chimpanzees are closer to humans than they are to the gorilla. This extreme similarity makes the chimpanzee a "sibling species" to humans. Researchers have compared nine amino-acid chains between chimpanzees and humans and found only 5 (0.4%) differences out of a total of 1,271 amino-acid positions. This means that, immunologically speaking, humans are 99.6% chimpanzee and vice versa.

Even more striking is that hemoglobin in chimpanzee and human blood is virtually identical. Human immunological defenses would recognize chimpanzee hemoglobin as "self" and not reject it. For example, when human type A red blood cells have been transferred into chimpanzees, they survive.

Genetic similarities have been directly examined as well. Here the findings confirm the extreme similarity noted in the immunological and blood characteristics, namely, that humans share 98.4% of their genes with chimpanzees. Another striking finding is that gorillas are 2.3% different from both humans and chimpanzees, and the orangutan is 3.6% different from humans and chimpanzees. Human similarity to chimpanzees shows that the environment humans evolved in can have a greater effect on their appearances than on their genes. Gorillas and orangutans are large impressive primates, and therefore we have assumed that they were our closer kin rather than the comical chimpanzees. Beyond the similarities already mentioned, chimpanzees in the wild are also similar to nontechnological humans. They live and hunt in communities, they form strong social bonds with their friends and families, and they make tools. They display a tremendous amount of cultural diversity in regard to toolmaking, tool use, and food preferences and even show evidence of self-medication. For example, some communities of chimpanzees use tools, others do not, and different communities use different tools.

Chimpanzees can suffer emotional and physical pain* just as humans do, and often for the same reasons. (Some have argued that human awareness of chimpanzees' ability to know and experience emotions similar to those of humans and humans' ability to empathize with them in their suffering means that humans have a responsibility to treat them with compassion and respect.)

Chimpanzees' cognitive abilities are as striking as their cultural similarities to humans. It has been demonstrated that chimpanzees can acquire and communicate with American Sign Language and that they can pass their signing skills on to the next generation. In addition, they use their signs to spontaneously converse with each other when no humans are present whatsoever, they sign to themselves, they use their signs during imaginary play, and much more (see ANIMAL COMMUNICATION, Language Debates).

The scientific evidence noted here clearly demonstrates that the difference between chimpanzees and humans is one of degree, just as it is with all of our fellow animals. This evidence is consistent with the Darwinian (see DARWIN, CHARLES) notion of continuity* that we are all relatives. The chimpanzee just happens to be our next of kin in our phylogenetic family. This scientific evidence contradicts the Dark Ages view that "Man" is different in kind from his fellow animals, which has been used to justify nonhuman-animal exploitation. Ironically, this extreme similarity of chimpanzees to humans has also worked against their welfare. For example, the biomedical community has used chimpanzees in research on the AIDS virus, organ-transplant research, hepatitis research, and even brain-injury research. The biomedical community justifies this research because the chimpanzee's physiology and biology are so similar to those of humans. Yet at the same time they ignore the ethical and moral responsibility for the damage they do to chimpanzees by relying on the view that humans are different in kind. (See also GREAT APE PROJECT.)

Selected Bibliography. Cavalieri, P., and P. Singer (Eds.), *The Great Ape Project* (London: Fourth Estate, 1993); Diamond, J., *The Third Chimpanzee* (New York: HarperPerennial, 1992); Fouts, R., and S. Mills, *Next of Kin* (New York: Wm. Morrow, 1997); Gardner, R. A., B. T. Gardner, and T. Van Cantfort (Eds.), *Teaching Sign Language to Chimpanzees* (Albany: State University of New York Press, 1989); Goodall, Jane, *The Chimpanzees of Gombe: Patterns of Behavior* (Cambridge, MA: Belknap Press of Harvard University Press, 1986).

ROGER FOUTS AND DEBORAH FOUTS

CHRISTIANITY. *See* RELIGION AND ANIMALS.

CIRCUSES AND CIRCUS ELEPHANTS

Circuses in North America were originally small shows that were based on equestrian acts and stayed in one location near large cities. In the late 1860s, circuses started to travel when they realized that they could attract much larger audiences in communities that had not recently experienced a circus. The shows added more animals and acts and grew in size into the huge shows of the 1920s. As large arenas appeared in cities, some circuses began to play in arenas, while others remained under tents.

Elephants are considered by many circuses to be the most important asset in drawing spectators. Using elephants in circuses, however, has become highly controversial in recent years. This issue involves the ethics of using animals for entertainment, the hesitation of some people to deprive a large, majestic species of living free in its natural habitat, the suspicion of abusive training methods, and the fear that many circus elephants are on the verge of going crazy.

Defenders of circuses cite that because of habitat depletion and poaching, Asian elephants are an endangered species* and African elephants are listed as threatened. They believe that there is a need to maintain a diverse genetic base if both species are to survive and that circuses can make a significant contribution to that gene pool. Defenders also think that it is unfair to condemn all trainers and circuses because of some isolated, highly publicized abuse cases, some of the most notorious of which occurred at zoos* and/or involved male elephants. Another defense is that elephants are very expensive to maintain in captivity, especially in sizable numbers, unless they can be used to generate income. Because of this expense, even some elephant "sanctuaries" will offer elephant rides to the public.

Estimates vary, but there are approximately 675 elephants in the United States, of which approximately 125 are owned by circuses, 250 are owned by zoos, and the remainder are under private ownership (e.g., sanctuaries, small exhibitors). In a traditional tented circus that moves to a new location each day, the elephants will spend 2 to 4 hours per day in a trailer while being transported 30 to 150 miles to the new lot and while the picket lines are being set up. They may also spend additional time in the trailers due to cold or inclement weather.

The picket line is the traditional method of restraining elephants in circuses (and is also used occasionally in zoos and other facilities) when they are not performing, giving rides, going for walks or baths, or putting up and taking down the tents. Picketing involves chaining one front and the diagonal rear leg to parallel picket lines made of cable or chain. The elephants can take only about one step forward and one backward, but can readily contact and interact with their neighbors if any are present. Normally, elephants spend 50% to 80% of their time on the picket line each day, although they may spend much more time on the line during rare occasions when performances are not scheduled or when the circus does not move to a new location.

In a study of picketed elephants, the single most common stereotypic (*see* STEREOTYPIES) behavior observed was weaving (rocking), which occupied up to 25% or more of some elephants' time, whereas others with the same circus did not weave during the same three 24-hour periods. The weaving of elephants is very different from that observed in other species, such as other zoo animals or stalled horses, because elephants are often performing a wide range of other behaviors while also weaving. For example, while weaving, elephants will frequently throw hay or dust on their backs, groom a neighbor with their trunks, or eat grain or hay. The frequency of weaving increases in apparent anticipation of being fed or performing, and especially prior to being watered.

Some circuses in North America are starting to use electric fences to create portable pens into which the elephants are released for varying periods of time when conditions permit. Many European circuses regularly use such electric pens. Electric pens appear to reduce the incidence of stereotypic behavior and offer increased opportunities for elephants to satisfy their behavioral needs. Anecdotal evidence also suggests that regular use of electric pens and taking elephants for walks and baths improve their "attitude," but this has yet to be confirmed. Until circuses have more experience with electric pens and there is a consensus regarding their reliability and security, electric pens will not replace picketing as the predominant method of restraint.

Most circuses use only female elephants because they usually are easier to handle than males. Also, most circus elephants are Asian because they are considered to be more reliable, more tolerant, and easier to handle than African elephants. Elephants will bond to good trainers, although trainers must always be able to assert dominance and a high degree of control over circus elephants because their size makes them potentially dangerous to humans. Good trainers know their individual elephants, and many will trade off or sell difficult-to-train or potentially dangerous elephants. Hence most circuses prefer to breed their own elephants rather than risk purchasing or leasing a potentially difficult or dangerous adult. Elephants are known to attack (usually a specific person) when they have been confused, scared, or pushed too far by a poor trainer. According to anecdotal information, the most common form of attack is sending someone flying with a slap of the trunk. People concerned about the welfare of elephants with a particular circus or other exhibitor of elephants in the United States can contact their regional office of the U.S. Department of Agriculture Animal and Plant Health Inspection Service (USDA-APHIS) to find out whether a circus or exhibitor is in compliance with federal regulations regarding the housing, training, and health care of elephants.

Selected Bibliography. Friend, T., and D. Bushong, Stereotypic Behaviour in Circus Elephants and the Effect of "Anticipation" of Feeding, Watering, and Performing, in *Proceedings of the 30th International Congress of the International Society for Applied Ethology, 14–17 August, 1996, Guelph, Ontario, Canada* (1996), 30; Hediger, H., *The Psychology and Behaviour of Animals in Zoos and Circuses* (New York: Dover Publications, 1968); Kiley-Worthington, M., *Animals in Circuses and Zoos: Chiron's World?* (Basildon, Essex, England: Little Eco-Farms, 1990); Schmid, J., Keeping Circus Elephants Temporarily in Paddocks: The Effects on Their Behaviour, *Animal Welfare* 4 (1995): 87–101; U.S. Department of Interior and U.S. Fish and Wildlife Service, Endangered and Threatened Wildlife and Plants, 50 CFR 17.11 and 17.12 (1991): 1–37.

TED FRIEND

COBBE, FRANCES POWER

Frances Power Cobbe (1822–1904) was an Irish-born social reformer, feminist, educationalist, and ardent antivivisectionist (*see* ANTIVIVISECTIONISM). In 1863, reports of cruelty to animals in a French veterinary school caught her attention. Subsequently, she went to Florence, where she discovered the work of Moritz Schiff. She organized the 1875 memorial (petition) signed by more than 600 leading intellectuals to the Royal Society for the Prevention of Cruelty to Animals (RSPCA)* in favor of restricting vivisection that in her own words "practically started the anti-vivisection movement in England" (note to Cobbe and Lloyd, *Memorial*, 1). When the RSPCA failed to act, she founded, with Lord Shaftesbury,* Cardinal Henry Manning,* and George Hoggan, the Victoria Street Society for the Protection of Animals from Vivisection in 1875. When the society (later named the National Anti-Vivisection Society) abandoned its insistence on immediate and total abolition, she founded the British Union for the Abolition of Vivisection (BUAV) in 1898. Cobbe's view that vivisection was evil derived from the theological conviction that the infliction of suffering* on animals was a denial of the God-given moral order. It was more important that "tender and just and compassionate feelings should grow and abound than that a cure should be found for any corporeal disease" (*Controversy in a Nutshell*, 5). While the "relationship of the brutes to God" might be the "humblest," she maintained that this should "move us to an emotion the reverse of such callous contempt" as was represented by vivisection ("Rights of Man," 596). Although she died without seeing the advancement of the cause she most loved, her personality and thought vastly influenced the entire movement. Among the many testimonials to her is a memorial in Manchester College, Oxford.

Selected Bibliography. Cobbe, Frances Power, *A Controversy in a Nutshell* (London: Victoria Street Society, 1889); Cobbe, Frances Power, *Life of Frances Power Cobbe by Herself*, 3rd ed. (London: Richard Bentley and Son, 1894); Cobbe, Frances Power, The Rights of Man and the Claims of Brutes, *Fraser's Magazine* 1 (November 1863), 586–602; Cobbe, Frances Power, and Miss Lloyd, *Memorial to the RSPCA on Vivisection, Presented on 23 January 1875* (London: RSPCA Records, vol. 1, 1875); Hume, E. Douglas, *The Mind-Changers*, foreword by HRH Prince Christopher of Greece and an introduction by George Arliss (London: Michael Joseph, 1939); Kramer, Molly Beer, and Andrew Linzey, Vivisection, in Paul Barry Clarke and Andrew Linzey (Eds.), *Dictionary of Ethics, Theology, and Society* (London and New York: Routledge, 1996), 870–874; Turner, James, *Reckoning with the Beast: Animals, Pain, and Humanity in the Victorian Mind* (Baltimore: Johns Hopkins University Press, 1980); Vyvyan, John, *In Pity and in Anger: A Study of the Use of Animals in Science* (London: Michael Joseph, 1969).

ANDREW LINZEY

COLLECTORS. *See* ANIMAL COLLECTORS.

COMMUNITY OF EQUALS

The term "community of equals" originally comes from the Declaration on Great Apes in the book *The Great Ape Project: Equality beyond Humanity*, edited by Paola Cavalieri and Peter Singer. It refers to a community that grants all its members equal moral protections, which are enforceable by law. Members of this community are regarded as moral equals in that they are all morally entitled to the same respect for their basic interests and needs; that is, they all have an equally justified claim to the same protection of their life, liberty, and freedom from deliberately inflicted harm.

"Equals" does not refer to any specific actual likeness, but to equal moral consideration without respect to morally irrelevant characteristics. The ability to understand or to undertake moral duties or responsibilities is regarded as a sufficient but not a necessary criterion for inclusion in this moral community, as is shown by the case of young children or severely mentally disabled humans. Though possibly not moral agents themselves (*see* MORAL

AGENCY AND ANIMALS), all members are regarded as moral equals, in that each of them is equally protected by and from the moral agents in this community.

At present, in both public opinion and national or international law, all and only humans are accepted as members of the community of equals. With the rise of the animal liberation* and animal rights* ethics over the last two decades there has been growing dissatisfaction among concerned people with the current boundaries of the community of equals. These boundaries are increasingly regarded as being an unjustified anthropocentric (*see* ANTHROPOCENTRISM) exclusion of nonhumans.

The speciesist (*see* SPECIESISM) nature of the exclusion of nonhumans has prompted steps to define the scope of equal moral concern on less biased grounds than just species membership, and to extend it beyond the human species. One major attack on the current boundaries of the community of equals is being mounted by the Great Ape Project,* which seeks to have all nonhuman great apes recognized as the moral equals of humans.

Selected Bibliography. Cavalieri, P., and P. Singer (Eds.), *The Great Ape Project* (London: Fourth Estate, 1993).

KARIN KARCHER

COMPANION ANIMALS AND PETS

Although often used as a synonym for "pets," the term "companion animals" refers primarily to those animals kept for companionship. "Pets" is a broader category than "companion animals" and includes animals kept for decorative purposes (for example, ornamental fish or birds), those kept for competitive or sporting activities (dog* shows, obedience trials, racing), and those kept to satisfy the interests of hobbyists (specialist animal collecting and breeding). In practice, of course, any particular pet may overlap two or more of these subcategories.

The practice of keeping animals primarily for companionship is certainly very ancient and may have contributed to the process of animal domestication at least 12,000 years ago (*see* DOMESTICATION). Recent hunter-gatherers and incipient agriculturalists are well known for their habit of capturing and taming wild mammals and birds and treating them with affection and concern for their well-being.

The existence of pet keeping in hunter-gatherer societies raises questions about the function of this activity. Until recently, it was widely assumed that the keeping of pet animals for companionship was a largely Western pastime associated with unusually high levels of monetary wealth. Viewed from this perspective, pet keeping tended to be categorized as an unnecessary luxury. Within the last 20 years, however, medical evidence has slowly accumulated suggesting that companion animals may contribute to their owners' mental and physical health. It is now known that close and supportive human relationships can exert a protective influence against many common life-threatening diseases, probably by buffering people from the negative health effects of chronic life stress. It appears that companion animals may serve a similar function (*see* ANIMAL-ASSISTED THERAPY). This would suggest that companion animals provide a means of augmenting the social support people receive from each other, and that this role may be as important in hunter-gatherer societies as it is in our own.

Despite the apparent contribution of pets to human well-being, some pet owners seem to have scant regard for the welfare of their animal companions. Welfare problems in companion animals arise from several sources. Most pets are restrained in various ways and are not permitted to express their full repertoire of behavior. These restrictions may cause some degree of distress* and frustration. The global trade in exotic pets, especially wild birds, reptiles,* amphibians,* and fish,* has seriously depleted some wild populations, as well as caused unestimated suffering* and death* during capture, handling, and transport. Since the middle of the 19th century, companion-animal breeders have created a wide range of hereditary breed defects, especially in dogs, while pursuing their own arbitrary standards of beauty. Many of these defects condemn the animals to lifetimes of distress and discomfort, and some require corrective surgery. Painful cosmetic "mutilations," such as tail docking* and ear cropping, and elective surgical procedures, such as declawing and debarking, designed to eliminate behavior problems, are widely performed, particularly in North America. The fate of unwanted pets is also a cause for concern.

These darker aspects of pet keeping have prompted some animal advocates to argue that the entire phenomenon constitutes a violation of animals' rights* and interests,* and that pet keeping should be abolished alongside other forms of animal exploitation. This position ignores the fact that at least some human–companion-animal relationships appear to be mutually beneficial and rewarding to both the human and animal participants. It also tends to discount the potentially positive effect of these relationships on our perceptions of animals in general.

Selected Bibliography. Katcher, A. H., and A. M. Beck (Eds.), *New Perspectives on Our Lives with Companion Animals* (Philadelphia: University of Pennsylvania Press, 1983); Manning, A., and J. A. Serpell (Eds.), *Animals and Human Society: Changing Perspectives* (London: Routledge, 1994); Robinson, I. (Ed.), *The Waltham Book of Human-Animal Interaction: Benefits and Responsibilities of Pet Ownership* (Oxford: Pergamon Press, 1995); Serpell, J. A., *In the Company of Animals*, 2nd ed. (Cambridge: Cambridge University Press, 1996).

JAMES A. SERPELL

Domesticated Companion Animals

Throughout the history of humanity, animals have had a place in human social communities and have been valued as guardians, work partners, and companions by individuals and families. Keeping animals such as dogs* and cats* as companions is so familiar to us that generally people do not regard it as an animal rights* issue so long as an animal is well cared for by the humans who are responsible for the animal's well-being.* Companion-animal keeping, however, is controversial among advocates of animal rights. The issues involved can be divided into two general categories: the harm or benefit to individual companion animals and the harm or benefit to populations of animals from which companion animals come.

Those who argue for the practice of keeping domesticated animals maintain that companions bring mutual benefit to both the animal and the human companion if the animal's needs* are valued and accommodated. These needs, most would agree, include not only adequate food and shelter, but also the needs that arise from the inherent nature of the particular species. A companion dog, for example, requires sufficient exercise and sensory stimulation, social contact with both humans and other dogs, and exposure to the outdoors in order to be psychologically as well as physically healthy. Keeping a companion dog in a small enclosure, chaining it in a yard, or isolating it from others of his or her kind are generally acknowledged to be examples of abusive practices. It is also commonly agreed among animal rights activists that it is not appropriate or ethical to keep animals such as birds, mice,* or hamsters, for example, as companions, because caging these animals violates their need for freedom of movement. When a companion animal's needs are met, this argument goes, the relationship between person and animal is reciprocally beneficial, as is evidenced by the genuine affection that exists between many people and their animal companions.

Those who argue against keeping animals as companions argue that the practice is motivated by a selfish human need to dominate and control members of other species. To support their argument, opponents of companion-animal keeping point to the many instances where people treat their animals in a patronizing or controlling manner, substitute animal affection for human affection, or use an animal as a surrogate child. That many animals are abused and/or neglected by their "owners" is a documented fact. The view that keeping animals as companions may violate the right of animals is further supported by the fact that the laws in most societies regard animals as chattel property (people are considered to be the "owners" of their animal companions; *see* LAW AND ANIMALS). The manner in which animals are kept and treated is considered to be largely within the discretion of the "owner," and there are few legal limits placed on how humans treat the animals they "own." Opponents of companion-animal keeping doubt that statutory or educational changes will adequately protect the majority of companion animals.

The latter view also addresses the significant harm that has been done to populations of animals through artificially selective breeding and the practice of inbreeding for looks and behavioral characteristics that are regarded as desirable in particular breeds of dogs and cats. Irresponsible breeders such as the puppy mills that sell dogs to "pet" stores cause great suffering* to both the animals that they use as breeding stock and to the puppies that are not adequately cared for or socialized when young.

There is a solution to these problems, advocates of companion-animal keeping would argue. Again, the answers according to this view lie in education and legal protection. Adequate legislation would prohibit or greatly reduce breeding by unscrupulous people who sacrifice the quality of animals' lives to their own greed for profit. If people were educated to the cruel breeding practices that produce "pet"-store animals and to the harm done through artificial selection for particular traits, it is argued, consumer demand for the involved animals would be eliminated and the harm would cease. Opponents of companion keeping find this approach unrealistic and believe that it is companion keeping itself that is the root of the problem. The extreme view in opposition to companion-animal keeping advocates that all companion animals be sterilized and that these domestic animals be allowed to become extinct. (*See also* DOMESTICATION.)

Selected Bibliography. Fox, Michael W., *Inhumane Society* (New York: St. Martin's Press, 1990); Francione, Gary L., *Animals, Property, and the Law* (Philadelphia: Temple University Press, 1995); Rollin, Bernard E., *Animals Rights and Human Morality* (Buffalo, NY: Prometheus Books, 1981); Ryder, Richard D., *Animal Revolution* (Oxford: Basil Blackwell, 1989); Serpell, James, *In the Company of Animals* (Oxford: Basil Blackwell, 1986).

ANN B. WOLFE

Exotic Companion Animals

An "exotic" animal is an individual member of any species that is not domesticated, that is, an animal who has not evolved either artificially or naturally to share a close living environment with humans (*see* DOMESTICATION). Some examples of exotic animals that are frequently kept as companions are parrots, iguanas, ferrets, Vietnamese pot-bellied pigs, snakes, and monkeys. People with a great deal of money sometimes acquire animals directly from their natural habitats by legal or illegal means. There is virtually no limit to the kind and number of exotic animals that may be procured through poachers and animal smugglers: bears, ocelots, panthers, and even elephants are some of the animals that can be obtained illegally. In many countries, the keeping of nondomesticated animals by private individuals is illegal.

There is virtually no disagreement among animal rights* advocates that it is both inappropriate and unethical to keep exotic animals as companions. The practice of keeping exotic animals as companions emerges from a view that all nonhuman animals should be at the disposal of humans for whatever purposes humans might choose.

Many of the exotic animals that are kept privately by animal collectors* and other people are members of endangered species.* Sometimes, the fact that a particular animal is rare makes it attractive as a companion. Taking these animals out of their natural environments and separating them from their conspecifics (other individuals of the same species) further endanger a species by reducing the chances of reproduction and therefore the renewal or survival of the involved species. Those who support human intervention to try and save endangered species advocate intervention by professionals through strategies that are well researched. The keeping or breeding of these animals in private settings is unlikely to yield results that will benefit a species as a whole.

Exotic animals are entitled to live in a way that allows for the nature (*telos*) of the particular animal to be accommodated. Since exotic animals by their very nature do not live in close proximity to humans, taming them and keeping them in captivity is a violation of that nature. Often exotic animals suffer ill health in captivity and have far shorter life spans than they might in a natural setting. From the perspective of both moderate and radical animal rights advocates, a decision as to the ethics of keeping any nonhuman animal as a companion must rely on the principle that an animal should be kept as a companion to humans only if the animal's nature can be fully accommodated. This principle would seem to be violated in the case of most, if not all, exotic animals.

ANN B. WOLFE

Animals in Public

The animals who live in our homes and with whom we share our lives frequently accompany us when we go out into everyday public settings. In these situations, companion animals often act as what sociologists call "social facilitators"; they provide a shared focus of attention and offer a reason for strangers to interact with each other. Being in the company of a companion animal in public not only gives strangers something to talk about, it also helps make people seem less threatening. Those with dogs* or other animals can be identified as "dog (animal) lovers." This public identity helps break down the suspicion we often feel for people we do not know while providing an acceptable reason for starting conversations with strangers.

This function appears to be particularly important for people with physical disabilities who are accompanied by service dogs. These people often feel that their special physical conditions make the "normals" they meet uncomfortable. Studies demonstrate that those who are visually handicapped or confined to wheelchairs have more frequent and friendly conversations with able-bodied people when they are with service dogs. Companion animals can act as more than simply the focus of brief public interactions among strangers.

The public interactions between people with animals and others are not always smooth and friendly, however. Like adults in the company of young children, people with animals typically are held responsible when the animal misbehaves in public. A study of the various ways of handling public misbehavior problems described "excusing tactics." Owner responses included such responses as blaming the unwanted behavior on the understandable stress* the dog feels in the situation, redefining the dog's actions as "cute" or normally doglike rather than "bad," and overtly punishing the misbehaving animal in order to make amends for the dog's violation of the rules.

Selected Bibliography. Adell-Bath, M., A. Krook, G. Sanqvist, and K. Skantze, *Do We Need Dogs? A Study of Dogs' Social Significance to Man* (Gothenburg: University of Gothenburg Press, 1979); Hart, Lynette, Benjamin Hart, and Bonita Bergin, Socializing Effects of Service Dogs for People with Disabilities, *Anthrozoös* 1(1) (1987): 41–44; Messent, Peter, Social Facilitation of Contact with Other People by Pet Dogs, in Aaron Katcher and Alan Beck (Eds.), *New Perspectives on Our Lives with Companion Animals* (Philadelphia: University of Pennsylvania Press, 1983), 37–46; Robins, Douglas, Clinton Sanders, and Spencer Cahill, Dogs and Their People: Pet-facilitated Interaction in a Public Setting, *Journal of Contemporary Ethnography* 20(1) (1991): 3–25; Sanders, Clinton R., Excusing Tactics: Social Responses to the Public Misbehavior of Companion Animals, *Anthrozoös* 4(2) (1990): 82–90.

CLINTON R. SANDERS

CONSCIOUSNESS AND THINKING IN ANIMALS. *See* ANIMAL COGNITION.

CONTENT, VALUE, AND RICHNESS OF ANIMAL LIFE

By "content," philosophers and others today refer to the subjective experiences of nonhuman animals, especially the "higher" animals (*see* ANIMAL COGNITION, Conscious Experience, Consciousness and Thinking, Subjectivity of Animals). That the "higher" animals have experiences, that they live experimental lives, is today widely accepted. The nature of these experiences and of the lives that contain them have become important for two reasons, moral standing* and value of life.

Increasingly on all sides today, quality of life, not life itself, is what matters essentially; the value of a life is determined by the quality of the life being lived. Debate exists over how to determine quality of life, not least over whether the issue is primarily a subjective or an objective one. One of the central difficulties with objective accounts is that while by objective criteria a life could be going well, by subjective criteria it might be going badly. A person might have all the calories needed to function well yet still not think that his or her life is going well. The subjective element is about how the life looks from the point of view of the creature living it, which requires some account of the subjective experiences of the creature in order to be properly understood. What we want to know, in essence, is how rich a life is from this individual's point of view, where "richness" means such things as the variety, depth, and extensiveness or kinds of experiences.

To hold that we have absolutely no access to the interior lives of animals seems false, at least if we take scientific work by ethologists, biologists, and others seriously. To hold that we cannot know *exactly* what these interior lives are like does not mean that we cannot know a good deal about them and so make some very provisional judgments about them. Playing fetch with a dog illustrates the point.

In discussions of the richness of animal lives, we must not apply criteria appropriate to judging richness in the human case as if they applied straightforwardly, without further defense, to the animal case. This would be a second-order form of speciesism.* Yet something here does set a kind of presumption of where both empirical science and argument must occur, for it does seem clear that richness of content in our lives is tied in large part to our capacities for enrichment: where these capacities are impaired or missing, as with the loss of a sense, a life appears less rich than an ordinary adult life that contains those kinds of experiences that that capacity makes possible. This does not mean that another capacity for richness cannot compensate for this loss, but it does mean that we should have to be convinced of this.

Thus, at the end of life, when we look back and say of a human that she or he lived a "rich" and "full" life, we refer to the array of kinds of experiences that characterize the lives of normal adult humans. At this level, we consider that we mean something far beyond what we would mean were we to say this of the life of a dog, for we think that we have capacities for enrichment that far outstrip anything the dog has. Nothing is settled, of course, by this presumption of argument; it simply means that something must be said in the dog's case, by way of compensation, to make us think that the richness of its life approaches that of the normal adult human. Again, nothing is prejudged; perhaps one can compare features of one of the dog's capacities that transforms its life through that single dimension to what is conferred on our lives by all our various capacities. If one thinks only of the role of culture or marriage or accomplishment of chosen ends in our lives, however, those who wish to contend that the dog's life is as rich as the lives of normal adult humans have a case to make.

Everything here is cast in terms of "normal adult" humans for the reason that it is false that all humans live lives of equal richness. Some human lives are so wanting in richness and scope for enrichment that we strive to avoid some paths for ourselves and our families; we do not appear to hold that all human lives are equally valuable. Rather, a quality-of-life view of the value of life commits us to another view: if human lives are not (approximately) equally rich, they are not of equal quality, and if they are not of (approximately) equal quality, they are not of equal value. In fact, what such a view appears committed to holding is that some animal lives can be of a richness and quality higher than those of some human lives, such as the brain-dead and anencephalic infants, and so can be of greater value.

Empirical work on the subjective lives of animals can be held to be necessary for these reasons. It must fit in with a philosophy of mind that makes intelligible to us ways of understanding and appreciating animal experience and with a moral philosophy that enables us to fit animal experience into our account of the value of a life. (*See also* WELL-BEING OF ANIMALS.)

Selected Bibliography. Bekoff, M., and D. Jamieson (Eds.), *Readings in Animal Cognition* (Cambridge, MA: MIT Press, 1996); Dawkins, Marian S., *Through Our Eyes Only: The Search for Animal Consciousness* (San Francisco: W. H. Freeman, 1993); Frey, R. G., Medicine, Animal Experimentation, and the Moral Problem of Unfortunate Humans, *Social Philosophy and Policy* 13 (1996): 181–211; Frey, R. G., *Rights, Killing, and Suffering* (Oxford: Basil Blackwell, 1983); Griffin, D., *Animal Minds* (Chicago: University of Chicago Press, 1992); Walker, S., *Animal Thought* (London: Routledge and Kegan Paul, 1983).

R. G. FREY

CONTINUITY

Less than 150 years after Charles Darwin's* *On the Origin of Species* we have not yet fully assimilated the meaning of evolutionary continuity. What Darwin proposed and then proved with many different arguments and examples is that all living organisms are *relatives*. All animals are related by common descent. For example, zebras and horses evolved from a common ancestor, as did chimpanzees* and humans, and wasps and ants; their common ancestors existed in a bygone time. All six of these animal species, however, also evolved from a common ancestor, only that ancestor existed and became extinct even further back in time. Species emerged like branches growing off other branches on a single tree, all originating from the same root.

Before the Darwinian revolution, it was believed that animals were organized according to a hierarchy called the "great chain of being." At the top of that hierarchy people put mammals and, at the very top, human beings. Then came birds, reptiles,* and amphibians*—that is, vertebrates, animals that have a backbone, like human beings. At the bottom of the scale came the invertebrates, among which are the insects. Instead of having a skeleton inside their bodies, as we do, insects wear their skeleton on the outside, almost like armor. This is only one of the ways that insects are different from us. Other ways in which they differ are that they are much smaller, they sense the world in totally unfamiliar ways (for example, bees see ultraviolet light), they communicate in ways we find hard to imagine (for example, using chemicals), and they look totally alien. Despite their minuscule size, fear of insects is not uncommon. Insects were placed at the bottom of this imagined hierarchical ladder because the less an animal resembled human beings, the lower its position. However, Darwin showed that the reason animals can be very unlike one another is not because they have different "essences," but because they are adapted to different conditions; because of common descent the core is the same, and only the manifest forms vary.

With his understanding of evolution as a process of descent from common ancestors, with new species shaped through encountering novel conditions, Darwin destroyed the self-promoting idea of the great chain of being. In its place he gave us a world in which there are no discontinuous leaps between species, for all animals are bound together by the single, very long story of life. Darwin went to great lengths to demonstrate this unbroken continuity at every level, not only in anatomy and physiology, but also in behavior and mental characteristics.

Despite the dismantling of the hierarchical great chain of being, in our practices and ideas we continue to uphold a radical break between vertebrates and invertebrates. We resist the idea, for example, that insects may feel pain* or suffering.* More deeply, we deny that insects lead a life that they experience from their perspective. Yet the impersonal and flawless reasoning of the evolutionary perspective would teach us that a discontinuous break between vertebrates and invertebrates is arbitrary and anthropocentric (*see* ANTHROPOCENTRISM).

Selected Bibliography. Darwin, Charles, *The Formation of Vegetable Mould through the Action of Worms with Observations on Their Habits* (1881; Chicago: Chicago University Press, 1985); Darwin, Charles, *On the Origin of Species*, facsimile of the first edition (1859) (Cambridge, MA: Harvard University Press, 1964); Fabre, Jean Henri, *The Insect World of J. Henri Fabre*, ed. E. Teale (Boston: Beacon Press, 1991); Frisch, Karl von, *Bees: Their Vision, Chemical Senses, and Language* (Ithaca, NY: Cornell University Press, 1950); Griffin, Donald, *Animal Minds* (Chicago: University of Chicago Press, 1992).

EILEEN CRIST

People and Animals

Treatment given to animals depends upon people's perceptions of and attitudes toward those animals. The most important determinant of human attitudes toward animals is the degree of similarity or difference that is believed to exist between people and nonhumans. Throughout history and among different cultures, ideas about human-animal differences have varied greatly from animals being seen as possessing greater powers and capacities than people and being regarded as gods to being viewed as lesser beings having nothing in common with the human species. Assignments of superiority or inferiority are generally based upon similarities or differences, qualities that have determined the appropriateness of exploiting various species for human purposes. Currently, this issue is particularly related to the use of animals for research. If research is intended to benefit the human species, then the animals must resemble humans enough so that results will be directly applicable to people. Yet in order to do the research, animals must be regarded as in some way different enough from humans to justify their use as subjects for experimentation.

Perceived differences and similarities between people and animals play a prominent role in current animals rights* controversies. Opposing views of animal rights advocates and their opponents rest partly on contrasting beliefs concerning the human place in nature. Supporters of animal rights typically see no great gap between humankind and animals, whereas those who oppose animal rights see a significant gap between the two. Both sides acknowledge some differences between people and animals, but those who favor animal rights see these as only quantitative (differences in degree), whereas their opponents believe that there are both quantitative and qualitative differences (differences in kind). The important question is whether the differences between animals and people are basic and significant enough to be the basis for excluding animals' interests and consideration of their welfare (*see* ANIMAL WELFARE) whenever there is a conflict between their interests and our own. The main issue in this debate is whether animals possess intrinsic (their own) value, regardless of what they provide or accomplish for the improvement of human life, or whether the significance of animals is only instrumental, dependent upon their usefulness for human ends.

Many criteria have been proposed for definitively distinguishing humankind from animals. Historically, the most profound separation between people and animals was delineated by the 17th-century philosopher and mathematician René Descartes.* Until fairly recently the idea that animals possess no symbolic cognitive (thought) process (*see* ANIMAL COGNITION) was widely accepted by Western science. Animals were generally assumed to be incapable of acting apart from instinctual motivation. Studies undertaken by Donald Griffin and other cognitive ethologists, however, lead to the conclusion that many species of animals do possess cognition. The scientific community, and to a lesser extent the public, often resist the idea that animals possess any degree of cognition, the trait that has long kept humans at the pinnacle of creation. As convincing evidence of animal awareness builds up, there is a trend away from denying animals any thoughts to claiming that their thoughts are different from ours.

Some other criteria that have been used to distinguish humankind from animals are tool use, toolmaking, teaching of cultural traditions, enactment of rituals with social significance, possession of individuality as opposed to an exclusively communal identity, awareness of death, converting nature to culture by building structures, creation of art, altruism, and the use of language. However, exceptions to these criteria have been convincingly demonstrated. Examples of tool use are Galapagos finches who use cactus spines to probe for insects and sea otters who use rocks to open abalones and obtain meat. Jane Goodall found that chimpanzees* actually make tools with which to obtain termites, using premeditation and planning. Termite fishing is an acquired skill passed on as part of cultural learning from older to younger individuals. Japanese macaques also have learned food washing as a behavior invented by one individual and taught to others. Wolves and chimpanzees perform rituals with social consequences. Individualized behavior has now been documented among many species, including birds, who previously were held to be automatons incapable of varying from specific inborn repertoires. Suggestions that elephants, baboons, and chimpanzees have some kind of awareness of death seem valid. Beavers, birds, and insects build structures that may not always be just the result of inflexible patterns of behavior. Chimpanzees and elephants create art. Dolphins, monkeys, and wolves and other canids demonstrate altruism. Possession of language, long considered the last bastion of human uniqueness, has been demonstrated in chimpanzees, gorillas, and orangutans who have learned sign language and have used it spontaneously and creatively. Studies also show that parrots can understand the language they speak, and that dolphins may be able to use language to communicate with people. Critics of these studies, however, claim that the animals in question have not mastered the syntactic structure characteristic of true language. As our knowledge about animals grows, the gap between the human and nonhuman worlds narrows.

Selected Bibliography. Beck, Benjamin B., *Animal Tool Behavior* (New York: Garland STPM, 1980); Griffin, Donald R., *Animal Minds* (Chicago: University of Chicago Press, 1992); Linden, Eugene, *Apes, Men, and Language* (New York: Saturday Review/Dutton, 1974); Rachels, James, Do Animals Have a Right to Liberty? in Tom Regan and Peter Singer (Eds.), *Animal Rights and Human Obligations* (Englewood Cliffs, NJ: Prentice-Hall, 1976); Ristau, Carolyn (Ed.), *Cognitive Ethology: The Minds of Other Animals* (Hillsdale, NJ: Erlbaum, 1991).

ELIZABETH ATWOOD LAWRENCE

COWHERD, WILLIAM

William Cowherd (1763–1816) was a minister and founder of the Bible Christian Church, a vegetarian sect that launched the world vegetarian movement. In 1800, Cowherd, then associated with the New Church of Emanuel Swedenborg, founded, together with Joseph Brotherton, Salford's first member of Parliament, a church at Salford near Manchester that would have an incalculable impact on the spread of vegetarianism* worldwide. Based on the biblical injunction to be vegetarian (Genesis 1:29–30), the main conditions of membership were vegetarianism and temperance. Moral considerations about the treatment of animals and a strong sense of respect for the whole created order complemented Cowherd's conviction that the consumption of animal flesh was prohibited by the Bible. The English Vegetarian Society was a direct offshoot of the Bible Christian Church when it was founded in 1847. Cowherd's influence was extended to the United States by his disciple William Metcalfe and other successors in the Bible Christian theology, such as Henry Clubb.

Selected Bibliography. [Maintenance Committee], *History of the Philadelphia Bible-Christian Church for the First Century of Its Existence from 1817–1917* (Philadelphia: J. B. Lippincott, 1922); Unti, Bernard, Vegetarian Roots, *Vegetarian Times*, April 1990, 52–57, 82; Williams, J. Howard, *The Ethics of Diet: A Catena of Authorities Deprecatory of the Practice of Flesh-Eating* (London: F. Pitman, 1883), 258–260.

BERNARD UNTI

CRUELTY TOWARD ANIMALS AND HUMAN VIOLENCE

The belief that one's treatment of animals is closely associated with the treatment of fellow humans has a long history, but despite the long history and popular acceptance of this concept, until recently there have been few attempts to systematically study the relationship between the treatment of animals and humans. In the early 1900s case studies by Richard Krafft-Ebbing and Sandor Ferenczi began to explore sadistic behavior toward animals associated with other forms of cruelty. However, single case histories do not provide much insight into the origins of animal abuse and its connections to other violent behavior. In 1966 Daniel Hellman and Nathan Blackman published one of the first formal studies of animal cruelty and violence. Their analysis of life histories of 84 prison inmates showed that 75% of those charged with violent crimes had an early history of cruelty to animals, fire setting, and persistent bed wetting. Several subsequent studies looked for this "triad" of symptoms in other violent criminals, with mixed results, but animal cruelty remained one of the strongest correlates of later violent behavior.

The concept became more widely appreciated within law-enforcement circles following a number of studies of criminal populations. FBI interviews of serial killers and other sexual homicide criminals initiated in the 1970s by Robert Ressler and his colleagues found that 36% of these violent criminals described instances of participating in animal mutilation and torture as children, and 46% described such activities in adolescence. Prevalence rates of early animal cruelty of 25% to 50% have been described in studies of aggressive prison inmates, assault offenders who are women, convicted rapists, and convicted child molesters. Questions regarding animal maltreatment have now become standardized in many investigations of violent crime and juvenile fire setting.

In the 1980s additional attention began to be given to instances of animal cruelty as part of the dynamics of child abuse and domestic violence. A review in one community in England of 23 families with a history of animal abuse indicated that 83% had also been identified by human social service agencies as having children at risk of abuse or neglect. A report on 53 pet-owning families in New Jersey being treated for child abuse or neglect indicated that at least one person had abused animals in 88% of the families with physical abuse. In two-thirds of these cases the pet abuser was the abusive parent. Recently, several studies have examined the incidence of animal cruelty in families of women seeking protection in shelters for battered partners. In one such survey in Utah, Frank Ascione found that 71% of the women with pets who sought shelter reported that their male partner had threatened to kill or had actually killed one or more of their pets.

Recognition of the significance of the interconnections between violence against animals and violence against people has led to a number of significant changes. A growing number of states have escalated extreme forms of intentional animal cruelty from misdemeanor to felony offenses. Larger fines, longer jail terms, and/or required counseling have become more commonplace in animal cruelty cases. Many areas have begun to train animal care and control officers in the recognition and reporting of child abuse, and some animal shelters* have begun to work closely with women's shelters to provide emergency housing for the pets of women and children at risk. Many advocates for animals and others hope that a better understanding of how cruelty to animals is related to other forms of violence may help in developing tools for prevention and intervention.

Selected Bibliography. Ascione, F. R., Children Who Are Cruel to Animals: A Review of Research and Implications for Developmental Psychopathology, *Anthrozoös* 6(4) (1993): 226–246; DeViney, E., J. Dickert, and R. Lockwood, The Care of Pets within Child Abusing Families, *International Journal for the Study of Animal Problems* 4(4) (1983): 321–329; Felthous, A. R., and S. R. Kellert, Violence against Animals and People: Is Aggression against Living Creatures Generalized? *Bulletin of the American Academy of Psychiatry and the Law* 14 (1986): 55–69; Hellman, D. S., and Nathan Blackman, Enuresis, Firesetting and Cruelty to Animals: A Triad Predictive of Adult Crime, *American Journal of Psychiatry* 122 (1966): 1431–1435; Lockwood, R., and G. R. Hodge, The Tangled Web of Animal Abuse, *Humane Society News* 31(3) (1986): 10–15; Ressler, R. K., A. W. Burgess, C. R. Hartman, J. E. Douglas, and A. McCormack, Murderers Who Rape and Mutilate, *Journal of Interpersonal Violence* 1 (1986): 273–287.

RANDALL LOCKWOOD

D

DAIRY INDUSTRY. *See* FACTORY FARMING; TRANSPORTATION AND SLAUGHTER; VEAL CALVES.

DARWIN, CHARLES

Charles Darwin's (1809–1882) theory of evolution by natural selection completely changed our understanding of the relation between humans and other species. Darwin was born on February 12, 1809, in Shrewsbury, England, the son of a prosperous doctor. As a young man he enrolled at Cambridge, intending to prepare for a life as a clergyman. But Darwin was already an accomplished amateur naturalist, and while he was at Cambridge, his talents were recognized by the science faculty, one of whom, John Henslow, recommended him for a post on HMS *Beagle*, which was about to embark on a five-year voyage around the world. The voyage would change Darwin's life.

Sometime during the *Beagle* voyage Darwin became an evolutionist. He did not invent the idea of evolution; others had already speculated that life might have evolved. But the idea was not taken seriously by most scientists because no one could think of a convincing mechanism by which evolutionary changes could occur. Darwin's contribution was to supply such a mechanism. In 1838, three years after returning to England, he devised the theory of natural selection to explain how evolutionary change takes place.

Darwin delayed publishing his theory for more than twenty years, partly because he dreaded the scandal it was sure to cause. In 1859 the publication of *On the Origin of Species* created just the sort of sensation Darwin had feared. Many readers were convinced, but many more were not. Evolution was resisted on religious and moral grounds—it was contrary to human dignity to imagine that man is kin to the apes. Darwin, however, went out of his way to emphasize the kinship. The resemblance, he said, is more than merely physical. Other animals also have social, mental, and moral lives similar to our own. In *The Descent of Man* (1871) he wrote, "There is no fundamental difference between man and the higher mammals in their mental faculties."

Darwin realized that such thoughts have ethical implications. In *The Descent of Man* he wrote that "humanity to the lower animals" is "one of the noblest virtues with which man is endowed" and represents the final stage in the development of the moral sentiments. It is only when our concern has been "extended to all sentient beings," he said, that our morality will have risen to its highest level.

Darwin's own feelings about the mistreatment of animals were unusually strong. Numerous stories illustrate the intensity of his feelings. Although he was generally mild mannered and disliked public confrontation, Darwin could fly into a rage when he saw animals being abused. In 1863 he wrote an article for the *Gardener's Chronicle*, a popular monthly magazine, with the title "Vermin and Traps." Using arguments that would not seem out of place in an animal rights* magazine today, he contended that "the setting of steel traps for catching vermin" is too cruel a business for civilized people to tolerate.

But as a man of science Darwin found his moral views about animals put to a severe test. In the 1870s antivivisection (*see* ANTIVIVISECTIONISM) agitation came to a boil in England, and Darwin was drawn into the controversy. His humanitarian impulse collided with his desire to see science advance, and he was uncomfortably caught in the middle. As he explained to one of his daughters, "I have long thought physiology one of the greatest of sciences, sure sooner, or more probably later, greatly to benefit mankind; but, judging from all other sciences, the benefits will accrue only indirectly in the search for abstract truth. It is certain that physiology can progress only by experiments on living animals." But on the other hand, as he wrote to a different correspondent, "You ask about my opinion on vivisection. I quite agree that it is justifiable for real investigations on physiology; but not for mere damnable and detestable curiosity. It is a subject which makes me sick with horror, so I will not say another word about it, else I shall not sleep tonight." So Darwin sought a compromise. In 1875 he testified before the Royal Commission on Vivisection and took the lead, lobbying the home secretary, in trying to have a bill passed that would "protect animals, and at the same time not injure physiology." But a more radical bill was passed that went further in protecting animals than Darwin thought wise.

Although Darwin saw himself as seeking middle ground, the animal rights advocates of the day regarded him more as a champion of the other side.

Frances Power Cobbe,* who had organized the National Anti-Vivisection Society, noted that "Mr. Darwin eventually became the centre of an adoring clique of vivisectors who (as his biography shows) plied him incessantly with encouragement to uphold their practice, till the deplorable spectacle was exhibited of a man who would not allow a fly to bite a pony's neck, standing forth before all Europe as the advocate of vivisection." Nonetheless, the long-term implications of Darwin's revolution seem to favor the animal advocates. These implications were summarized in 1880, two years before Darwin's death, by Asa Gray, the Harvard professor of botany who was Darwin's chief defender in America: "We are sharers with the higher brute animals in common instincts and feelings and affections. It seems to me that there is a meanness in the wish to ignore the tie. I fancy that human beings may be more humane when they realize that, as their dependent associates live a life in which man has a share, so they have rights which man is bound to respect."

Selected Bibliography. Browne, E. J., *Charles Darwin: A Biography* (Princeton, NJ: Princeton University Press, 1995); Darwin, Charles, *The Descent of Man, and Selection in Relation to Sex* (London: John Murray, 1871); Darwin, Charles, *On the Origin of Species* (London: John Murray, 1859); Gray, Asa, *Natural Science and Religion* (New York: Scribner's, 1880); Rachels, James, *Created from Animals: The Moral Implications of Darwinism* (New York: Oxford University Press, 1990).

JAMES RACHELS

DEATH OF ANIMALS

Animals are routinely killed for food, for leather and fur, for sport, and for education* and research. Society is comfortable with these killings as long as we believe that the animals do not suffer during the process (*see* SUFFERING OF ANIMALS). That is why there are training sessions, regulations, and inspections to ensure that the slaughter of animals for these purposes is "humane," meaning that it is done as quickly and painlessly as possible (*see* TRANSPORTATION AND SLAUGHTER).

People commonly believe that killing human beings is ordinarily immoral, even if the people being killed experience no pain during the process. Philosophers who agree with this difference in our attitudes toward killing animals and killing people defend it in two ways. First, humans have the ability to understand death and to value life itself. Humans can fear death itself, even when the dying process is painless. Animals, these philosophers contend, are not capable of understanding death and, consequently, are incapable of fearing death. Thus there is a kind of distress* and loss involved in the killing of humans that is not involved in the killing of animals.

Second, these philosophers stress that humans are capable of making long-range plans. When humans are killed, their long-range projects are frustrated. Animals, however, do not form such projects. Consequently, when they are killed, this does not frustrate long-term plans.

Some animal rights* advocates accept this common distinction between killing humans and killing animals. These animal rights advocates are particularly concerned with the suffering that animals endure and with how we can reduce that suffering. Utilitarian philosophers (*see* UTILITARIANISM), such as Peter Singer, who believe that moral concern should be focused on minimizing suffering and maximizing happiness in the world, take this position. However, they go on to emphasize that when large numbers of animals are routinely killed, it is not a painless process. The animals who are slaughtered for food or fur are not killed painlessly. They feel pain* when being shot, cut, gassed, or clubbed to death. Animals also smell the blood of the animals killed just before them, and this frightens them, causing them distress.

Animal rights philosophers also make the further point that most of our killing of animals is avoidable. Killing animals for sport is unnecessary. Most people can live healthy, happy lives without wearing fur or eating meat. Many of the animals killed in education and research are sacrificed for trivial information. Consequently, these philosophers contend that even if the killing of animals lacks the moral dimensions of killing humans, the pain we inflict on animals when we kill them is unnecessary, and as a result, our routine killing of great numbers of animals is morally objectionable.

Other animal rights philosophers emphasize that even if animals cannot value life itself or form long-range plans, killing them is ordinarily morally objectionable, even if it is done painlessly. When animals are killed, they suffer the loss of the rest of their lives. If those lives are lives the animals could have enjoyed living, they have suffered a great loss in being deprived of those lives. This deprivation makes killing animals morally objectionable independent of the issue of pain suffered during the process of killing.

Some animal rights advocates also question the assumption that animals cannot understand and fear death itself. It is difficult to tell whether an animal threatened with death fears death or the pain that ordinarily accompanies dying. Many animal rights advocates also question the assumption that animals cannot form long-range plans. They claim that even if animals cannot formulate plans that are as detailed and long term as human plans, they are not without plans altogether. Birds building nests, beavers building dams, squirrels storing nuts, and dogs* waiting for their human companions to come home at the usual time seem to be examples of animal planning. If these animal rights advocates are correct, the reasons cited for the common difference in the moral significance of killing animals and killing humans are more a matter of degree than a difference in kind, that is, a difference based on something humans have but animals lack.

Selected Bibliography. Cigman, Ruth, Death, Misfortune, and Species Inequality, *Philosophy and Public Affairs* 10 (1981): 47–64; Frey, R. G., *Rights, Killing, and Suffering*

(Oxford: Basil Blackwell, 1983); Johnson, Edward, Life, Death, and Animals, in H. Miller and W. Williams (Eds.), *Ethics and Animals* (Clifton, NJ: Humana Press, 1983); Sapontzis, S. F., *Morals, Reason, and Animals* (Philadelphia: Temple University Press, 1987); Singer, Peter, *Practical Ethics* (Cambridge: Cambridge University Press, 1979).

STEVE F. SAPONTZIS

DEEP ECOLOGY

The notion of deep ecology was first suggested by the Norwegian philosopher Arne Naess in 1973. He strongly sought terms and methods by which human beings might develop "a wider self," an unselfish self that identified wholly with the biosphere. In his contrasting of so-called shallow with deep ecology, Naess characterized what he viewed to be an ethical battlefield. All living beings in their natural diversity are viewed as resources, useful for humans but not in and of themselves, according to the shallow view; sustaining our way of life and our individual habits of mind are basic elements of a democratic society that should be tolerated and sustained, even if it entails our being cruel, our polluting the biosphere, our driving to extinction other life forms, and our declining quality of life. Deep ecology, however, disagrees with this view, faulting cruelty, respecting other life forms for their intrinsic worth irrespective of their potential usefulness to humans, and arguing that through such respect and nurturance, our own lives will be greatly enriched, made more meaningful, and assured of a better chance at survival.

In Asian tradition, what Westerners have only recently called "deep ecology" has always been understood. Throughout China and Japan, all animals and plants are capable of becoming Buddha (*see* RELIGION AND ANIMALS). In India, Hindu tradition has always been steeped in the sacredness of nature, best exemplified in the tradition of *pancavati*, or five sacred groves. Trees all over the country were traditionally revered, a fact now mourned by most Indians who have seen their forests mowed down to the extent that less than 9% of standing canopies remain in India. Among the Jains, Todas, and Bishnoi of India, the tenets of so-called deep ecology are at the very wellspring of their living spiritual traditions. In the future, deep ecology may well evolve into the science of what Marc Bekoff calls "deep ethology."*

Selected Bibliography. Banwari, *Pancavati: Indian Approach to Environment*, trans. Asha Vohra (New Delhi: Shri Vinayaka Publications, 1992); Chapple, Christopher Key, *Nonviolence to Animals, Earth, and Self in Asian Traditions* (Albany: State University of New York Press, 1993); Naess, A., *Ekologi, Samhalle, och Livsstil: Utkast til en Ekosofi* (Stockholm: LTs forlag, 1981); Naess, A., Identification as a Source of Deep Ecological Attitudes, in Michael Tobias (Ed.), *Deep Ecology* (San Diego, CA: Avant Books, 1984), 256–270; Naess, A., The Shallow and the Deep, Long Range Ecology Movement, *Inquiry* 16 (1973): 95–100; Rolston, H., III, *Environmental Eth-*

ics: Duties to and Values in the Natural World (Philadelphia: Temple University Press, 1988).

MICHAEL TOBIAS

DEEP ETHOLOGY

The term "deep ethology" carries some of the same general meaning that underlies the term "deep ecology,"* in which it is asked that people recognize not only that they are an important part of nature, but also that they have unique responsibilities to nature as moral agents (*see* MORAL AGENCY AND ANIMALS). Deep ethological research will pursue a detailed (and compassionate) understanding of the unique worlds of nonhuman animals themselves in order to learn more about their own points of view—how they live and how they experience pain* and suffering.* The development of what are called species-fair tests that take into account the different sensory worlds and abilities of animals will allow humans to learn more about how all animals deal with their social and nonsocial environments, including pleasurable and painful or stressful (*see* STRESS) stimuli.

Selected Bibliography. Bekoff, M., Deep Ethology, in M. Tobias and K. Solisti (Eds.), *Kinship with the Animals* (Portland, OR: Beyond Words, 1998).

MARC BEKOFF

DEROGATION OF ANIMALS

Derogation is the depreciation of others by means of symbolic (*see* ANIMAL SYMBOLISM) expression, usually through language (*see* OBJECTIFICATION OF ANIMALS), such as with racial slurs. Derogation is conducive to hostile actions, which is why it has become a social taboo and even illegal in many societies. Derogation of animals is ingrained in the European and other languages. In one of its derived meanings, the noun "animal" is frequently used to describe the cruelest, most heinous criminals (*see* BEASTLINESS). Aside from individual dislikes and propaganda from those in trades that depend on the use of animals as commodities, the two main universal reasons for the derogation of animals are cultural tradition and the psychological reaction of blaming an innocent victim (*see* VICTIMIZATION OF ANIMALS).

Derogation of animals has deep cultural (including religious) roots and goes back at least to the beginning of farming societies when domestication* of farm animals led to their demystification. It may go even further back in human prehistory. Contrary to common beliefs, not all hunter-gatherer peoples show any respect, even symbolic, for animals (for example, the Mbuti pygmies of the African equatorial forest mock and denigrate hunted animals that are dying of wounds). In Judeo-Christian and Moslem religions (*see*

RELIGION AND ANIMALS), animals are denied a hereafter and thus must be derogated if this belief in a perfectly just world is to be sustained. In the everyday language, the most derogated are pigs* and cows, which are the main source of meat and leather (and, in the industrial societies, almost never have a chance to be seen by the public as individual animals).

Less obvious than the direct, verbal derogation is the indirect, contextual derogation through public displays (especially on the part of known personalities) that disregard the life and suffering* of those deemed inferior. This type of derogation can be achieved by portraying acts of abuse as neutral or positive acts. Advertising cartoons of meat and dairy products commonly show pigs, cows, and other animals happy to be exploited and killed (*see* DISNEYFICATION).

Selected Bibliography. Lerner, M., *The Belief in a Just World: A Fundamental Delusion* (New York: Plenum, 1980); Mason, J., *An Unnatural Order* (New York: Simon and Schuster, 1993); Ryan, W., *Blaming the Victim* (New York: Pantheon Books, 1971); Sorrentino, R. M., Derogation of an Innocently Suffering Victim: So Who's the "Good Guy"? in J. P. Rushton and R. M. Sorrentino (Eds.), *Altruism and Helping Behavior: Social, Personality, and Developmental Perspectives* (Hillsdale, NJ: L. Erlbaum, 1981), 257–283.

ANDRZEJ ELZANOWSKI

DESCARTES, RENÉ

René Descartes (1596–1650) was a French philosopher who is also known as one of the fathers of modern science and mathematics. A dualist, he believed that only two kinds of substance exist in the universe: mental substance and corporeal, or bodily, substance. Human beings, he thought, are composed of mind (which he equated with the soul) and body. Nonhuman animals, however, he saw as mindless automata or machines. The traditional interpretation is that he even denied that animals have feelings.*

Descartes himself not only influenced the formation of the scientific method, but also engaged in various studies of his own, including, apparently, vivisection. In a little-cited passage from his *Description of the Human Body*, Descartes took issue with William Harvey's theory of blood circulation by cutting off part of the heart of a live dog and feeling the length of the pulse in various parts. He was an avid observer of animal bodies by his own account, stating in a letter of 1639, "I have spent much time on dissection during the last eleven years, and I doubt whether there is a doctor who has made more detailed observations than I." He inspired generations of scientists after him to dissect live animals without inhibition, since after all these living machines are without feeling—or so Descartes believed.

However, John Cottingham, who translated the philosophical works of Descartes, claimed that Descartes did think that nonhuman animals have conscious feelings, just not self-conscious awareness of feelings. Supposedly, this interpretation would mean that an animal can feel, but has no sense that the feelings are associated with that animal's own self (*see* ANIMAL COGNITION, Consciousness and Thinking). Some philosophers have said that animal feelings have no significance if animals lack self-consciousness, even though an animal can still be hurt on this theory.

Cottingham refers to letters of Descartes in which animals are said to feel joy, anger, and fear,* for example. Tom Regan has explained this apparent inconsistency in Descartes by reference to a distinction that the latter makes between three different types of sensation. According to Descartes, animals can have three different grades of sensation: physical, conscious, and self-conscious. Descartes indicated that we have only the first in common with nonhuman creatures. His denial that animals have minds prevents animals from having either conscious or self-conscious souls. Regan's interpretation is more consistent with what Descartes actually wrote. Animals only "felt joy" and other emotions in the first grade of sensation, which is a very unfamiliar sense of "feeling": the animals, in response to a physical stimulus, would mechanically respond by dancing about, appearing happy, or the like, even though the "animal machines" would not *consciously* feel anything. Thus Descartes actually wrote that animals do not feel "pain in the strict sense," since they lack an understanding or a mind, and also that they are not aware of any thing. This appears to rule out the view that animals have conscious feelings according to Descartes. This view did not go uncontested even in Descartes's own time. Voltaire (1694–1778) famously wrote a generation later: "Answer me, machinist, has nature arranged all the means of feeling in this animal, so that it may not feel?"

There are a few people who still hold to Cartesianism. Bernard Rollin found the animals-feel-no-pain thesis expressed in the *Bulletin of the National Society for Medical Research*, a U.S. lobby group that tries "to block legislation that would in any way place restrictions on biomedical research." Peter Harrison, a philosopher, defends Cartesianism based largely on the view that we cannot absolutely prove that animals feel pain* (his argument is much more detailed, however). The criticism of Descartes's view of animals stems from its conflict with commonsense experience of animals and also its being at odds with a variety of considerations in favor of holding that animals can suffer.

Selected Bibliography. Cottingham, John, "A Brute to the Brutes?": Descartes' Treatment of Animals, *Philosophy* 63 (1988): 175–183; Descartes, René, *The Philosophical Writings of Descartes*, 3 vols., trans. John Cottingham, Robert Stoothoff, and Dugald Murdoch (Cambridge: Cambridge University Press, 1991); Regan, Tom, *The Case for Animal Rights* (Berkeley: University of California Press, 1983); Rollin, Bernard, *The Unheeded Cry: Animal Consciousness, Animal Pain, and Science* (New York: Oxford University Press, 1989); Singer, Peter, *Animal Liberation* (New York: New

York Review of Books, 1990); Williams, Bernard, *Descartes: The Project of Pure Enquiry* (Harmondsworth, England: Penguin Books, 1978).

                                                                    DAVID SZTYBEL

DEVIANCE AND ANIMALS

Social scientists typically understand deviant behavior in two ways. Deviance, on the one hand, is a characteristic of how people act. If the behavior violates social norms—the basic guidelines for behavior that are known and obeyed by well-socialized members of a society—then it is, by definition, deviance. In contrast, some sociologists speak of deviance as a subjective or personal phenomenon. From this view a behavior is deviant or not depending on who does it, for what reason, and who finds out about it.

Deviant animals are usually displayed in the media in much the same way as are deviant humans. At times they are threatening and dangerous because they are innately evil, like, for example, the shark in *Jaws*. At other times, animals are presented in the media as behaving in deviant ways because they are mad (e.g., the dogs* in *Cujo* and *Man's Best Friend*) or because they have been trained by humans to do evil things (e.g., the rats in *Ben* or the guard dog in *White Dog*). Like the human deviants portrayed in the media, deviant animals are easy to recognize because they are slimy, foam at the mouth, bare their teeth, or in other ways physically display their malevolence. It is likely that the fear that many people have for pit bull terriers, bats, snakes, and other definably ugly animals has its roots in our cultural connection of appearance and deviance.

Another common connection between animals and deviance is seen in the tendency for animal terms to be used in most, if not all, cultures as labels that diminish the importance of the person so labeled (*see* BEASTLINESS; DEROGATION OF ANIMALS). In our society, for example, a person can be degraded by calling him or her such things as an "animal," "pig," "chicken," "snake," or "dirty dog." These animal labels are intended to demonstrate that those to whom they are applied are less than "real" human beings.

From the Middle Ages until the 18th century, it was common in Europe for nonhuman animals to be seen as being able to choose how they behaved. This meant that animals were often put on trial for such things as murder, assault, and destruction of property. If they were judged guilty, the animal defendants were usually executed. One writer recorded 191 judicial proceedings involving such animal defendants as bulls, horses, pigs,* dogs,* turtledoves, field mice,* flies, caterpillars, and bees.

Bestiality* is one type of behavior involving people and animals that is seen as a serious norm violation. A far more common and less controversial example of the relationship of animals and deviance is seen in the everyday lives we share with companion animals.* In some ways, "training" a dog or "breaking" a horse may be seen as forms of socialization. We typically teach animals to abide by certain rules—not to relieve themselves in our homes, not to jump up on visitors, not to make unnecessary noise, and so forth. As is the case with humans, animal companions often break the rules we would like them to obey. When this happens, their misbehavior is usually either ignored or steps are taken to control the "deviant" animals.

One study by Sanders (1994) focused on how doctors in a veterinary clinic defined and responded to violations by their animal patients. Typically, the misbehavior of animals was not seen as being "their fault" but as being caused by the stress* of being in the clinic or the pain* they were experiencing. While patients' unruliness usually was not understood as due to their moral failings, veterinarians* were rarely as charitable in their evaluations of owners. The "bad" behavior of patients was commonly seen as the fault of "bad" (ignorant, weak, overly permissive) clients.

Selected Bibliography. Dekkers, Midas, *Dearest Pet: On Bestiality* (London: Verso, 1994); Evans, E. P., *The Criminal Prosecution and Capital Punishment of Animals* (1906; Boston: Faber and Faber, 1987); Hearne, Vickie, *Bandit: Dossier of a Dangerous Dog* (New York: HarperCollins, 1991); Laurent, Erick, Definition and Cultural Representation of the Ethnocategory *Mushi* in Japanese Culture, *Society and Animals* 3(1) (1995): 61–77; Sanders, Clinton, Biting the Hand That Heals You: Encounters with Problematic Patients in a General Veterinary Practice, *Society and Animals* 1(3) (1994): 47–66.

CLINTON R. SANDERS

DISENSOULMENT. *See* RELIGION AND ANIMALS.

DISNEYFICATION

Disneyfication of animals refers to the assignment of some human characteristics (*see* ANTHROPOMORPHISM) and cultural stereotypes onto the animals. Although this practice is best shown by the way cartoon characters and animals are pictured in Walt Disney movies, it is not restricted to the Disney Corporation but is widespread as a marketing strategy.

The most noticeable human characteristic projected onto animals is that they can talk in human language. Physically, animal cartoon characters (and toys made after them) are also most often deformed in such a way as to resemble humans. This is achieved by showing them with humanlike facial features (eyebrows, expressive lips) and altered forelimbs to resemble human hands (although with a smaller number of fingers). In more recent animated movies the trend has been to depict the animals in a more "natural" way. However, they still use their limbs like human hands (for example, lions can pick up and lift small objects with one paw), and they still talk with an appropriate facial expression. A general strategy that is used to make the animal characters more emotionally appealing, both to children and adults, is to give them enlarged and distorted childlike features.

Probably the most significant aspect of Disneyfication of animals is the projection of cultural stereotypes onto animal behavior. The members of the animal "kingdom" are often used as a means for presenting male-dominated societies with stereotypical gender roles. Racist attitudes are subtly conveyed not only through the choice of physical characteristics of "bad" animal characters, but also through the use of language with accents and characteristic expressions indicative of racial or ethnic background. In Disney's 1994 best-selling *The Lion King* the members of the royal family speak with British accents, whereas the voices of hyenas resemble the ones of urban black and Latino populations.

Disneyfication is widely used in popular visual culture, including everything from video games, television, and film to amusement parks and shopping malls. Its effects on the formation of individual and collective identities of children and youth are not yet fully understood. One of the direct effects of misrepresentation of animals is that animals and their behavior tend to be misinterpreted by children, sometimes with tragic consequences. Objectification* of animals promotes the pet industry and the view of animals as goods to be bought. This strategy may lead to formation of adult personalities incapable of functioning outside of stereotypical frameworks modeled after their childhood experiences.

Selected Bibliography. Complete Details on Disney's Animal Kingdom, *Orlando Sentinel*, June 21, 1995, A1, A6; Giroux, H. A., Animating Youth: The Disneyfication of Children's Culture, *Socialist Review* 24(3) (1994): 23–55; Mike, O., Report on the Potentially Dangerous Dog Program: Multnomah County, Oregon, *Anthrozoös* 4(4) (1991): 247–254; Noske, B., *Humans and Other Animals* (London: Pluto Press, 1989); Thompson, W. I., Disney's World: The American Replacement of Culture, in *The American Replacement of Nature* (New York: Doubleday, 1991).

SLAVOLJUB MILEKIC

DISSECTION. *See* EDUCATION AND THE USE OF ANIMALS.

DISTRESS IN ANIMALS

Distress denotes mental suffering* and may be reflected in a change in molecular receptor binding in the central nervous system (e.g., benzodiazepine, opioid, serotonin, noradrenalin) (*see* DYSTRESS). It may be an integral part of other aspects of suffering. An animal in pain* from a broken leg may be fearful of being moved or touched, as well as being distressed by its inability to move normally. Such changes in receptor binding in the CNS may lead to stereotypic behaviors (*see* STEREOTYPIES IN ANIMALS).

DAVID B. MORTON

DOCKING

Docking refers to the removal of varying amounts of the tail. Docking is done for reasons of fashion (dogs,* horses), protection of some animals from diseases where other preventative measures are impracticable (lambs, hill farming of sheep), or convenience of the stockperson (cattle). Sometimes it is done therapeutically for the benefit of the animal.

DAVID B. MORTON

DOGS

Over the years, dogs have been widely used in biomedical research to investigate heart disease, bone injury, hearing loss, blindness, lung disorders, infectious diseases, the effects of lethal poisons, and other conditions that have relevance to human health. In the 20th century, as the volume of animal experiments increased, researchers found a ready supply of dogs and cats* for their work from shelters* and pounds. Shelters (and pounds) are places where lost, stray, and abandoned animals are temporarily housed. By law, shelters must retain animals in their care for a certain number of days in order that owners have an opportunity to reclaim their pets or, alternatively, that adoptive homes be sought. If a suitable home is not found, the dogs are painlessly killed. There is an enormous overpopulation of dogs in the United States due to irresponsible overbreeding, and currently approximately six to eight million animals have to be killed in shelters each year because no home can be found for them. Humane societies have worked hard to educate the public about spaying and neutering their pets, but the overpopulation persists.

In 1945, a lobbying group for animal researchers was formed whose primary purpose was to work for passage of state laws to permit researchers to have access to unwanted and unclaimed animals in shelters. These efforts persist to this day. However, these efforts are strongly resisted by members of the animal welfare* and animal rights* movements who hold that shelter animals should not be used for research. Leading humane societies, including the Animal Welfare Institute, the Humane Society of the United States,* the American Humane Association, and others, have been involved. Currently, state laws are mixed. Some laws specifically require shelters to hand over their animals to research, while others prohibit this practice. In states where there is no law, shelters operated by humane societies usually will not permit their dogs or cats to go to research. But city pounds, whose responsibility is to keep stray animals off the streets, do not share the same compunctions about the eventual fate of one-time pets and so are often glad to sell dogs to labs.

Rationales for these opposing viewpoints of researchers and members of the humane movement are as follows: Researchers argue that shelter animals are unwanted and are doomed to die anyhow, so why not use them for a socially useful purpose? Also, the animals are less expensive than animals bred specially for the purpose of research, thus saving research dollars. The animal welfare/rights view is that human beings have a profound moral responsibility to domesticated (*see* DOMESTICATION) animals that cannot be forsaken at any point in those animals' lives. Shelters should be sanctuaries for animals, not a supply line for biomedical researchers. From a dog's viewpoint, a humane death may be a better choice than a longer life being a subject of a painful experiment. Animal welfarists hold that overpopulation of pet animals should not be exploited for the benefit of researchers. Animals for research should be a different population of animals than those that were one-time pets.

This clash of viewpoints has been somewhat lessened by the fact that since the 1980s, commercial breeders for laboratory dogs have become well established. It is a profitable business. Commercial breeders can supply animals who are healthy, are of known age and genetic makeup, and are more reliable experimental subjects than so-called random-source dogs from shelters. Increasingly, researchers are turning to this source of supply. Currently, researchers obtain about half their dogs from commercial suppliers and the other half from shelters. Increasingly, researchers are finding that so-called purpose-bred animals are scientifically preferable to random-source animals, and less public criticism is encountered.

Selected Bibliography. Festing, M., Bad Animals Mean Bad Science, *New Scientist* 73(1035) (1977): 130–131; Giannelli, M. A., The Decline and Fall of Pound Seizure, *Animals' Agenda*, July/August 1986, 10–13, 36; National Association for Biomedical Research, *The Use of Dogs and Cats in Research and Education*, NABR Issue Update (Washington, DC: National Association for Biomedical Research, 1994); Orlans, F. B., *In the Name of Science: Issues in Responsible Animal Experimentation* (New York: Oxford University Press, 1993), 209–220.

F. BARBARA ORLANS

DOMESTICATION

Domestication is a process rather than an event, and it is hard to define the point at which a tame or captive wild animal can be classed as domesticated. In general, truly domesticated animals exhibit some obvious genetic divergence from the ancestral "wild type" due to the effects of artificial (human) selection over many generations.

The first species to undergo the change from wild to domestic life was probably the wolf (*Canis lupus*), the ancestor of the dog.* The oldest known archaeological remains of a probable domesticated wolf come from a 14,000-year-old site in central Europe. Wild sheep (*Ovis orientalis*) and goats (*Capra aegagrus*) appear to have been domesticated more or less simultaneously in the Near East around 11,000 years ago, while remains of domesticated cattle (*Bos primigenius*) and pigs* (*Sus scrofa*) first occur in the archaeological record around 8,000–9,000 years B.P. (before present) at various sites in Asia. Horses (*Equus ferus*), asses (*Equus africanus*), camels (*Camelus* spp.), water buffalo (*Bubalus bubalis*) and chickens* (*Gallus gallus*) all appear to have been domesticated in different parts of Asia and North Africa between 7,000 and 5,000 years ago, while the first domesticated cats* (*Felis silvestris*) appeared in ancient Egypt between 4,000 and 5,000 years ago. Meanwhile, in the New World, llamas, alpacas (*Lama* spp.), guinea pigs (*Cavia* sp.) and turkeys (*Meleagris gallopavo*) were undergoing a similar process at various times and locations.

It is unlikely that Paleolithic and Neolithic peoples consciously domesticated animals for specific economic or practical purposes. It appears more probable, given the record of human history, that domestication, at least in its early stages, was a largely unconscious process in which tame or semitame wild animals were gradually brought under increasing levels of human control. Nor should we assume that the different species were necessarily domesticated in the same ways. For example, species that could be herded, such as the wild ancestors of domestic sheep, goats, or llamas, may originally have been followed and hunted by nomadic human groups long before people began to play an active role in guiding the movements of the animals, protecting them from predators, or interfering selectively in their reproduction. Large and potentially dangerous animals such as wild cattle may have been coaxed gradually into semidependent relationships with humans by the provision of salt licks close to villages. Other species, such as dogs, pigs, cats, and poultry, may have been captured or collected when young and then kept primarily as pets (*see* COMPANION ANIMALS AND PETS). This form of pet keeping is known to be extremely widespread among living or recent hunter-gatherer societies, and there is no obvious reason to believe that Paleolithic hunters were any different. Although these pets are regarded with affection and are not usually killed or eaten, moral inhibitions about slaughtering them for food are sometimes overcome by the demands of hunger.

The relationship between humans and domesticated animals is sometimes pictured as a successful "adaptive strategy" in the evolutionary sense, a kind of "symbiosis" in which both the humans and the animals have benefited. This idea is often used to justify the continued exploitation of domestic animals by humans and is based on the simple observation that most domestic species are more numerous nowadays than their wild ancestors (some of which are extinct). While the argument carries some force for species, such as the cat or dog, that have increased hugely in numbers and range at relatively little cost in terms of loss of individual freedom, it becomes more difficult to sustain in relation to intensively reared food or research animals, such as pigs, broiler chickens,* or laboratory mice.*

Selected Bibliography. Budiansky, S., *The Covenant of the Wild: Why Animals Chose Domestication* (New York: Morrow, 1992); Clutton-Brock, J., *Domesticated Animals from Early Times* (London: British Museum [Natural History] and Heinemann, 1981); Cohen, M. N., *The Food Crisis in Prehistory: Overpopulation and the Origins of Agriculture* (New Haven: Yale University Press, 1977); Serpell, J. A., *In the Company of Animals*, 2nd ed. (Cambridge: Cambridge University Press, 1996); Zeuner, F. E., *A History of Domesticated Animals* (London: Hutchinson, 1963).

JAMES A. SERPELL

DOMINIONISM

According to one dictionary, the word "dominion" means "a supremacy in determining and directing the actions of others . . . the exercise of such supremacy." Dominionism is the West's basic ideology, one that views the world and all its life forms as God-given property to serve human needs and whims. Dominionism drives science and technology to take ever-increasing power and control over the living world so that human beings—some, at least—may have safety, comfort, convenience, longer lives, and other benefits.

Dominionism is older than the Judeo-Christian (*see* RELIGION AND ANIMALS, Christianity, Judaism). As farmers, humans stepped up ways to use some plants and animals while they subdued the competition—the plants and animals of the natural world. As farmers, humans learned to take the laws of nature into their own hands. In time, agrarian peoples regarded the living world less as a divinity and more as an enemy. Nature was not to be held in awe; it was to be subdued, outwitted, and controlled. Animals, who had long been regarded as the souls and powers of the mysterious living world, became tools, goods, and pests. With their inferior status, the much older sense of kinship and continuity with the living world broke up, and the agrarian sense of superiority and alienation set in.

Selected Bibliography. Collard, Andree, and Joyce Contrucci, *Rape of the Wild* (Bloomington: Indiana University Press, 1989); Eisler, Riane, *The Chalice and the Blade* (San Francisco: Harper and Row, 1987); Shepard, Paul, *Man in the Landscape*, 2nd ed. (College Station: Texas A&M University Press, 1991); Thomas, Keith, *Man* *and the Natural World: A History of the Modern Sensibility* (New York: Pantheon Books, 1983).

JIM MASON

DOWDING, LADY MURIEL

Lady Muriel Dowding (1908–1981), a leading British humanitarian, vegetarian, and antivivisectionist, was the founder in 1959 and later chairperson of Beauty without Cruelty, the organization that led the way in the commercial production of synthetic alternatives to fur and cruelty-free cosmetics. She was a longtime president of the National Anti-Vivisection Society. In 1969, she cofounded the International Association against Painful Experiments on Animals (IAAPEA) and remained a patron until her death. She was the wife of Air-Chief Marshall the Lord Dowding, former commander-in-chief of the British Fighter Command, who died in 1970. Together, they shared a lifelong interest in spiritualism that informed their ethical concern for animals.

Selected Bibliography. Berry, Rynn, Interview with Lady Dowding, in *The New Vegetarians* (New York: Pythagorean Publishers, 1993), 137–152; Brophy, Brigid, The Darwinist's Dilemma, in David A. Paterson and Richard D. Ryder (Eds.), *Animals' Rights: A Symposium* (London: Centaur Press, 1979), 63–72; Muriel, the Lady Dowding, Furs and Cosmetics: Too High a Price? in Stanley Godlovitch, Roslind Godlovitch, and John Harris (Eds.), *Animals, Men, and Morals: An Enquiry into the Maltreatment of Non-Humans* (London: Gollancz, 1971), 25–40.

ANDREW LINZEY

DRAIZE TEST. *See* ACTIVISM FOR ANIMALS; LABORATORY ANIMAL USE.

DUCK STAMP ACT. *See* HUNTING.

DYSTRESS

Dystress is taken from the Greek root *dus* (bad), which has a notion of hard or bad or unlucky and removes the good sense of a word or increases its bad sense (e.g., dyspepsia, dysentery). Dystress means stress* with which the animal cannot cope (*see* ANIMAL WELFARE, Coping) and is usually a result of long-term (chronic) stress. It is to be differentiated from stress with which an animal can cope, sometimes referred to as *eustress*. It often involves activation of the hypothalamus with its connections to the pituitary gland, which controls many of the endocrine glands in the body. The adrenal cortex is often involved, and this leads to a rise in circulating corticosteroids. On other occasions, compromised functioning of the other endocrine glands can lead, for example, to poor weight gain and reproductive failure. Dystress may be an integral part of other aspects of suffering—an animal in pain* from a broken leg may be fearful of being moved or touched, as well as distressed by its inability to move.

DAVID B. MORTON

E

EAR CROPPING. *See* VETERINARIANS.

ECOFEMINISTS. *See* ANIMAL RIGHTS.

EDUCATION AND THE USE OF ANIMALS

Animals teach us about the world around us. Sometimes this learning is informal when we encounter animals in film, stories, and legends; as political or sports mascots; or even in everyday language. Other times it is formal in classrooms, laboratories, or museums or zoos.* The roles that animals play in education may be formal or informal, direct or indirect. Informal and indirect learning helps us form our attitudes about nonhuman animals. Formal and direct learning provides us with information about them. Together, our attitudes and information become the foundation of the principles and practice of animal welfare.*

Much of the use of animals in education is formal and indirect. The lessons serve primarily as examples of broader knowledge beyond the classroom or lab. This focus may tend to minimize animal welfare considerations related to teaching with animals due to the primary emphasis given to the teaching and learning objectives of these lessons.

From an animal rights* and welfarist perspective, each proposed educational use of animals must begin by evaluating its contribution to learning. Does the use actively engage the senses as we expect? Is it appropriate to the developmental stage of the students? Does it directly support the goals of the lesson(s)? How does it contribute to a base of knowledge that is necessary for the student's continued learning? Together, these questions serve to establish an educational justification for a specific proposal to use animals.

Even if educational objectives can clearly be met, it is still important to consider the animal welfare objectives. Every educational use of animals has an impact on the animals being studied. Therefore, a conscientious educator seeks to learn as much as possible about the natural way of life of the proposed animal subject. For living animals, this includes the normal social life, life-cycle needs (nutrition, growth, development, reproduction, survival), habitat needs (physical and social environment), and normal behaviors.

However, even for projects that use animals' body parts or remains, a concern for animal welfare dictates that we learn as much as possible about where and how the animals are acquired, how they were treated before they came to the classroom, and whether their choice was dictated by specific educational objectives or was based on availability, mass purchases, or other noneducational and nonwelfare conditions. Furthermore, the appropriate educational use of animals requires that we consider what happens to the animals after the educational objectives have been served.

The challenge for humane educational use of animals is to make the lesson reflect not only the specific materials and concepts to be learned by the students, but also to make the students aware that this use is a choice that must be made actively. One approach to humane education is to develop lists of allowed and prohibited activities to regulate the educational use of animals. The chief appeal of this approach is its ease of understanding and enforcement. However, the lessons of humane education may not be reinforced when students and teachers only follow a list of rules.

A second approach involves students in the choices to be made. The student learns about the natural life, the environmental setting, and the costs of the capture and study of the animal. In this approach, nonhuman animals become more than just the "material" for the lesson. The students and teacher take on the responsibility of deciding actively whether and how animals ought to be introduced into the curriculum. Humane education becomes a process of learning about animals in a new way and about the impact of their inclusion in educational activities. Furthermore, teachers and students who take the responsibility for asking about the roles that nonhuman animals will play in the curriculum engage in a process that is the essence of humane education, because it requires students and teachers to confront the issues of animal use explicitly. For the whole learning community (students, teachers, administrators, parents), the process of considering issues of acquisition, care and use, and disposition of nonhuman animals in educational activities is the foundation of a humane education ethic. In concert with clearly defined educational objectives, this process promotes a thorough, multifaceted understanding of animal welfare.

ANDREW J. PETTO

Policies in the United States

Policies on the use of animals in biology education in high schools and colleges have been developed by a number of organizations, but there is no national agreement. Many students contend that they can learn all the biology they need by studying the natural activities of animals without resorting to harming or killing them. However, there is a tradition in the United States, not found in several European countries, that permits high-school and undergraduate college students to conduct animal experiments that cause pain* and suffering.* The primary policy issues to be addressed are to what extent, if any, beginning biology students should be allowed to conduct experiments that involve inflicting painful conditions on sentient (*see* SENTIENTISM) animals, and whether dissections of frogs (*see* AMPHIBIANS) and other vertebrate species should be permitted or phased out.

The humane movement has sought to prohibit educational projects involving live vertebrate animals that cause either pain or death.* It has also opposed frog and other animal dissections. Recommended policies include provisions that student projects should not interfere with the animals' health or cause any pain, suffering, or death.

Historically, inhumane animal use has been most common in science fairs. These are extracurricular competitions in which junior- and senior-high-school students exhibit their projects. Humane standards at these fairs have been lacking in the past. In the 1960s and 1970s, the Animal Welfare Institute (AWI) reported on many student projects that treated animals cruelly. Typical were projects that involved failed attempts of surgical procedures on monkeys, rabbits, and other species, guinea pigs who were forced to inhale nicotine fumes until they died, and mice* who were given known toxic agents (such as cleaning fluid) to demonstrate their death. Frequently these teenagers worked in their home basements or garages, and they often won prizes for their efforts. What was particularly troubling was that these students were becoming insensitive to animal suffering and learning all the wrong lessons about how to treat animals from these school-sanctioned activities.

Today, humane standards in science competitions are much better than they were. Monkey surgery is no longer encountered, and at least some restrictions have been placed on harming and killing animals. Permissive policies and practices have been revised, and supervision and oversight of student experiments are now improved. But the rules of the largest science fair still permit the infliction of pain and death on vertebrate animals.

Even so, animal experiments conducted by students, either at high school or in colleges, are still largely unregulated. The federal Laboratory Animal Welfare Act* does not include the use of animals by elementary and secondary students. Policies prepared by various professional biology teachers' organizations for elementary and secondary schools are of a voluntary nature (with no force of law) and, in any case, do not always prohibit the infliction of animal pain.

At the college level, only a small fraction of animal experiments (those involving cats,* dogs,* hamsters, rabbits, guinea pigs, and farm animals) are legally subject to oversight review by institutional animal care and use committees* (IACUCs) under provisions of the federal Laboratory Animal Welfare Act. The many colleges that use only rats or mice or birds for student instruction (species not covered by the Laboratory Animal Welfare Act) do not have required oversight committees. This omission acts as a serious detriment to humane standards.

Policies on dissection vary but on the whole tend to favor the continuation of dissection in elementary and high schools and in colleges. When the National Association of Biology Teachers issued a policy in 1989 that recommended alternatives to dissection, some teachers objected. The policy was revised in 1990, and some of the 1989 provisions were reversed. Much work still needs to be done to improve the laws and policies governing student use of animals.

Selected Bibliography. [Documentation of improper high-school student science-fair projects,] *Information Report* (Animal Welfare Institute, Washington, DC) 17(2) (1968), 17(4) (1968), 18(2) (1969), 19(1) (1970), 19(2) (1970), 21(3) (1972), 21(4) (1972); National Association of Biology Teachers, Policy Statement, The Responsible Use of Animals in Biology Classrooms, Including Alternatives to Dissection (Reston, VA: NABT, 1989, revised 1990); Orlans, F. B., *In the Name of Science: Issues in Responsible Animal Experimentation* (New York: Oxford University Press, 1993); Westinghouse Science Talent Search, Facts and Official Entry Form, Science Service, 1719 N Street, N.W., Washington, D.C. 20036; Youth Science Foundation, Canada-Wide Science Fair Rules and Regulations, Youth Science Foundation, 151 Slater St., Suite 904, Ottawa KIP 5H3, Canada.

F. BARBARA ORLANS

Dissection and Vivisection Laws

Efforts to reform the use of animals in education have included the passage of laws and the adoption of policies concerning dissection and vivisection. Currently, four states in the United States have dissection-choice laws: Florida (enacted 1985), California (1988), Pennsylvania (1992), and New York (1994). These laws provide students below college or university level the option not to dissect an animal in lieu of another exercise not harmful to animals. Similar legislation has been introduced in Massachusetts, Maine, Illinois, New Jersey, Rhode Island, Louisiana, and Maryland, but has thus far failed to pass.

These laws are not ideal for a student conscientious objector. They apply only to students at the level of kindergarten through 12th grade; college students are not covered. Three of the four laws (Florida, California, and

New York) require written consent from the student's parent or guardian, a provision that limits the student's power to choose. Only two of the four laws (California and Pennsylvania) require that students be notified of their parent's choice. Teacher discretion not to exempt students is also granted in two of the states (California and New York). Definitions of "animal" are narrow in the Florida and New York laws, whose wording implies that choice applies only to the dissection of mammals and birds; only the California law appears to include invertebrates. Finally, private schools are exempted in all but the Pennsylvania law.

In the absence of legislation, progressive policies have been enacted. The Maine Department of Education in 1989 adopted a bill titled "Student's Right to Refrain from Harmful or Destructive Use of Animals" that had failed to pass. The Louisiana legislature issued a similar resolution in 1992. The Chicago Public School System in 1993 implemented a policy allowing students choice in dissection. Many other U.S. schools and school boards have similar policies. Various organizations and professional societies (e.g., the Humane Society of the United States,* the National Association of Biology Teachers, and the National Science Teachers Association) publish guidelines for animal use in education; these are advisory and have no legislative authority.

The great majority of schools in the United States, however, continue to operate without dissection policies. As the number of students who object to classroom practices harmful to animals grows, so do conflicts. The Dissection Hotline (800-922-FROG), operated by the National Anti-Vivisection Society, has received more than 100,000 calls since it was started in 1989 by the Animal Legal Defense Fund. A significant percentage of these calls come from the four states with dissection-choice laws, suggesting that these laws do not fully resolve the dissection issue for conscientious objectors. The National Association for Biomedical Research (202-857-0540) issues frequent reports concerning legislation dealing with animals in research, testing, and education. Classroom vivisection (harmful experiments on living animals) has also been addressed by U.S. laws. California, Florida, Illinois, Massachusetts, New Hampshire, New York, Maine, and Pennsylvania have laws prohibiting the infliction of painful and invasive experiments on animals in precollege education.

Internationally, the past decade has witnessed some changes. In 1987, animal vivisection and dissection were banned from all teaching establishments in Argentina. In 1993, the Italian parliament enacted a law recognizing the right of any person to refuse to participate in animal experimentation and dissection. In 1995, animal dissection was banned in all primary and secondary schools in the Slovak Republic. In June 1996, a committee of the High Court of Delhi issued a ruling banning animal dissection in Indian schools, where an estimated 60 million animals have been dissected annually.

Europe has seen probably the most changes regarding animal use in the schools. In England, pupils who object to dissection are allowed to use suitable alternatives without penalty, classroom procedures likely to harm vertebrate animals are disallowed, and living animals may not be used by surgeons or others merely to perfect their techniques, with the exception of microsurgery training. Several countries, including Germany, the Czech Republic, Norway, and Holland, prohibit the use of live animals in education when viable alternative methods exist; the rigor of such policies is questionable due to the discretion an educator could exercise regarding what does or does not qualify as a viable alternative. In 1995, the Karl-Franzens University in Graz, Austria, declared its intention to become completely animal free. A well-run activist organization called the European Network of Individuals and Campaigns for Humane Education (EuroNICHE) has spawned ongoing campaigns in Germany, Denmark, Belgium, the Czech Republic, Austria, Portugal, Poland, Bulgaria, Estonia, Holland, Ireland, and Romania.

At more advanced levels of education, such as medical and veterinary training, growing sensitivity to animal protection, increasing costs of animals, and improved alternatives technology are generating more animal-friendly approaches. According to the Physicians Committee for Responsible Medicine (202-686-2210), at least 34 medical schools in the United States now use no animals in their curricula. Of the 20 respondents to a 1995 survey of 31 veterinary schools in the United States and Canada conducted by the Association of Veterinarians for Animal Rights (AVAR),* 16 schools (80%) have implemented curriculum changes to accommodate students who do not wish to harm healthy animals. A 1994 survey found that 25 of 37 U.S. medical institutions (68%) no longer use cats* and kittens in intubation training.

Clearly, trends are emerging concerning the use of animals in education. Conscientious objection to dissection is increasing. Practices that cause harm to animals, while still common, appear to be declining. The number of laws granting the rights of citizens to choose humane alternatives is rising. Some nations have acknowledged the many problems surrounding classroom dissection and vivisection by banning such practices altogether. As activism gains momentum, new laws are being enacted.

Selected Bibliography. Animal Welfare Institute, *Animals and Their Legal Rights: A Survey of American Laws from 1641 to 1990*, 4th ed. (Washington, DC: Animal Welfare Institution, 1990); Balcombe, J. P., Dissection and the Law, *AV Magazine* 105(3) (1996): 18–21; Downie, R., and J. Meadows, Experience with a Dissection Opt-out Scheme in University Level Biology, *Journal of Biological Education* 29(3) (1995): 187–194; Francione, G. L., and A. E. Charlton, *Vivisection and Dissection in the Classroom: A Guide to Conscientious Objection* (Jenkintown, PA: American Anti-Vivisection Society, 1992); Sapontzis, S. F., We Should Not Allow Dissection of Animals, *Journal of Agricultural and Environmental Ethics* 8(2) (1995): 181–189.

JONATHAN BALCOMBE

Student Objection to Dissection

Increasingly, students have been objecting to the practice of dissection in the classroom on ethical grounds and demanding the "student rights option," a policy that guarantees the right of a student to an alternative educational exercise. As a legal issue, their objections pit the rights of students to freedom of religion or, more broadly, of conscience, under the First Amendment of the federal Constitution against teachers' rights to academic freedom (*see* EDUCATION AND THE USE OF ANIMALS, Student Rights and the First Amendment). The claim against dissection is based on the civil liberties of a human animal (the student) and only indirectly concerns a claim to rights for animals other than humans. To date, in several cases, the courts appear to be sympathetic to student claims.

A second issue raised by dissection in the classroom is whether using animals in laboratory exercises is an effective way of teaching anatomy, medicine, and behavior, for example. The few available studies comparing educational effectiveness suggest that the use of alternatives (*see* ALTERNATIVES TO ANIMAL EXPERIMENTS), such as computer software, models, and transparencies, is at least as likely to achieve the instructional goals.

Supporters of dissection frequently argue that "hands-on" experience is essential to the student's education. There is no evidence supporting this claim. Further, the term must be redefined to reflect current practices. Increasingly, as techniques of observation and intervention become more sophisticated, both for scientists and surgeons, hands-on experience is coming to refer more to the microscope, computer, and television monitor than to direct observation and manipulation of organs and body parts.

A number of studies have explored the impact of the experience of dissection on student attitudes and psychology. In a study of adults formerly involved in classes involving dissection, it was found that most people remember their first laboratory dissection vividly, with strong associated feeling, and many consider it an important experience of their childhood or adolescence. For a minority of these, the memory has some features of a traumatic event: it is easily remembered and negatively emotionally loaded. Interviews with these adults and with students currently involved in classroom dissection suggest several reasons why this experience is emotionally loaded for most individuals, and negatively so for a minority. (1) Unresolved issues around the early exploration of death by young people in this culture are part of what gives emotional loading to the experience of dissection. Whereas children are exposed to death and violence graphically through television and other media every day, often they are shielded from direct exposure to serious illness, dying, and death when it strikes loved ones. For this reason, the killing, dying, and death of a frog or rat in the classroom tends to assume significant psychological importance (*see* DEATH OF ANIMALS). (2) Dissection teaches lessons that are strikingly at odds with the constructive adolescent self-discovery process. Instead of being associated with individuality, integrity, and privacy, the body is objectified, reduced to internal workings, and publicly displayed. (3) In dissection, there is public encouragement and sanction of the otherwise censured impulse to kill and/or mutilate. This likely arouses a developmentally early form of evil called "defilement"—a common childhood experience exemplified by pulling the wings off a butterfly or tormenting other small animals. The impulse to defile is a mixture of disgust and fascination at the suffering* of another individual.

Selected Bibliography. Francione, G., and A. Charlton, *Vivisection and Dissection in the Classroom: A Guide to Conscientious Objection* (Jenkintown, PA: American Anti-Vivisection Society, 1992); Hepner, L., *Animals in Education: The Facts, Issues, and Implications* (Albuquerque, NM: Richmond, 1994); Kelly, J., Alternatives to Aversive Procedures with Animals in the Psychology Teaching Setting, in M. Fox and L. Mickley (Eds.), *Advances in Animal Welfare Science 1985/86* (Washington, DC: Humane Society of the United States, 1985), 165–184; Shapiro, K., The Psychology of Dissection, *Animals' Agenda* 11(9) (November 1991): 20–21.

KENNETH J. SHAPIRO

Student Rights and the First Amendment

The free-exercise clause of the First Amendment to the U.S. Constitution provides that "Congress shall make no law . . . prohibiting the free exercise" of religion. Although the U.S. Supreme Court has not yet had an opportunity to interpret this First Amendment guarantee in the precise context of a student objection to dissection and vivisection in the classroom, the Court has guaranteed First Amendment protection in cases that are relevant to the issue.

The Supreme Court has long drawn a distinction between *belief* and *conduct* in the context of interpreting the constitutional guarantee of freedom of religion. In *Cantwell v. Connecticut* (1940) the Court held that the free-exercise clause "embraces two concepts—the freedom to believe and the freedom to act. The first is absolute but, in the nature of things, the second cannot be. Conduct remains subject to regulation for the protection of society." That is, government cannot regulate religious belief and can only regulate religious conduct, a notion that was upheld in *Thomas v. Review Board* (1981) and reaffirmed with the Religious Freedom Restoration Act (1993).

The legal framework established by the Court and Congress involves six elements for evaluating the suitability of the regulation of conduct that is claimed to be protected by the free-exercise clause of the First Amendment. First, the regulation must constitute *state action*. The reason for this requirement is that with certain exceptions not relevant here, the U.S. Constitution protects us only from the action of some branch of government. Although there may be other federal and state laws that apply to the actions of private institutions, a claim under the First Amendment requires that the student show that there is a legally relevant relationship between the government (state, federal, local) and the challenged regulation so that the regulation may be treated as an act of the state itself. For example, a requirement to vivisect or dissect imposed by a state university would constitute state action. The same requirement imposed by a private school, even one that receives state money, may not qualify as state action depending on the relationship of the private institution to the government.

Second, the First Amendment guarantee of freedom of religion protects only *religious* or *spiritual* beliefs and does not protect bare "ethical" beliefs. It is important to understand, however, that the Supreme Court has held quite clearly that the religious belief need not be "theistic" or based on faith in a "God" or "Supreme Being," and that the claimant need not be a member of an organized religion. So, for example, a person who accepts "reverence for life" as a spiritual belief, but who does not believe in "God" per se, would qualify for First Amendment protection. Finally, it is not necessary that the belief be recognized as legitimate by others who claim to be adherents of a religious or spiritual doctrine. So, for example, it is not relevant to a claim that the killing of animals is contrary to Christian belief that others who identify themselves as Christians feel that animals have no rights and should not be the subject of moral concern.

Third, the student who asserts a First Amendment right must be *sincere*. If, for example, a student objects to vivisection on the ground that it violates the student's belief in the sanctity of all life, the fact that the student eats meat, wears leather, and trains fighting dogs for a hobby may indicate that the student's asserted concern for the sanctity of all life is insincere and should not be protected.

Fourth, the state action must actually *burden* the religious belief. This requirement is not usually a problem in the context of student rights to oppose animal exploitation because in most cases the state is conditioning the receipt of a benefit—an education—on the performance of an act (vivisection or dissection) that is proscribed by the student's religious belief system.

Fifth, once it is determined that the state is placing a burden on a sincerely held religious or spiritual belief, then the state has the burden to prove that the regulation serves a *compelling state interest*. That is, the state must prove that there is a very important reason for the regulation. Normally, schools argue that the state has a compelling interest in establishing educational standards. That may very well be true, but if the school has allowed other students not to vivisect or dissect because they happened to be ill on the day of the lab, then the claim that the state has a compelling interest in particular educational standards has less force.

Sixth, the state must show that the requirement is the *least restrictive means* of satisfying any state interests. For example, if there are educationally sound nonanimal alternatives to the vivisection/dissection requirement, then the state must allow such alternatives.

In addition to the protection afforded the free exercise of religious and spiritual beliefs (broadly defined) protected by the First Amendment, there may be other federal and state laws that are relevant to the student's claim depending on the particular case. Other relevant federal laws concern freedom of speech and association, due process and equal protection, procedural due process, and civil rights. Other relevant state laws include state (as opposed to federal) constitutional guarantees, as well as laws concerning contract, tort, and discrimination within educational institutions. Several states (California, Florida, Pennsylvania, New York) have provided for a limited statutory right to object to vivisection and dissection.

Selected Bibliography. *Cantwell v. Connecticut*, 310 U.S. 296 (1940); *Church of the Lukumi Babalu Aye, Inc. v. City of Hialeah*, 508 U.S. 520 (1993); Francione, Gary L., and Anna E. Charlton, *Vivisection and Dissection in the Classroom: A Guide to Conscientious Objection* (Jenkintown, PA: American Anti-Vivisection Society, 1992); The Religious Freedom Restoration Act, 42 U.S.C. 2000bb (1993); *Thomas v. Review Board*, 450 U.S. 707 (1981).

ANNA E. CHARLTON

Field Studies

Field studies of many animals contribute information on the complexity and richness of animal lives that has been, and is, very useful to those interested in animal rights* and animal welfare.* Students of behavior want to be able to identify individuals, assign gender, know how old animals are, follow them as they move about, and possibly record various physiological measurements, including heart rate and body temperature. Animals living under field conditions are generally more difficult to study than individuals living under more confined conditions, and various methods are often used to make them more accessible to study. These include activities such as (1) handling, (2) trapping,* using various sorts of mechanical devices that might include luring using live animals as bait, (3) marking individuals using colored tags or bands, and (4) fitting individuals with various sorts of devices that transmit physiological and behavioral information telemetrically (radio collars, other instruments that are placed on an animal, or devices that are implanted).

Trapping is often used to restrain animals while they are marked, fitted with tags that can be used to identify them as individuals, or equipped with radiotelemetric devices that allow researchers to follow them or to record physiological measurements. However, the trapping and handling of wild animals are not the only ways in which their lives can be affected, for just "being there" and watching or filming them can influence their lives—what seem to be minor intrusions can really be major intrusions. Here are some examples:

1. Magpies who are not habituated to human presence spend so much time avoiding humans that this takes time away from essential activities such as feeding.
2. Adélie penguins exposed to aircraft and directly to humans showed profound changes in behavior, including deviation from a direct course back to a nest and increased nest abandonment.
3. The foraging behavior of little penguins (average mass of 1,100 grams) is influenced by their carrying a small device (about 60 grams) that measures the speed and depth of their dives. The small attachments result in decreased foraging efficiency. Changes in behavior such as these are called the "instrument effect."
4. Mate choice in zebra finches is influenced by the color of the leg band used to mark individuals, and there may be all sorts of other influences that have not been documented.
5. The weight of radio collars can influence dominance relationships in adult female meadow voles. When voles wore a collar that was greater than 10% of their live body mass, there was a significant loss of dominance.
6. Helicopter surveys of mountain sheep that are conducted to learn more about these mammals disturb them (as well as other animals) and greatly influence how they use their habitat, increase their susceptibility to predation,* and also increase nutritional stress.*

While there are many problems that are encountered both in laboratory and field research, the consequences for wild animals may be different from and greater than those experienced by captive animals, whose lives are already changed by the conditions under which they live. This is so for different types of experiments that do not have to involve trapping, handling, or marking individuals. Consider experimental procedures that include (1) visiting the home ranges, territories, or dens of animals, (2) manipulating food supply, (3) changing the size and composition of groups by removing or adding individuals, (4) playing back vocalizations, (5) depositing scents (odors), (6) distorting body features, (7) using dummies, and (8) manipulating the gene pool. All of these manipulations can change the behavior of individuals, including movement patterns, how space is used, the amount of time that is devoted to various activities such as hunting, antipredatory behavior, and various types of social interactions such as caregiving, social play, and dominance interactions. These changes can also influence the behavior of groups as a whole, including group hunting or foraging patterns, caregiving behavior, and dominance relationships, and can influence nontarget individuals as well. There are also individual differences in responses to human intrusion.

Although we often cannot know about various aspects of the behavior of animals before we arrive in the field, our presence does seem to influence what animals do when we enter into their worlds. What appear to be relatively small changes at the individual level can have wide-ranging effects in both the short and long terms. On-the-spot decisions often need to be made, and knowledge of what these changes will mean to the lives of the animals who are involved deserves serious attention. A guiding principle should be that wild animals whom we are privileged to study should be respected, and when we are unsure about how our activities will influence the lives of the animals being studied, we should err on the side of the animals and not engage in these practices until we know the consequences of our acts.

Selected Bibliography. Bekoff, M., and D. Jamieson, Ethics and the Study of Carnivores, in J. L. Gittleman (Ed.), *Carnivore Behavior, Ecology, and Evolution*, vol. 2 (Ithaca, NY: Cornell University Press, 1996), 15–45; Cooper, N. S., and R. C. J. Carling (Eds.), *Ecologists and Ethical Judgements* (New York: Chapman and Hall, 1996); Farnsworth, E. J., and J. Rosovsky, The Ethics of Ecological Field Experimentation, *Conservation Biology* 7 (1993): 463–472; Jamieson, D., and M. Bekoff, Ethics and the Study of Animal Cognition, in M. Bekoff and D. Jamieson (Eds.), *Readings in Animal Cognition* (Cambridge, MA: MIT Press, 1996), 359–371; Kirkwood J. K., A. W. Sainsbury, and P. M. Bennett, The Welfare of Free-living Wild Animals: Methods of Assessment, *Animal Welfare* 3 (1994): 257–273; Laurenson, M. K., and T. M. Caro, Monitoring the Effects of Non-trivial Handling in Free-living Cheetahs, *Animal Behaviour* 47 (1994): 547–557.

MARC BEKOFF

EMBRYO RESEARCH

The study of nonhuman animal embryos has provided a wealth of information about normal embryonic development. This basic research has important clinical relevance. For example, the research on fertilization in sea urchins and mice* has provided the data needed to develop methods for in vitro fertilization. Studies of the development of the nervous system in frogs (*see* Amphibians) have permitted researchers to identify the processes involved in a major birth defect, spina bifida, in which the spinal cord does not form normally. Limb development is another developmental process that has been extensively studied in nonhuman animal models (*see* ANIMAL MODELS, Biomedical and Behavioral Science). Basic research on chicken embryos first identified the importance of retinoic acid in limb formation. These studies made it clear that drugs containing forms of retinoic acid, often used in formulations designed to treat acne and wrinkling of the skin, are potentially dangerous to the unborn fetus.

The choice of animal model for a particular embryological question depends on several factors. For example, fruit flies are an excellent model for examining how genes control the formation of the basic body plan, and for asking questions such as where the head will be and where dorsal and ventral will be located. On the other hand, sea urchins have been widely used for studies of fertilization because the processes are easily visualized. The advantage of using invertebrates such as fruit flies and sea urchins is that they are available in large numbers, at low cost, and are small in size and relatively easy to house in a laboratory. On the other hand, the disadvantage is that the relevance of the mechanisms used in invertebrate embryonic development to those used in humans is not always immediately clear. The use of vertebrates, and particularly mammals such as mice and primates, has the advantage that the results are likely to be more directly relevant to human development. However, smaller numbers of embryos are typically available, and they are larger in size and cost more to maintain. As a result, research is often first carried out in animals that are less closely related to humans. Once mechanisms are understood there, then more targeted research can be carried out on vertebrates and finally mammals.

The ethics of using nonhuman animal embryos in research has not been widely discussed. This is most likely because the vast majority of embryonic research takes place in the newly fertilized egg and early embryo. The stages studied most often occur before the nervous system is functional, so that neither pain* nor consciousness (*see* ANIMAL COGNITION, Consciousness and Thinking) is an issue. In contrast, the question of whether human embryos should ever be used in research has generated a great deal of controversy. However, even here, most people agree that prior to neural tube closure, even human embryos are "too rudimentary to have interests or rights and thus cannot be harmed when used in research" (Robertson 1995).

Selected Bibliography. Robertson, J. A., Symbolic Issues in Embryo Research, *Hastings Center Report* 25 (1995): 37–38; What Research? Which Embryos? *Hastings Center Report* 25 (1995): 36–46.

ANNE C. BEKOFF

EMPATHY FOR ANIMALS

"Empathy" is a term used to describe the tendency that most people have to be emotionally affected by witnessing the emotion (e.g., suffering* or distress*) of another person. On the whole, the more empathic we are, the more likely we are to show compassion and concern and to offer help to someone in distress.

Psychologists studying empathy have tended to assume that people who are strongly emotionally affected by the distress of a human being will also be strongly emotionally affected by the distress of a nonhuman animal. Indeed, a recent questionnaire study has shown that empathy with people and empathy with animals do seem to be correlated, that is, people who reported greater emotional concerns about humans were also more likely to report greater concerns about animals, but this association was not as strong as might have been expected. There were still plenty of people who showed high empathy with humans but low empathy with animals, and others who were very concerned about animals but showed concern no greater than average about people. So, although there does appear to be some association, feeling empathy with or compassion for animals seems to be a process that is not entirely the same as feeling empathy with or compassion for people.

From a developmental perspective, it has long been a popular belief that children who are brought up to love and care for pet animals will develop into people who also love and care for people. The notion seems to be that caring for something smaller, weaker, and more dependent than oneself during childhood will instill an enhanced sense of empathy or compassion in adulthood that can be applied to the weaker and more dependent individuals in society. However, the mere existence of a few well-known tyrants and mass murderers (for example, Hitler) who were also pet lovers seems to weaken the idea that keeping pet animals leads inevitably to empathy with humans.

A recent study has found that childhood pet keeping is indeed associated with higher levels of human-directed empathy in adulthood. When university students were asked to report on the pets they (and their immediate families) had kept during childhood, it was found that those who had grown up with more pets, and those who had been more attached to those pets, tended to obtain higher scores on questionnaire measures of empathy with humans. But even more striking was the finding that students who had grown up with pets were more likely to show concern and compassion for the welfare of nonhuman animals. Thus it seems that childhood pet keeping is related to adult empathy with humans, but it is even more closely related to adult empathy with animals. (*See also* ATTITUDES TOWARD ANIMALS, Attitudes among Children, Attitudes among Students; COMPANION ANIMALS AND PETS; CRUELTY TOWARD ANIMALS AND HUMAN VIOLENCE.)

Selected Bibliography. Eisenberg, N., and J. Strayer, *Empathy and Its Development* (Cambridge: Cambridge University Press, 1987); Mehrabian, A., and N. Epstein, A Measure of Emotional Empathy, *Journal of Personality* 40 (4) (1972): 525–543; Paul, E. S., and J. A. Serpell, Childhood Pet Keeping and Humane Attitudes in Young Adulthood, *Animal Welfare* 2 (1993): 321–337.

ELIZABETH PAUL

ENDANGERED SPECIES

Few persons doubt that humans have obligations concerning endangered species. Persons are helped or hurt by the condition of their environment, which includes a wealth of wild species, many of which are currently under threat of extinction. Whether humans have duties directly to endangered species is a deeper question, part of the larger issue of biodiversity conservation. Many believe that humans have such duties. The United Nations World Charter for Nature states, "Every form of life is unique, warranting respect regardless of its worth to man." The Biodiversity Convention affirms "the intrinsic value of biological diversity." Both are signed by over a hundred nations.

Many endangered species have no resource value, nor are they particularly important for the usual humanistic reasons: medicine, industry, agricultural resources, scientific study, recreation, ecosystem stability, and so on. Many environmental ethicists (*see* ENVIRONMENTAL ETHICS) believe that species are good in their own right, whether or not they are good for anything. The duties-to-persons-only line of argument leaves deeper reasons untouched.

There are two levels of questions: facts (a scientific issue, about species) and values (an ethical issue, involving duties). Sometimes species can seem made up, since some biologists regularly change their classifications as they attempt to understand and classify nature's complexity. On a more realist account, a biological species is a living historical form (Latin *species*), an ongoing lineage expressed in organisms and encoded in the flow of genes. In this sense, species are objectively there—found, not made up.

Responsibility to species differs from that to individuals, although species are always exemplified in individuals. When an individual dies, another replaces it. Tracking its environment, the species is conserved and modified. Extinction shuts down the generative processes, a kind of superkilling.

A species lacks moral agency,* reflective self-awareness, sentience, or organic individuality. An ethic that features humans or sentient animals may hold that specific-level processes cannot count morally. But each ongoing species defends a form of life, and these forms are, on the whole, good kinds.

Humans are shutting down the life stream. One argument is that humans ought not to play the role of murderers, superkillers. The duty to species can be overridden, for example, with pests or disease organisms. But a prima facie duty (presumably, unless there are reasons to the contrary) stands. Increasingly, humans have a vital role in whether these species continue. The duties that such power generates no longer attach simply to individuals or persons but are duties to the species lines, kept in ecosystems, because these are the more fundamental living systems, the wholes of which individual organisms are the essential parts. On this view the appropriate survival unit is the appropriate level of moral concern.

It might seem that for humans to terminate species now and again is quite natural. Species become extinct all the time. But there are important theoretical and practical differences between natural and anthropogenic (human-generated) extinctions. In natural extinction, a species dies when it has become unfit in habitat, and other species appear in its place, a normal turnover. By contrast, artificial extinction shuts down speciation. One opens doors, the other closes them. Humans generate and regenerate nothing; they dead-end these lines. Relevant differences make the two as morally distinct as death by natural causes is from murder.

Selected Bibliography. Gunn, Alastair S., Preserving Rare Species, in Tom Regan (Ed.), *Earthbound: New Introductory Essays in Environmental Ethics* (New York: Random House, 1984); Norton, Bryan G. (Ed.), *The Preservation of Species* (Princeton, NJ: Princeton University Press, 1986); Rolston, Holmes, *Conserving Natural Value* (New York: Columbia University Press, 1994); Rolston, Holmes, *Environmental Ethics* (Philadelphia: Temple University Press, 1988); Wilson, E. O., *The Diversity of Life* (Cambridge, MA: Belknap Press of Harvard University Press, 1992).

HOLMES ROLSTON III

ENRICHMENT FOR ANIMALS

During the past 25 years the recognition that captive wild animals are in need of richer environments than those traditionally afforded them has become the accepted norm. Often this recognition has spurred the production of more beneficial behavioral conditions for animals in our care, but in some cases it has resulted in richer-appearing environments that please humans, but do little or nothing to improve the animals' well-being.* The term "enrichment" might better be limited to those circumstances in which there is measurable improvement in the behavioral and physiological well-being of the animal.

Historically, there were distinctions between "behavioral enrichment" and "environmental enrichment." These were based on suggestions that there were two radically different approaches to improving the lot of captive animals. The behavioral enrichment approach focused on engineering environments that provided opportunities that were likely to elicit species-typical behaviors. For example, occasionally producing the sounds of crickets in an otter exhibit and providing means by which the otters could hunt and capture crickets resulted in considerable display of species-typical behaviors. Supporters of environmental enrichment suggested that providing a rich-enough environment precluded the need for engineering artificial "hunts" or other apparatus that rewarded animals for particular responses. For example, if a captive forest with sufficient food was provided for chimpanzees,* this might be sufficient to encourage significant amounts of species-typical behaviors. Today, the term "environmental enrichment" is typically used to refer to all efforts to improve the circumstances of captive animals (*see also* ZOOS). Methods of providing more stimulating environments for animals obviously depend on the species involved, but some examples of simple and inexpensive methods that will serve for many animals include the following:

1. Provide interesting ways for them to hunt for food. Hide their provisions in trees or behind objects in ever-varying ways so that they may have the joy of discovering them.

2. Simple objects such as balls can be rotated with other toys, and where possible, some possibility for their animation can be provided. A ball suspended tetherball style will often lead to greater interaction and entertainment for animals than one simply thrown in their living space.

3. Most young animals love to explore new situations. A trip to the local toy store may yield giant, durably made building elements that may be assembled and reassembled into ever-changing steps to climb and holes to dive into.

4. A simple switch or motion detector can be used to allow animals to control various parts of the environment. The range of opportunities is limited only by imagination and budget. Inexpensive suggestions include allowing animals to control the dimming or brightening of lights in their room; to control radios, televisions, or video recorders (perhaps even with motion pictures of their favorite companions to entertain themselves while humans are at work); to rotate a wheel or perform other exercise to deliver food treats; or to turn on showers or mists in which to play.

Selected Bibliography. Markowitz, H., *Behavioral Enrichment in the Zoo* (New York: Van Nostrand Reinhold, 1982); Markowitz, H., and C. Aday, Power for Captive Animals: Contingencies and Nature, in D. Shepherdson, J. Mellen, and M. Hutchins (Eds.), *Second Nature: Environmental Enrichment for Captive Animals* (Washington, DC: Smithsonian Institution Press, 1998); Markowitz, H., C. Aday, and A. Gavassi, Effectiveness of Acoustic "Prey": Environmental Enrichment for a Captive African Leopard (*Panthera pardus*), *Zoo Biology* 14 (1995): 371–379; Markowitz, H., and A. Gavassi, Eleven Principles for Improving the Quality of Captive Animal Life, *Lab Animal* 24 (1995): 30–33; Markowitz, H., and S. W. Line, The Need for Responsive Environments, in B. E. Rollin and M. L. Kesel (Eds.), *The Experimental Animal in Biomedical Research*, vol. 1 (Boca Raton, FL: CRC Press, 1990), 153–170.

HAL MARKOWITZ

Enrichment and Research

Changes in the conditions under which animals are kept that appear superficially to improve animals' lives do not always have the desired effect. Such contradictory results have most often been found when animals are kept in large numbers under standardized conditions on farms or in laboratories. To measure effects of proposed improvements in living conditions on the welfare of large numbers of animals usually requires carefully designed experiments. If you want to know whether changing the diet of 1,000 rats in a laboratory colony improves their health, you have to keep careful records of the animals' condition before and after the diet change to see if the new diet really improves the health of colony members.

"Enrichment" has potential costs as well as potential benefits. On the surface, it seems likely that an animal living with others or in an interesting environment would be happier than an animal that spends its entire life alone in a standard laboratory cage. But consider the Norway rat, a common laboratory animal. When placed together, groups of male rats will engage in a series of fights and form a dominance hierarchy with one or more males dominant over the rest. Subordinate individuals are continually harassed by dominant animals, and within the confines of a laboratory cage, subordinate rats are forced into constant contact with their superiors.

Enhancement of the physical environment can also have undesired side effects. Consider the Mongolian gerbil. Gerbils are easy to handle and do not appear stressed by interaction with humans. However, if you provide a breeding pair of gerbils with an environment where they are free to dig tunnels (as they do in nature) and allow them to rear their young in the underground nest chambers they construct, such young behave strangely when they are grown. They flee when you attempt to pick them up. When captured, they frequently have seizures. Here, enrichment seems to decrease, not increase, the well-being* of animals who are going to spend their lives interacting with humans.

Other attempts to improve the well-being of caged animals may have similar paradoxical effects, not because of the nature of the animals, but because of the economics of animal maintenance. Most people seem to believe that the larger the enclosure in which an animal is kept, the better off the animal will be. However, rats in nature spend most of their lives in burrows consisting of small nest chambers connected by even smaller tunnels. Perhaps rats like to be kept in closely confined spaces. In fact, when given a choice between tall cages and short ones, rats are nonresponsive. Similarly, researchers at Oxford University in England have found that domesticated hens raised in the cramped "battery cages" (*see* CHICKENS) used for commercial egg production show no preference when given the choice between a large pen and a battery cage.

Existing standards for animal maintenance have evolved over the years with revisions based on professional judgment and personal evaluations. Such informal development of standards for animal maintenance does not inspire confidence that the procedures in use today are optimal. On the other hand, the equally personal basis for many proposed changes in maintenance procedures suggests that such changes may not have the desired result of enhancing the well-being of animals. Paradoxical consequences of alterations in maintenance conditions intended to improve the well-being of animals in laboratories and on farms are likely. More research on consequences for animals of proposed changes in living conditions is needed. (*See also* LABORATORY ANIMAL USE.)

Selected Bibliography. Clark, M. M., and B. G. Galef, Jr., Effects of Rearing Environment on Adrenal Weights, Sexual Development, and Behavior in Gerbils: An Examination of Richter's Domestication Hypothesis, *Journal of Comparative and Physiological Psychology* 94 (1980): 857–863; Dawkins, M. S., Do Hens Suffer in Battery Cages? Environmental Preferences and Welfare, *Animal Behaviour* 25 (1977): 1034–1046; Dawkins, M. S., From an Animal's Point of View: Motivation, Fitness, and Animal Welfare, *Behavioral and Brain Sciences* 13 (1990): 1–61; Galef, B. G., Jr., and

P. Durlach, Should Large Rats Be Housed in Large Cages? An Empirical Issue, *Canadian Psychology* 34 (1993): 203–207.

BENNETT G. GALEF, JR.

ENVIRONMENTAL ETHICS

Anthropocentric (human-centered) environmental ethics bases concern for the nonhuman natural environment (including animals) on the benefits it provides humans. It treats only humans as of direct and intrinsic moral concern. Taking care of a pet (*see* COMPANION ANIMALS AND PETS) or a park is done solely because they are useful to us. Anthropocentrism* is often defended by appeals to biblical passages that give humans "dominion over . . . every living thing that moves upon the earth" (Genesis 1:28). In contrast, nonanthropocentric environmental ethics bases the protection of the environment on its intrinsic value. It conceives of nonhuman nature as important in ways that surpass its instrumental (or use) value to humans.

A sentiocentric (sentience-centered; *see* SENTIENTISM) environmental ethic holds that sentient creatures—those who can feel and perceive—are morally important in their own right. Some of the best-known defenders of animals accept this ethic, including Peter Singer. Because it is likely that only vertebrate animals—mammals, birds, fish,* amphibians,* and reptiles*— consciously feel and perceive, a sentiocentric environmental ethic treats invertebrate nature as solely of instrumental value for sentient creatures. Such an ethic protects trees and ecosystems, for example, not for their own sake, but because they provide a habitat for sentient creatures.

Sentiocentrism ruptures the boundary of the traditional human-only moral club and may have radical implications for animal agriculture, animal experimentation, and hunting.* Nonetheless, from the perspective of broader environmental ethics, sentiocentrism is but a small modification of the traditional ethic. It extends moral concern beyond humans only to our closest cousins, the sentient animals, and denies direct moral concern to 99% of living beings on the planet, as well as species and ecosystems. Sentiocentrists respond that it makes no sense to care directly about trees or ecosystems and that the idea of owing obligations to bacteria is foolish.

Biocentric (life-centered) environmental ethics views all living beings as worthy of direct moral concern. Biocentrists contend that although plants and invertebrate animals do not have preferences, they nonetheless have goods of their own that we should morally consider. Though a tree does not care if its roots are crushed by a bulldozer, crushed roots are still bad for the tree and not just for the homeowner who wants its shade. Insentient living beings have a welfare of their own that should be part of direct environmental concern. Albert Schweitzer's* reverence-for-life ethic is an example of biocentrism.

Ecocentric (environment-centered) environmental ethics holds that entire species and ecosystems are morally important in their own right. Ecocentrists reject the idea that only individuals (a particular animal or plant) are appropriate objects of direct moral concern. They believe that whole ecosystems and species are intrinsically valuable and not simply the individuals in them. Aldo Leopold's concern to preserve the integrity, stability, and beauty of the biotic community is an example of an ecocentric ethic.

These broader environmental ethics view concern for animals as only a first step toward extending moral concern beyond humans to include the natural, nonhuman environment. This broadening of concern creates conflict. For example, hunters and fishers can show great ecocentric concern for the perpetuation of species and ecosystems while placing low (or no) moral value on the lives and welfare of individual animals. Conversely, defenders of sentient animals can have great concern for the well-being* of individual animals while placing low (or no) moral value on the protection of plants, the perpetuation of species, or the preservation of ecosystems.

These conflicts are not simply theoretical. Feral goats (*see* FERAL ANIMALS) have been shot to protect rare plants. Conservation of endangered species,* like the California condor, often involves captive breeding programs that harm the individual for the sake of the species. Preservation of ecosystems calls for the elimination of "exotics," as when lake trout introduced into Yellowstone Lake are poisoned to protect the integrity of the ecosystem. Restoration of ecosystems sometimes involves bringing back predators. This not only disrupts the lives of the predators, but puts responsibility for the suffering of their prey in the hands of humans.

Some defenders of animals say that only human-induced suffering* and death* are bad things that should be prevented. It is human violation of animal rights* that needs to be prevented, not natural suffering and death in the wild. However, if one believes that animal rights are logically analogous to human rights, then humans are responsible for failing to assist an animal in distress,* just as we are culpable when we fail to assist a human in distress. The worry that a consistent commitment to protect the lives and welfare of animals would involve massive human intervention into natural systems has led some to claim that defenders of animals cannot be environmentalists.

Selected Bibliography. Callicott, J. Baird, Animal Liberation: A Triangular Affair, in *In Defense of the Land Ethic* (Albany: State University of New York Press, 1989); Hettinger, Ned, Valuing Predation in Rolston's Environmental Ethics: Bambi Lovers versus Tree Huggers, *Environmental Ethics* 16(1) (Spring 1994): 3–20; Rolston, Holmes, III, *Environmental Ethics* (Philadelphia: Temple University Press, 1988); Sagoff, Mark, Animal Liberation and Environmental Ethics: Bad Marriage, Quick Divorce, *Osgoode Hall Law Journal* 22(2) (Summer 1984): 297–307; Varner, Gary, Can Animal Rights Activists Be Environmentalists? in Christine Pierce and Donald

Van DeVeer (Eds.), *People, Penguins, and Plastic Trees*, 2nd ed. (Belmont, CA: Wadsworth, 1995).

NED HETTINGER

Environmental Ethics versus Animal Rights

The modern animal rights* and environmental ethics movements have much in common. They both became popular in the 1970s. They are both opposed to anthropocentrism,* the belief that the only value things have is the value they have for fulfilling human needs and desires. Both movements insist that wild animals are sources of value independent of human needs and desires. Environmentalists and animal rights advocates have frequently been on the same side of public controversies. They both favor preserving the forests spotted owls need to survive, rather than allowing loggers to cut down these forests, and they both favor banning whaling and fishing techniques that kill dolphins.

Nevertheless, there are four significant differences between animal rights philosophy and environmental ethics. First, animal rights philosophy is directly concerned with the condition of animals used in agriculture, education,* and entertainment. Animal rights principles apply as directly to the suffering of farm animals kept in small cages and stalls and to the burning, cutting, and killing of animals in research centers as to the plight of spotted owls in the wild and dolphins in the oceans. However, except where they are allowed to graze on open range, an increasingly rare practice, animals that are bred and controlled by humans are not part of a natural environment. Consequently, their well-being* is not of direct concern to environmental ethics, although the impact of massive animal husbandry on the environment is of concern.

Second, animal rights principles refer directly only to sentient beings (*see* SENTIENTISM), beings capable of feeling pleasure and pain.* These are all animals; none are plants or nonliving things. Plants and nonliving things, like the redwood forests in which spotted owls live and the clean water fish* need to live, are morally important in animal rights philosophy only as supports for animal life. However, in environmental ethics, plants, rivers, the atmosphere, species, and ecosystems are frequently objects of moral concern for their own sake. Environmental ethicists have even acknowledged that they would support killing animals such as deer if that were the only way to preserve a species of plants the animals were eating.

Third, animal rights principles tend to focus on pain and death* as moral evils to be minimized and postponed when they cannot be avoided altogether. Environmental ethicists have criticized animal rights supporters for this. They advocate recognizing pain and death as essential elements of the life process and of the order of nature. They focus on the elimination of elements of nature, as when species are driven to extinction, and on the destruction of natural systems, such as pollution of lakes, as the evils that we need to oppose.

Fourth, animal rights principles focus on the well-being of individuals. On the other hand, much environmental ethics is holistic. This means that the object of primary moral concern is a system or structure—a whole—rather than the individuals found in the system. Some animal rights philosophers have criticized the willingness of environmental ethicists to sacrifice individual well-being if that is what it takes to preserve a species or ecosystem.

Selected Bibliography. Callicott, J. Baird, Animal Liberation: A Triangular Affair, in *In Defense of the Land Ethic* (Albany: State University of New York Press, 1989); Rodman, John, The Liberation of Nature, *Inquiry* 20 (1977): 83–145; Russow, Lilly-Marlene, Why Do Species Matter? *Environmental Ethics* 3 (1981): 106–112; Sapontzis, S. F., Plants and Things, in *Morals, Reason, and Animals* (Philadelphia: Temple University Press, 1987); Warren, Mary Anne, The Rights of the Nonhuman World, in Robert Elliot and Arran Gare (Eds.), *Environmental Philosophy: A Collection of Readings* (University Park: Pennsylvania State University Press, 1983).

STEVE F. SAPONTZIS

EQUAL CONSIDERATION

Equal consideration, whether for humans or animals, means in some way giving equal moral weight to the relevantly similar interests of different individuals. By itself this is very vague and abstract, yet it is extremely important. Aristocratic, feudalistic, Nazi, and other elitist worldviews have often denied that human beings are subject to any sort of basic moral equality. Moreover, to extend equal consideration (on any reasonable interpretation of this idea) to animals would represent a major departure from common thinking and practice throughout the world.

At an abstract level, equal consideration for animals would rule out a general discounting of animals' interests, an across-the-board devaluing of their interests relative to ours. An example of such devaluing would be the judgment that a monkey's interest in avoiding pain* of some amount is intrinsically less important than a human's interest in avoiding pain of the same amount. At a practical level, equal consideration for animals would rule out the routine overriding of animals' interests in the name of human benefit. While equal consideration is in agreement with numerous ethical theories, it is not in agreement, if extended to animals, with any view that sees animals as essentially resources for human use and amusement.

Assuming that humans are entitled to equal consideration, then unequal consideration for animals is justified only if there is some morally relevant difference between humans and animals. Peter Singer has argued that there is no such difference between all humans and all animals, so that denying equal consideration to animals is speciesism.*

Among leading philosophical arguments for a crucial moral difference between humans and animals are the following. Contract theories typically argue that only those who have the capacities needed for forming contracts are entitled to full (equal) consideration; such theories are often motivated by the belief that morality is constructed by humans primarily for human benefit. A somewhat related view is that only moral agents (*see* MORAL AGENCY)—those who can have moral obligations—are entitled to equal consideration. In these views, only humans qualify as potential contractors and moral agents. A different approach appeals to social relations: How much moral consideration one is due depends on how closely or distantly moral agents are socially related to one. As bond-forming creatures, we moral agents (humans) are much closer to other humans than to animals. Yet another argument appeals to the comparative value of human and animal lives. Equal consideration would require giving equal moral weight to the relevantly similar interests of humans and animals. According to the argument, a dog's life and a human's life are relevantly similar (equally important to the dog and human, respectively), so equal consideration implies that a dog's life is as morally valuable as a human's. A final argument appeals to the alleged authority of moral tradition: Because our moral tradition, the only source of moral authority, has always given animals' interests a subordinate place, there is no compelling reason to grant animals equal consideration.

The debate over equal consideration remains open because the issues are complex. Two points deserve mention. First, defenders of equal consideration generally deny that this principle means that human and animal lives are of equal value, but their supporting arguments have been incomplete at best. Second, defenders of unequal consideration for animals need to contend with the so-called problem of marginal cases*: Any criterion that supposedly marks a relevant difference between humans and animals (e.g., moral agency) will seemingly fail to apply to all humans, with the apparent suggestion that the exceptional humans are not due equal consideration.

Selected Bibliography. Carruthers, Peter, *The Animals Issues: Moral Theory in Practice* (Cambridge: Cambridge University Press, 1992); DeGrazia, David, *Taking Animals Seriously: Mental Life and Moral Status* (Cambridge: Cambridge University Press, 1996); Midgley, Mary, *Animals and Why They Matter* (Athens: University of Georgia Press, 1983); Regan, Tom, *The Case for Animal Rights* (Berkeley: University of California Press, 1983); Singer, Peter, *Animal Liberation* (New York: New York Review of Books, 1990).

DAVID D. DeGRAZIA

ERSKINE, THOMAS

Thomas Erskine (1750–1823) was a one-time lord chancellor and Whig politician who introduced an anticruelty bill in Parliament on May 15, 1809.

The bill passed the Lords but was lost in the Commons. Erskine later supported and helped to secure passage of Richard Martin's* anticruelty bill in 1822. It was the first national law of its type.

Selected Bibliography. Erskine, Thomas, *Cruelty to Animals: The Speech of Lord Erskine in the House of Peers on the Second Reading of the Bill Preventing Malicious and Wanton Cruelty to Animals, Taken in Short Hand* (London: Richard Phillips, 1809); Herbert, J. A., Thomas Erskine, in Leslie Stephen and Sidney Lee (Eds.), *Dictionary of National Biography*, vol. 6 (Oxford: Oxford University Press, 1921), 853–863; Moss, Arthur W., *Valiant Crusade: The History of the R.S.P.C.A.* (London: Cassell, 1961).

BERNARD UNTI

EUTHANASIA

"Euthanasia" comes from two Greek words: *eu* (good, well) and *thanatos* (death). Euthanasia is a central concern in animal welfare* studies because several million animals are euthanized (or *euthanatized*; both forms are used) in the United States each year in animal shelters,* veterinary clinics, and research laboratories. The definition of euthanasia differs slightly in veterinary medicine and human medicine. In human medicine, the term is restricted to "mercy killing"—killing a patient when death is a welcome relief from a life that has become too painful or no longer worth living. The definition is broader in veterinary medicine, however, including as well the euthanasia of healthy animals for owner convenience, for reasons of overpopulation, for behavior problems, or as donors of tissues for research.

Occasionally, other terms, such as "put down," "put to sleep," "sacrifice," or "destroy," are substituted as euphemisms. These terms can lead to confusion or may carry particular connotations. In particular, veterinarians* may describe either euthanasia or anesthesia as putting an animal to sleep, needlessly confusing clients and their children about what is to be done to their pet.

The American Veterinary Medical Association (AVMA) first published guidelines for animal euthanasia in 1963 and updated them in 1972, 1978, 1986, and 1993. Primary criteria for the evaluation of euthanasia techniques are the physical pain* and psychological distress* experienced by the animal. Other criteria include the emotional effect on humans who are present; the availability of appropriate drugs; and compatibility with the subsequent examination or use of the animal's body and tissues. The veterinary guidelines only cover methods of euthanasia, not issues of why, when, or whether specific animals should be euthanized.

The best euthanasia method for any particular animal will vary with his or her species, age, size, health, and temperament. Intravenous injection of a barbiturate anesthetic is currently preferred when an experienced operator must euthanize a calm, friendly dog. Smaller animals, frightened or wild animals, sick animals with low blood pressure, and many species of animals are not as easily injected. In these cases, additional tranquilizers or sedatives may be necessary, or alternative euthanasia methods such as carbon dioxide inhalation may be used.

The AVMA panel divides euthanasia techniques into three categories: those that directly destroy the conscious centers of the brain, those that interrupt the supply of blood or oxygen to the brain, and those that anesthetize the brain. The animal's behavior is not always a reliable guide. For example, succinylcholine paralyzes animal muscles. Paralysis of the respiratory muscles causes death, while general body paralysis prevents struggling or other obvious behavioral signs of distress. Since the drug does not anesthetize or sedate the animal before respiratory paralysis occurs, the use of succinylcholine is considered stressful and painful and is condemned by the AVMA panel. By contrast, barbiturates cause deep anesthesia and unconsciousness before leading to respiratory failure, making barbiturate euthanasia a preferred method. Other euthanasia methods that have been studied scientifically remain controversial.

In many jurisdictions, only veterinarians or their assistants may legally euthanize privately owned companion animals.* Other professionals, such as animal shelter workers, also euthanize animals, often in large numbers. In the past, humane associations have hailed decompression and carbon monoxide chambers, which allowed untrained workers to quickly kill large numbers of stray animals, as humane innovations over gunshot, drowning, or strychnine. Carbon monoxide chambers are still in use, though many shelters have turned to barbiturate injections as a more reliable and personal way of providing a smooth and pain-free euthanasia. Decompression chambers, which induce death by simulating the sudden loss of pressure that can occur in airplanes at high altitudes, are now illegal in many states. Studies on human volunteers showed that there can be chest and abdominal pain before unconsciousness, and shelter workers have reported dog behaviors in these chambers that could indicate severe distress.

Pet owners often grieve the euthanasia of a loved animal just as we grieve the death of our loved human friends and family. Social workers and therapists are increasingly recognizing this important response to animal death. They work to help people come to terms with this loss, rather than ridiculing it. Euthanasia training for shelter workers also includes seminars on dealing with the irony that responsible animal care sometimes includes killing animals.

Selected Bibliography. American Veterinary Medical Association, *Pet Loss and Human Emotion* (Schaumberg, IL: American Veterinary Medical Association, n.d.); Andrews, Edwin J., B. Taylor Bennett et al., Report of the AVMA Panel on Euthanasia, *Journal of the American Veterinary Medical Association* 202 (2) (1993): 229–249; Church, J. A., *Joy in a Woolly Coat: Living with, Loving, and Letting Go of Treasured Animal Friends* (Tiburon, CA: H. J. Kramer, 1987); Cohen, Susan P., and Carole E.

Fudin (Eds.), *Animal Illness and Human Emotion, Problems in Veterinary Medicine*, vol. 3, no. 1 (Philadelphia: J. B. Lippincott, 1991); Kay, William J., Susan P. Cohen et al. (Eds.), *Euthanasia of the Companion Animal: The Impact on Pet Owners, Veterinarians, and Society* (Philadelphia: Charles Press, 1988); Lagoni, Laurel, Carolyn Butler, and Suzanne Hetts, *The Human-Animal Bond and Grief* (Philadelphia: W. B. Saunders, 1994).

LARRY CARBONE

Social Attitudes

Interactions surrounding euthanasia in a large veterinary hospital showed ambivalence by clients, doctors, and veterinary staff when confronted by the decision to end an animal patient's life. From the veterinarian's perspective, the most legitimate reason for euthanizing an animal was if the patient was old, infirm, or suffering pain* because of serious illness or injury. Far less justifiable were those cases in which owners requested that their animals be "put down" because they were problematically aggressive, too expensive to keep, or simply more trouble than they were worth. When encountering these latter cases, veterinarians* regularly attempted to persuade the client to change his or her mind or simply refused to euthanize the animal. Veterinarians had a variety of ways of dealing with the emotions surrounding their involvement in putting an animal to death. They regularly offered clients advice about the decision, were overtly sympathetic, and sometimes recommended helpful readings or grief counselors. To safeguard their own emotional well-being, veterinarians commonly rationalized euthanasia as a necessary, if unfortunate, aspect of their work or, as is common in medical settings generally, employed humor as a protective device.

Research has also been done in two settings where animals are routinely euthanized: biomedical laboratories (*see* LABORATORY ANIMAL USE) and animal shelters.* Workers in these places feel ambivalence about regularly killing animals similar to that experienced by the veterinary personnel. Laboratory workers, for example, were torn between seeing the animals as experimental "objects" or transforming them into individual "pets." Objectification* helped provide some measure of emotional protection, while thinking of and treating some animals as pets (*see* COMPANION ANIMALS AND PETS) increased workers' emotional conflict. The informal culture and routines of the laboratory, such as referring to experimental animals by identifying numbers and speaking of killing animals as "sacrificing," helped provide workers with means of protecting their emotional health. On their part, workers in animal shelters coped with putting animals to death by using grim humor, focusing attention on the technical skills involved in performing euthanasia, defining their task as a humane response to animal suffering, and blaming negligent owners and pet overpopulation* for creating the necessity for euthanasia.

It is in places where causing the death of animals is, at least ostensibly, an act of mercy that killing presents the most moral and emotional problems. The ambivalence surrounding euthanasia commonly experienced by veterinarians, shelter workers, and laboratory technicians, as well as the intense grief typically felt by pet owners when confronted with euthanizing a beloved animal companion, speak to the power of our cultural inclination to regard some animals as individual persons.

Selected Bibliography. Arluke, Arnold, Coping with Euthanasia: A Case Study of Shelter Culture, *Journal of the American Veterinary Medical Association* 198(7) (1991): 1176–1180; Arluke, Arnold, Sacrificial Symbolism in Animal Experimentation: Object or Pet? *Anthrozoös* 2 (1988): 98–117; Kay, William J., Susan P. Cohen et al. (Eds.), *Euthanasia of the Companion Animal: The Impact on Pet Owners, Veterinarians, and Society* (Philadelphia: Charles Press, 1988); Sanders, Clinton, 1995, Killing with Kindness: Veterinary Euthanasia and the Social Construction of Personhood, *Sociological Forum* 10(2) (1995): 195–214; Shapiro, Kenneth, The Death of the Animal: Ontological Vulnerability, *Between the Species* 5(4) (1989): 183–193.

CLINTON R. SANDERS

EXOTIC COMPANION ANIMALS. *See* COMPANION ANIMALS AND PETS.

EXPERIMENTAL VACCINES. *See* MICE.

F

FACTORY FARMING

Factory farming, the mass production and daily slaughter of millions of other creatures for food in circumstances designed solely for cost and handling efficiency rather than the welfare of the animals involved, raises many concerns, not all of which are ethical in nature. Strong challenges to harsh confinement conditions have also been based on the consequences of such practices for humans in terms of health and environmental damage. Factory farming also has hidden costs that must be considered in evaluations of "success" measured only by short-term economic and production advantages.

While factory farms are common, they are usually socially invisible, and this, together with the accepted practice of using other animals as commodities and property (*see* LAW AND ANIMALS), accounts for the fact that they have seldom been subjected to basic ethical questions. The practice relies on certain moral perceptions of animals that are culturally rooted and socially reinforced through language (*see* OBJECTIFICATION OF ANIMALS) and daily practice.

The development of modern confinement methods first began in the 1950s with poultry and dairy livestock, who were moved from outdoor circumstances to confinement facilities; livestock kept for meat followed in the 1960s. Modern practices rely heavily on science for genetic control and design of animals and for remedies to combat the inevitable results of unnatural confinement such as virulent diseases, cannibalism, debilitating stress,* and stereotyped behavior (*see* STEREOTYPIES IN ANIMALS). Economic factors have led to technologically specialized intensive factory farms that are typically owned by large corporations that mass-produce and market food.

Conditions within factory farms vary considerably with the types of animals who are being intensively farmed. The dairy and chicken industries, often associated with the benign, idyllic images of "milk is good for you" and free-range chickens,* offer a good profile of modern factory farms. Modern dairy production is designed to maximize cows' output by manipulating their physiology. In order to allow as much milk to be produced as possible (which involves milking several times per day), each cow is kept in a constant state of pregnancy. A by-product of this process are the veal calves* that are, in turn, factory farmed in confined conditions and fed a diet that produces anemia. Dairy cows, which have a normal life expectancy of around twenty years, are also fed a special diet of chemicals, vitamins, and medicines designed to maximize production. Any individual cow is kept a limited number of years (usually only six or seven years of their maximum production), at the end of which they are sent to the slaughterhouse.

Poultry is factory farmed in even more intensive conditions. They are kept in crowded conditions and subjected to debeaking. Perhaps most symbolic is the discarding of all male chicks in the process of producing laying hens. These chicks are simply gassed or dumped alive into plastic sacks in which they suffocate. Female chicks are integrated into the extraordinarily deprived conditions of the modern battery-cage system.

It has been argued that factory farming has brought benefits to animals, such as safety from predators and a steady supply of balanced diet and sanitary water. The role of humans as predators is ignored, and the realities of the situation are best seen by the fact that this kind of argument could never be made with regard to humans being factory farmed. The intensive practices of factory farming were first introduced under the assumption that animals could be used without limit. There has since been an increase in awareness in many different ethical systems.

Selected Bibliography. Adams, Carol J., *The Sexual Politics of Meat: A Feminist-Vegetarian Critical Theory* (New York: Continuum, 1991); Curtis, S. E., The Case for Intensive Farming of Food Animals, in T. Regan and P. Singer (Eds.), *Animal Rights and Human Obligations*, 2nd ed. (Englewood Cliffs, NJ: Prentice-Hall, 1989); D'Silva, Joyce, and Peter Stevenson, *Modern Breeding Technologies and the Welfare of Farm Animals* (Petersfield, Hampshire: Compassion in World Farming Trust, 1995); Fiddes, Nick, *Meat: A Natural Symbol* (London and New York: Routledge, 1991); Fox, Michael, *Superpigs and Wondercorn* (New York: Lyons and Burford, 1992); Johnson, Andrew, *Factory Farming* (Oxford: Basil Blackwell, 1992); Kalechofsky, Roberta, *Autobiography of a Revolutionary: Essays on Animal and Human Rights* (Marblehead, MA: Micah Publications, 1991); Linzey, Andrew, *Animal Theology* (London: SCM Press; Urbana: University of Illinois Press, 1995); Mason, Jim, and Peter Singer, *Animal Factories* (New York: Crown, 1990); Swann Report, *Use of Antibiotics in Animal Husbandry and Veterinary Medicine*, Cmnd. 41990 (London: HMSO, 1969/1970).

PAUL WALDAU

FAIR CHASE. *See* Hunting.

FARM-ANIMAL WELFARE

Ruth Harrison's book *Animal Machines*, published in Britain in 1964, introduced the British public to a large-scale and highly intensified animal agriculture that was a far cry from their cherished image of the pastoral family farm. Harrison coined the term "factory farming"* to describe this new agriculture, which she viewed as being more concerned with profits than with animals.

The farming practices that Harrison described were the outcome of a number of scientific and technological advances. Two critical discoveries, that vitamin D could be provided artificially in the food rather than requiring sunlight for its synthesis and that antibiotics could be used to minimize the spread of certain infectious diseases, meant that large numbers of animals could be housed together indoors. Feeding, watering, and handling could then be more easily mechanized, decreasing labor requirements. Indoor environments also allowed better monitoring and control of nutrition, temperature, lighting, and animal health. Combined with more sophisticated methods of genetic selection for production traits, these factors resulted in a more economically efficient animal agriculture, but one in which many of the behaviors of the animals were prevented, and in which the contact between the animal and the human caretaker was minimized.

In response to the outcry stimulated by Harrison's book, the British government formed a committee of inquiry, the Brambell Committee, to listen to testimony and visit farms throughout Britain. The committee recommended that, at a minimum, every farm animal should have "sufficient freedom of movement to be able, without difficulty, to turn round, groom itself, get up, lie down and stretch its limbs" (*see* ANIMAL WELFARE, Freedom). These are referred to as the five freedoms. Twenty-five years later, the Farm Animal Welfare Committee modified these into a more expansive set of recommendations, the five new freedoms: freedom to display most normal patterns of behavior and freedom from hunger, thirst, or malnutrition; from inadequate comfort and shelter; from disease or injury; and from fear.*

Animal agriculture has continued to intensify in the years since the Brambell Committee issued its report, in the process creating some additional welfare problems. The primary criticisms directed against contemporary animal agriculture relate to (1) animal health, (2) pain* and distress,* and (3) restriction of movement and other behaviors of animals, including social behaviors, in husbandry systems that involve close confinement or are barren of stimulation. Coupled with these concerns are worries about the human health effects of drugs and antibiotics administered to animals, as well as the potentially negative effects of large-scale animal production on the environment.

The welfare of farm animals is now the subject of extensive regulation in many industrialized nations. In Europe, the Council of Europe and the European Community develop requirements for the care of farm animals that are translated into legislation in the different member countries. Other countries, like Canada and the United States, rely mainly on codes of practice or guidelines rather than legislation.

Farm-animal welfare may prove to be a particularly thorny issue to resolve because of its economic and social implications. Generally, systems that allow the animal more behavioral freedom are also associated with higher production costs and therefore with increased food prices. The role that this factor will play in determining the direction that animal agriculture takes in the future will depend on a complex interplay between attitudes toward animals, environmental and food safety concerns, economic forces, and the structure of urban and rural human communities.

Selected Bibliography. Brambell, F. W. R., Report of the Technical Committee to Enquire into the Welfare of Animals Kept under Intensive Livestock Husbandry Systems, Cmnd. Paper 28 36 (London: Her Majesty's Stationary Office, 1965); Fraser, A. F., and D. M. Broom, *Farm Animal Behaviour and Welfare*, 3rd ed. (London: Ballière Tindall, 1990); Harrison, R., *Animal Machines* (London: Vincent Stuart, 1964); Mench, J. A., and W. R. Stricklin (Eds.), An International Conference on Farm Animal Welfare: Ethical, Technological, Sociopolitical, and Scientific Perspectives, *Journal of Agricultural and Environmental Ethics* 6 (special supplements 1 and 2); Swanson, J. C., Farm Animal Well-Being and Intensive Production Systems, *Journal of Animal Science* 73 (1995): 2744–2751.

JOY A. MENCH

FEAR

Fear involves the perception of stressful environmental factors through an animal's senses such as smell (olfaction), sight (vision), and sound (hearing), activating the sympathetic part of the autonomic nervous system, which has direct neural connections to other parts of the body through sympathetic efferent (motor) nerves (for example, the skin, leading to raising the hairs or fur; the eyes, leading to dilatation of the pupils). There is also stimulation of the adrenal medulla leading to a release of catecholamines such as adrenalin and noradrenalin. As with pain,* there are neuronal connections with other areas of the central nervous system resulting in the expression of a variety of physiological responses. Fear may be an integral part of other aspects of suffering*—an animal with a broken leg may be fearful of being moved or touched, as well as being in pain and distressed by its inability to move normally.

DAVID B. MORTON

FEELINGS OF ANIMALS

Because of the difficulty in determining factors for animal welfare,* many have concluded that the important thing determining welfare is how an animal feels. Thus if an animal feels frightened or frustrated or in pain,* its welfare will be reduced; if it feels happy or contented, its welfare will be enhanced. The problem is that subjective or personal feelings, of human beings or of animals, are not directly available to scientific investigation. We can have a good idea of how other human beings feel because they are built like us, they have the same sensory and processing mechanisms as part of their nervous systems, and, moreover, they have language that enables them to describe how they feel. It is much more difficult with nonhuman animals; although there are similarities, their sensory information-processing mechanisms are different from ours. Also, we do not share a common language with them, so they cannot describe how they feel. However, we may be able to gain much information about animals' feelings from indirect evidence. It is not necessary to know exactly what the animal is feeling. Thus in the case of a dog* that we suspect is suffering from pain following tail docking,* it is not necessary to know whether the dog experiences something similar to a human being with a bad burn or toothache or a broken leg. If the dog behaves as if it is having a very negative experience, if its behavior becomes much more normal when it is given a painkiller, and if it strongly avoids anyone dressed like the veterinarian who performed the surgery, then we can conclude that it is suffering and that its welfare is reduced. A very crude measure of feelings such as how positive or negative they are is very helpful in assessing welfare.

One way to find out what an animal feels is to study various states of suffering* such as fear,* frustration (*see* ANIMAL BOREDOM), and pain. For example, we can say that if an animal has a strong tendency to behave in a particular way and we prevent that behavior, then the animal will be frustrated. Then we can find out how that animal behaves when it is frustrated in many different ways. This type of information is now being collected for fear, frustration, and pain. There is, of course, a moral dilemma; in order to understand the state of suffering in question, we have to subject the animal to that state. We also have to show that an animal exhibiting symptoms of fear or frustration or pain is actually suffering. A good demonstration that the animal is having an unpleasant experience would be if it avoids or works to escape from situations associated with suffering. A simple method used to "ask" the animal how it feels about the conditions and procedures to which it is subjected is the preference test (*see* PREFERENCE AND MOTIVATION TESTING), in which animals are allowed to choose various aspects of their environment. It is assumed that they will express at least some of their feelings in their actions and choose in the best interests of their welfare. As with any scientific method, there are pitfalls associated with preference testing, but when these are known, steps can be taken to avoid them. Preference tests must be used with other tests that measure how strong the preference is. This ensures that the choice being made is not a trivial choice, or that the animal is not choosing "the lesser of two evils." Tests that give some insight into the feelings of animals will eventually give more definitive answers about their welfare.

Selected Bibliography. Dawkins, M. S., From an Animal's Point of View: Motivation, Fitness, and Animal Welfare, *Behavioral and Brain Sciences* 13 (1990): 1–61; Dawkins, M. S., *Through Our Eyes Only: The Search for Animal Consciousness* (San Francisco: W. H. Freeman, 1993); Duncan, I. J. H., Animal Rights–Animal Welfare: A Scientist's Assessment, *Poultry Science* 60 (1981): 489–499; Duncan, I. J. H., Animal Welfare Defined in Terms of Feelings, *Acta Agriculturae Scandinavica, Section A, Animal Science*, Supplement 27 (1996): 29–35; Rushen, J., The Validity of Behavioural Measures of Aversion: A Review, *Applied Animal Behaviour Science* 16 (1986): 309–323.

IAN J. H. DUNCAN

FERAL ANIMALS

The term "feral animals" generally refers to those individuals who belong to domesticated species (*see* DOMESTICATION), such as dogs* and cats,* but who themselves have not adapted to domestic life or to interactions with humans. Many of these animals were born never knowing the care of humans, but some may have once, years ago, been family pets (*see* COMPANION ANIMALS AND PETS). These animals typically live on the fringes of society. They are not just strays—former pets who were recently lost or abandoned—but animals who have survived in the wild on wits and an occasional bowl of food left on the back porch, caught somewhere between a wild and domestic existence.

Feral cats are probably the most well known of the feral animals. Although it is impossible to say exactly how many exist, Alley Cat Allies (ACA), a feral-cat advocacy group, estimates that there are approximately 60 million feral cats living in the United States. Litters of unneutered cats allowed to roam outside and breed indiscriminately also contribute to the feral cat population. It is estimated that an unspayed female cat and her offspring can produce 420,000 cats in just seven years.

The existence of feral cats and what should be done about them is a controversial topic facing humane organizations, animal control agencies, and communities across the country. They present a wide variety of problems: concern for the animals themselves, disease transmission (to both animals and humans), property destruction, wildlife predation,* and a threat to owned pets allowed outdoors. Some feel that the cats should be trapped and euthanized (*see* EUTHANASIA). Others believe that sterilizing the cats and returning them to the wild is the only answer. Still others, such as the Humane Society of the United States (HSUS),* believe that communities must develop policies regarding feral cats in association with policies on cat ownership and control.

Those who believe that the cats should be trapped and euthanized feel that it is better that they die a quick, humane death rather than live a rough life on the streets. They believe that it is also the best way to keep the feral-cat population from increasing. Many cat owners who let their own cats roam freely outside oppose this type of program, fearing a sweeping roundup of any and all cats found outside.

Others suggest that the solution is to "trap, test, vaccinate, alter, and release" (TTVAR). Such programs were developed in the 1970s in Denmark and the United Kingdom and are endorsed by a number of humane organizations throughout the world, including the Royal Society for the Prevention of Cruelty to Animals (RSPCA),* which in 1977 founded the Feral Cat Working Party to study the feral-cat problem and possible solutions. In these programs, a colony of feral cats is watched over by a "caregiver" who provides food, water, and shelter and keeps an inventory of colony members. When a new member joins the colony, that cat is humanely trapped, transported to a clinic or shelter,* spayed or neutered, tested for disease, given vaccinations and some method of identification, and returned to the colony. Often these caregivers are registered with some kind of community animal welfare* organization that runs a structured feral-cat program complete with records, participation rules, and requirements. Promoters of this type of feral-cat control say that the result is a healthier and stable cat population. Such programs, however, are extremely time-consuming, require a long-term commitment from the caregivers, and may not be feasible in every community.

Feral dogs, whose numbers were once controlled through roundups and extermination beginning in the mid-to-late 1800s and then by dog control laws created and implemented in the 1940s, present a much greater threat to the safety of humans. Although wild dogs may be traditionally associated with rural areas, recent news stories report their existence in urban areas, such as New York City and Los Angeles. Some say that feral dogs form packs in which they become aggressive and more likely to attack humans or other animals.

Communities have found it difficult to devise an immediate solution to control feral dogs. Animal control, in some areas, has attempted to round up these animals. While this is the best solution, it is not always successful since animals are being abandoned by their owners daily and some animals inevitably escape capture and breed new litters. The existence and presence of feral animals may never be stopped, but fighting the problem at the root cause—irresponsible animal ownership—can decrease suffering.*

Selected Bibliography. Alley Cat Allies, PO Box 397, Mt. Rainier, MD 20712; Berkeley, Ellen Perry, Feral Cats, *Cat Fancy* 33(7) (July 1990): 20; Donald, Rhonda Lucas, Should Feral Cats Be Euthanized? *Shelter Sense* (Humane Society of the United States), May 1992, 3; Doris Day Animal League, Standards for Maintaining Feral Cat Colonies, 227 Massachusetts Ave, NE, Suite 100, Washington, DC, 20002; Searle, Milton C., Overpopulation: The Perennial Problem, in Robert D. Allen and William H. Westbrook (Eds.), *The Handbook of Animal Welfare* (New York: Garland STPM Press, 1979), 47–50.

CYNTHIA STITELY

FIELD STUDIES. *See* EDUCATION AND THE USE OF ANIMALS.

FISH

Nearly 500 million years ago, the earliest known fishes cast their shadow on the fossil record. They lacked both jaws and paired fins and are believed to have made their living from sucking the bottoms of prehistoric seas and lakes. Today, more than 21,000 species of fish have been identified. The vast majority are ray-finned fishes, which account for nearly half of the planet's known vertebrate species.

The use of fish for human consumption predates recorded history. The widely held attitude that fish exist primarily to benefit humans has led to a disregard for the welfare of individual animals (*see* ANIMAL WELFARE). There are few restrictions on the killing and eating of fish, and little thought is given to their treatment in the process. Those protective measures that do exist, such as the U.S. Magnuson Fisheries Conservation and Management Act (1976) and the United Nations Convention on the Law of the Sea (1982), are concerned with the regulation of fishery fleets such that catch and wealth are maximized while irreversible population effects are minimized.

Despite illusions to the contrary, fishes are sentient (*see* SENTIENTISM) beings with a proven capacity to feel discomfort and pain.* They share the basic biological processes of any living being and respond accordingly when these processes are disrupted.

Modern fishing technology has greatly enhanced our ability to catch fishes on a grand scale. Approximately 9,000 fish species are currently caught by marine fisheries, resulting in an annual global catch of more than 85 million metric tons. In the North Sea, for example, commercial fisheries remove between 30% and 40% of the biomass of fish each year. Over 82% of commercial fish stocks in U.S. waters have been classified as overexploited.

Recent trends in catch data indicate that fish populations are declining. The 1990s have seen a 5% decrease from the worldwide fish catch of the late 1980s, despite increases in fishery effort. Nearly 70% of known marine fish species are heavily exploited or depleted, and some populations have been reduced to dangerously low levels. While such trends may be interpreted as a response to the dietary needs of a growing human population, fish protein is actually becoming less accessible to many coastal communities who depend on it for adequate nutrition because the fish are exported.

The vast amount of fish consumed by humans is only part of the picture. At least one-third of the annual marine catch becomes animal feed for pets, livestock, and farm-raised fish. In addition, the United Nations estimates that between 18 and 40 million tons of unwanted (and usually fatally injured) fish are thrown back into the sea by commercial fishers. Shrimp trawlers are the worst offenders, in some cases discarding 15 tons of fish for every ton that is retained. Incidental to the capture of marine fishes are the deaths of many thousands of marine mammals, sea turtles, and seabirds that become entangled in the equipment used to catch fish.

The plight of the world's fishing industry has been met by the intensification of raising fish as crops. Aquaculture is a fast-growing producer of both freshwater and marine fishes, with its contribution to the world's food-fish supply increasing from 12% in 1984 to 17% in 1992. While regarded by many as a solution to overfishing, large-scale aquaculture invites its own host of problems. Reared in crowded pens, farmed fish are major sources of water pollution and are subject to parasites, infections, and disease. Animals who escape carry disease and exotic genes into the surrounding waters, as well as becoming resource competitors to local species. Marine aquaculture is a primary cause of coastal habitat destruction, with mangroves and wetlands being developed for farming at an ever-increasing rate.

The moral issues surrounding fish as food are numerous and complex and require further detailed attention. As currently managed, large-scale fisheries threaten the well-being of humans and nonhumans alike.

Selected Bibliography. *Ambio: A Journal of the Human Environment* (Royal Swedish Academy of Sciences); Moyle, P. B., *Fish: An Enthusiast's Guide* (Berkeley: University of California Press, 1993); Simoons, F. J., *Eat Not This Flesh: Food Avoidances from Prehistory to the Present*, 2nd ed. (Madison: University of Wisconsin Press, 1994), 253–296; Weber, P., Protecting Oceanic Fisheries and Jobs, in *State of the World 1995: A Worldwatch Institute Report on Progress toward a Sustainable Society* (New York: W. W. Norton, 1995), 21–37.

PAULA MACKAY

FISH AND WILDLIFE SERVICE. *See* HUNTING.

FIVE FREEDOMS. *See* FARM-ANIMAL WELFARE.

FRESHEL, M. R. L.

M. R. L. ("Emmarel") Freshel (1867–1948) was the founder of the Millennium Guild, the first American animal rights* organization. Founded in 1912, the guild published Freshel's *Golden Rule Cook Book* (first published in 1907) and *Selections from Three Essays by Richard Wagner with Comment on a Subject of Such Importance to the Moral Progress of Humanity That It Constitutes an Issue in Ethics and Religion* (1933), an impassioned attack on vivisection. An associate of Mary Baker Eddy, founder of the Christian Science Church, Freshel resigned from the Christian Science Church after it expressed support for the entry of the United States into World War I. Through the Millennium Guild, she promoted alternative fur fabrics and vegetarianism* and spoke out against all forms of animal exploitation. After her death, control of the Millennium Guild fell to her husband Curtis. After his death, the organization was directed by New York radio personality Pegeen Fitzgerald.

Selected Bibliography. Freshel, M. R. L. [M. R. L. Sharpe on title page], *The Golden Rule Cook Book* (Boston, MA: Little, Brown, 1910); Freshel, M. R. L., Greetings to the Congress, *Anti-Vivisection Review* 6 (July/August 1927): 171; Sharpe, M. R. L., *Selections from Three Essays by Richard Wagner with Comment on a Subject of Such Importance to the Moral Progress of Humanity That It Constitutes an Issue in Ethics and Religion* (New York: Millennium Guild, 1933); *Proceedings of the International Anti-Vivisection and Animal Protection Congress, Philadelphia, 1926* (Philadelphia: American Anti-Vivisection Society, 1926), 104–110, 149–154.

BERNARD UNTI

FROGS. *See* AMPHIBIANS; EDUCATION AND THE USE OF ANIMALS.

G

GAME MANAGEMENT. *See* HUNTING.

GANDHI, MOHANDAS KARAMCHAND

Mohandas Karamchand Gandhi (1869–1948) was a world statesman, pacifist, and vegetarian. Reading Henry Salt's* *A Plea for Vegetarianism* and Howard Williams's *The Ethics of Diet* reinforced his ethical vegetarianism* on his first visit to England in 1887. Thereafter Gandhi became a committed vegetarian "by choice," and this commitment was deepened through his conversion to the Hindu (*see* RELIGION, Hinduism) philosophy of *ahimsa*, nonviolence or noninjury, which became fundamental to his religious outlook and which especially informed his insistence upon nonviolent civil disobedience as a means of political struggle. During his first stay in London Gandhi became a member of the executive committee of the London Vegetarian Society. Gandhi made special arrangements to meet Salt during his trip to England for the Round Table Conference in 1931.

Selected Bibliography. Chapple Christopher Key, *Nonviolence to Animals, Earth, and Self in Asian Traditions* (Albany: State University of New York Press, 1993); Gandhi, Mohandas K., *Diet and Diet Reform* (Ahmedabad: Navajivan Publishing House, 1949); Gandhi, Mohandas K., *How to Serve the Cow* (Ahmedabad: Navajivan Publishing House, 1954); Gandhi, Mohandas K., *The Moral Basis of Vegetarianism* (Ahmedabad: Navajivan Publishing House, 1959); Gandhi, Mohandas Karamchand, *The Story of My Experiments with Truth: Autobiography*, ed. Mahadev Desai (Ahmedabad: Navajivan Press, 1927); Mehta, Ved, *Mahatma Gandhi and His Apostles* (London: Andre Deutsch, 1977).

ANDREW LINZEY AND BERNARD UNTI

GAP. *See* GREAT APE PROJECT.

GENEROSITY PARADIGM

The generosity paradigm maintains that humans owe animals not equal consideration,* or equality of treatment, but moral generosity, that is, more than equal treatment. According to the generosity paradigm, our obligations to animals and children (and to all beings who are vulnerable, unprotected, undefended, and morally innocent) are not exhausted by the language of rights and duties but require practical costly action to promote their well-being.* Such a notion is centered theologically in the notion of the generosity of God, who is disclosed in the self-sacrifice of Jesus Christ. Historically the idea was pioneered in many 18th- and 19th-century works of zoophily that celebrated a newly found sensitivity toward animals, including those of William Hamilton Drummond, Henry Crowe, John Hildrop,* and especially Humphry Primatt,* who argued that mercy and benevolent regard are foundational to Christian living (*see* RELIGION, Christianity). These pioneering works laid the foundations for a radical modern interpretation that insists that the weak should have "moral priority" over other competing claims. This interpretation maintains that the nature of human power is morally legitimate only when exercised in a self-costly sacrificial way. The generosity paradigm resists the flattening of all obligations into one catchall equality view; it recasts the debate about animal rights* not just in terms of moral limits but rather in terms of extending these limits beyond what is currently assumed even by most animal advocates.

Selected Bibliography. Crowe, Henry, *Zoophilos; or, Considerations on the Moral Treatment of Inferior Animals* (London, 1820); Drummond, William Hamilton, *Humanity to Animals the Christian's Duty* (London, 1830); Drummond, William Hamilton, *The Rights of Animals* (London, 1838); Hildrop, John, *Free Thoughts upon the Brute-Creation, or, an Examination of Father Bougeant's "Philosophical Amusement, &c. on the Language of Beasts"* (London: R. Minors, 1742); Linzey, Andrew, *Animal Theology* (London: SCM Press; Urbana: University of Illinois Press, 1995); Linzey, Andrew, The Moral Priority of the Weak: The Theological Basis of Animal Liberation, in *The Animal Kingdom and the Kingdom of God* (University of Edinburgh: Centre for Theology and Public Issues, 1991), 25–42; Primatt, Humphry, *The Duty of Mercy and the Sin of Cruelty to Brute Animals* (London: T. Cadell, 1776).

ANDREW LINZEY

GENETHICS

Genethics is the application of moral or social values to genetics. Within the last decade, the techniques of genetics have grown greatly, allowing us to pinpoint genes for cancer, mental illness, obesity, and a host of other traits and diseases. Although we can locate on the chromosome the gene(s) for such characters, our ability to treat them lags far behind. Genethics is typically applied to humans, but could also be applied to other animals.

Nonhuman animals are currently the experimental organisms of choice for research geneticists interested in human diseases and other traits. The reason is simple: the experimental work necessary to understand the genetic basis of a characteristic is often invasive and typically involves the rapid breeding of large numbers of offspring, procedures that cannot readily be applied to humans. For example, in research that focuses on the genetics of a behavior in mice* that may be similar to alcoholism in humans, it is necessary to inject mice with a standard dose of alcohol so that researchers can assess its effect on them. Animals also have to be euthanized (*see* EUTHANASIA) so that we can do necessary analyses.

There are three types of genetic research that involve animals. The first is the use of animal models* for human genetic diseases. These include diseases caused by abnormalities in single genes, such as cystic fibrosis, sickle-cell anemia, and Huntington's disease, as well as polygenic (many-gene) diseases such as cancer, heart disease, and alcoholism. Next come the genome projects that have as their goal the identification of all the genes of a given organism. Currently, genome projects are at work on several bacterial species and on yeast, nematode (a type of roundworm), mouse, rat, and human genomes. Finally, there is transgenic research, also known as recombinant DNA technology, which moves genes from one organism into another (see GENETIC ENGINEERING). This area of research initially allowed the insertion of human genes into bacteria, primarily for the purpose of production of the protein specified by the human gene (e.g., insulin). Now, many human genes are being moved into a variety of mammalian species both for production and for studying the function of the human gene. Currently, more scientists are beginning to work on mammals, particularly humans, to concentrate on human genetic conditions. As the potential to work directly on humans becomes more accessible, there may be a reduction in the use of animal subjects.

Selected Bibliography. Crisp, R., Making the World a Better Place: Genes and Ethics, *Science and Engineering Ethics* 1 (1995): 101–110; Hubbard, R., *Exploding the Gene Myth* (Boston: Beacon Press, 1993); Jones, S., *The Language of Genes* (New York: Doubleday, 1993); Kevles, D. J., *In the Name of Eugenics* (New York: Random House, 1985); Suzuki, D. T., and P. Knudtson, *Genethics*, revised and updated ed. (Cambridge, MA: Harvard University Press, 1990).

BETH BENNETT

GENETIC ENGINEERING

Although humans have always "genetically engineered" domesticated animals (*see* DOMESTICATION) to suit their uses of these animals, the only tool available to accomplish this in the past was to breed animals selected specifically for this purpose. This in turn required many generations of gradual change in order to produce significant changes in the animals and also limited manipulation of genes to those that could be introduced by normal reproduction. Since the late 1970s, however, the technology for inserting all manner of genes into an animal's genome, including radically foreign genes (for example, genes from human beings), has progressively developed in sophistication. This opens up a vast range of possibilities for manipulating animals' genetic makeup and thus their phenotypic traits. In 1989, the U.S. Patent Office announced that it had issued the first animal patent for a mouse (*see* MICE) that was genetically engineered to be highly susceptible to developing tumors, a trait rendering the animal extremely valuable for cancer research.

Genetic engineering and the potential for patenting the resulting animals have evoked strong negative criticism, largely from theologians and animal advocates. Theologians express concern that genetic engineering does not show proper respect for the gift of life and implies that humans are "playing God." Although such religiously based criticisms are perhaps meaningful within the context of a religious tradition, it is difficult to extract from them any ethical content that can be used to illuminate the issue of genetic engineering of animals in the context of social ethics. Animal advocates, on the other hand, express the concern that genetic engineering and animal patenting will result in increased animal suffering.*

It is certainly not necessarily the case that genetic engineering of animals must inevitably result in increased suffering for animals. Genetic engineering can, in principle, significantly reduce animal suffering by, for example, increasing animals' resistance to disease. This has already been accomplished in chickens* who have been genetically engineered to resist some cancers. Furthermore, genetic engineering could be employed to correct suffering created by traditional breeding, as in the case of the more than 400 genetic diseases in purebred dogs* that have been introduced into these animals by breeding them to fit aesthetic standards. Third, genetic engineering could be used to make animals more suited to the harsh environments in which we raise them, for example, hens kept in battery cages, though both common sense and common decency suggest that it makes more sense to change the environment to fit the animals than vice versa.

But animal advocates are correct in their concern that if current tendencies in animal use continue unchanged, they will favor genetic engineering being used in ways whose result, albeit unintended, will increase animal suffering. Consider animal agriculture (*see* GENETIC ENGINEERING, Genetic Engineering, Pesticides, and Agriculture). Traditional (pre–mid-20th-century) agriculture was based on animal husbandry, that is, caring for the animals, respecting their biological natures, and placing them into environments for which they would be optimally suited; the producer did well if and only if the animals did well. Animal suffering worked as much against the farmer's interest as against the animal's interest, and thus animal welfare* was closely connected with animal productivity. However, the advent of high-technology agriculture allowed farmers to put animals into environments that did not suit them biologically (e.g., battery cages), yet in which they could still be productive.

One major and legitimate concern is that genetic engineering not be used as yet another tool for augmenting productivity at the expense of animal welfare. Thus, for example, in the early 1980s, pigs* were genetically engineered to produce leaner meat, faster growth, and greater feed efficiency. While this was accomplished, the negative effects of the genetic engineering were unexpected and striking, with the animals suffering from kidney and liver problems, diabetes, lameness, gastric ulcers, joint disease, synovitis, heart disease, pneumonia, and other problems.

To prevent the use of genetic engineering as a tool enabling us to further erode animal welfare for the sake of efficiency, productivity, and profit, Bernard Rollin proposed the following morally based regulatory principle as a check on commercial use of genetic engineering of animals, the principle of conservation of welfare: Genetically engineered animals should be no worse off than the parent stock would be if they were not so engineered. Such a principle should serve to forestall new suffering based in genetic engineering for profit.

The second major source of suffering growing out of genetic engineering of animals comes from our increasing ability to create transgenic animal models (*see* ANIMAL MODELS, Biomedical and Behavioral Science) for human genetic disease. Genetic engineering gives researchers the capability of genetically creating animals who suffer from human genetic diseases. This means that vast numbers of defective animals will be created to research these human diseases. In many if not most cases of genetic disease, there is no way to control the painful symptoms, and reducing the animals' suffering through early euthanasia* is excluded, since researchers wish to study the long-term development of the disease. Thus this sort of genetic engineering creates a major problem of animal suffering. Thus far, neither the research community nor society in general has addressed this issue, despite society's 1985 expression in federal law of its ethical commitment to limit animal suffering in biomedical research (*see* LABORATORY ANIMAL WELFARE ACT OF 1966, Law [Federal] Governing Animal Research).

Selected Bibliography. Fox, Michael W., *Superpigs and Wondercorn* (New York: Lyons and Burford, 1992); Pursel, Vernon, et al., Genetic Engineering of Livestock, *Science* 244 (1989): 1281–1288; Rifkin, J., *Declaration of a Heretic* (Boston: Routledge and Kegan Paul, 1985); Rollin, Bernard E., *The Frankenstein Syndrome: Ethical and Social Issues in the Genetic Engineering of Animals* (New York: Cambridge University Press, 1995); Seidel, George E., Biotechnology in Animal Agriculture, in J. F. Mac-

Donald (Ed.), *National Agricultural Biotechnology at the Crossroads: Biological, Social, and Institutional Concerns*, NABC Report 3 (Ithaca, NY: NABC, 1991).

BERNARD E. ROLLIN

Genetic Engineering, Pesticides, and Agriculture

Industrial farming methods of food and fiber crop production that use various types of biotechnology to keep these methods operating are, in spite of political support, publicly unacceptable. We know very little about the risks of releasing genetically engineered biopesticides, as is proposed for the control of myriad insect pests, like the pine beauty moth and cotton boll weevil. Nor do we know the long-term ecological and economic risks and potential harm to ecosystems, wildlife, and natural biodiversity of releasing genetically engineered plants and animals into the environment, and their potentially harmful ecological consequences are legitimate concerns. For example, fish* such as trout and salmon containing the growth and antifreeze genes of other species are being developed for commercial fish farming. These animals could escape and breed with wild fish. Genetically engineered plants could transmit herbicide and insect resistance to other plants.

As of January 1995, the U.S. Department of Agriculture and the Environmental Protection Agency had approved over 2,000 releases of genetically engineered organisms into the environment for agricultural field tests. Several patented varieties of crops like corn and wheat have also been engineered to be resistant to patent holders' herbicides, the continued use of which may be economically unwise. Various crops have been given bacterial genes to produce insecticides like *Bacillus thuringiensis* in order to repel and kill pests, but this is a short-lived miracle since pests quickly develop resistance.

Conventional industrial agriculture has globally contaminated surface waters and groundwaters and the entire terrestrial and aquatic food chains with harmful agricultural chemicals. Industrial agriculture has also brought us intensive factory poultry and livestock production (*see* FACTORY FARMING). These have been shown to cause animals to suffer stress,* distress,* and disease. Surface groundwater pollution from animal wastes is a serious environmental and public health problem. Using biotechnology to correct these problems—so-called bioremediation—is of questionable value if no efforts are made to change the agricultural system and especially to raise farm animals under less intensive and more humane conditions, which can be done efficiently and profitably.

The widespread use of veterinary drugs and new genetically engineered vaccines to keep farm animals productive in the intensive confinement systems of factory farms that cause sickness and suffering,* as well as harm to the contract labor that cares for them, is ethically questionable. So is genetically engineering livestock to better resist stress and disease.

Animal-production scientists also continue to seek ways to make farm animals more efficient and productive. One product, genetically engineered recombinant bovine growth hormone (rBGH), which dairy farmers inject into cows to boost milk production, gives rise to a variety of animal health problems and potential consumer health risks. A more economic and ecologically sound alternative, rotational grazing, where cows are moved to fresh pasture at intervals, is seen as a major obstacle by agribusiness in its attempts to get dairy farmers to buy this new drug. Biotechnology companies have been testing and trying to market rBGH in developing countries, which would undermine traditional sustainable livestock and forage-production systems. The new field of "pharming" useful medical products from the milk of transgenic cows, sheep, and goats raises many ethical and regulatory questions, as does the genetic engineering of pigs* to be used as human organ donors.

Selected Bibliography. Fox, M. W., *Superpigs and Wondercorn: The Brave New World of Biotechnology and Where It All May Lead* (New York: Lyons and Burford, 1992); Gussow, Joan Dye, *Chicken Little, Tomato Sauce, and Agriculture: Who Will Produce Tomorrow's Food?* (New York: Bootstrap Press, 1991); Krimsky, S., *Biotechnics and Society: The Rise of Industrial Genetics* (New York: Praeger, 1991); Rollin, Bernard, *The Frankenstein Syndrome: Ethical and Social Issues in the Genetic Engineering of Animals* (New York: Cambridge University Press, 1995); Suzuki, David, and Peter Knudtson, *Genethics* (Cambridge, MA: Harvard University Press, 1989).

MICHAEL W. FOX

GOMPERTZ, LEWIS

Lewis Gompertz (1779–1861) was the second secretary of the organization now known as the Royal Society for the Prevention of Cruelty to Animals (RSPCA)* and sustained the society through financially troubled times with his personal contributions. Gompertz held tenaciously to his principles, abstaining from meat and avoiding the use of coaches because of the abuse inflicted on horses. He served as secretary until 1832, when religious prejudice resulted in his ouster. He then founded the Animals' Friend Society. For a time, the new organization commanded the allegiance of significant supporters of what was then called the Society for the Prevention of Cruelty to Animals (SPCA), including Richard Martin.* Eventually, however, the stability and respectable appearance of the SPCA won out, and Gompertz's remaining colleagues rejoined the parent group. An inventor, credited with the development of 38 devices, Gompertz was the author of *Moral Inquiries on the Situation of Man and of Brutes* (1824) and a collection of essays, *Fragments in Defence of Animals* (1852). He also edited the journal *Voice of Humanity*.

Selected Bibliography. Gompertz, Lewis, *Fragments in Defence of Animals, and Essays on Morals, Soul, and Future State* (London: W. Horsell, 1852); Gompertz,

Lewis, *Moral Inquiries on the Situation of Man and of Brutes: On the Crime of Committing Cruelty to Brutes, and of Sacrificing Them to the Purposes of Man* (London: Lewis Gompertz, 1824); Moss, Arthur W., *Valiant Crusade: The History of the R.S.P.C.A.* (London: Cassell, 1961); Turner, James, *Reckoning with the Beast: Animals, Pain, and Humanity in the Victorian Mind* (Baltimore: Johns Hopkins University Press, 1980).

BERNARD UNTI

GREAT APE PROJECT

Launched by scientists and scholars from various nations and disciplines, the Great Ape Project (GAP) seeks to extend the scope of three basic moral principles to include all members of what the GAP founders call the five great-ape species (humans, chimpanzees,* bonobos, gorillas, and orangutans). These principles are set out in the Declaration on Great Apes and include the right to life, the protection of individual liberty, and the prohibition of torture, all currently enjoyed only by humans. The GAP follows the tradition of animal liberation* ethics, which requests a fundamental change in the moral status of nonhumans and views the unequal ranking of equal interests and needs solely on grounds of nonmembership of the human species (*see* SPECIESISM) as ethically unjustified discrimination. The project's founders and first signatories to the declaration were Paola Cavalieri and Peter Singer, who also edited the book *The Great Ape Project: Equality beyond Humanity*, on which the GAP's challenge is founded.

The scientific basis for GAP is provided mostly by recent biological, ethological, and psychological findings that unanimously indicate that all the species of great apes have highly complex emotional lives, form long-lasting social relationships, are self-aware and thus see themselves as distinct from others, make at least short-term plans for the future, have memories and anticipation, and possess mental capacities comparable to those of two- to three-year-old human children. This redrawn picture of the other great apes is underpinned by recent taxonomic investigations that indicate that chimpanzees* share 98.4% of their DNA with humans. Studies using sign language have further revealed that some nonhuman great apes can comprehend, use, and pass on abstract symbols to communicate with humans and other group members or to talk to themselves.

The change in the moral and legal status of the other great apes envisaged by the Great Ape Project is to be seen as an extension of the community of equals* beyond the boundaries of the human species. As members of this community, nonhuman great apes are entitled to the same previously mentioned basic rights as humans. The GAP points to the contrast between this moral entitlement and the reality of the great apes' existence. They are frequently subjected to extended or lifelong imprisonment, to the destruction

The Great Ape Project seeks to extend the scope of three basic moral principles.
Source: Reprinted from *The Great Ape Project: Equality beyond Humanity*, edited by
Paola Cavalieri and Peter Singer, © 1993, published in the United Kingdom by
Fourth Estate, Ltd., and in the United States by St. Martin's Press.

of family or other important social bonds, and to grave physical and psychological injury and deprivation.

In practice, the inclusion of the nonhuman great apes into the community of equals requires that the declaration be contained in U.N. resolutions and national law. As particular models for concrete political measures, the Great Ape Project takes two already-existing protective devices afforded to powerless members of human societies. Nonhuman great apes still living in their natural habitats are to be protected by the establishment of U.N. trust territories, like those set up to protect weaker nations against stronger ones. As with young children and some mentally handicapped humans, approved guardians should be appointed to plead the cause of individual nonhuman great apes who are currently imprisoned. For individuals who cannot be reintroduced into the wild because of their long imprisonment, either as human surrogates in biomedical or psychological experiments or as objects of education* and amusement in zoos* or other forms of entertainment industry, the Great Ape Project proposes carefully considered resettlement in sanctuaries and reserves especially designed to meet their manifold physical, emotional, and social needs, where they can live their own lives among others of their kind.

The Great Ape Project conceives the case of the nonhuman great apes as the best example for demonstrating the arbitrariness that, within the conception of animal liberation ethics, underlies a speciesist discrimination. According to the policy of the GAP, the focus on great apes is to be regarded in the broader political context of tackling the moral and social problems due to prejudice in favor of one's own group.

Selected Bibliography. Cavalieri, Paola, and Peter Singer (Eds.) *The Great Ape Project: Equality beyond Humanity* (London: Fourth Estate, 1993); Goodall, Jane, *Through a Window: My Thirty Years with the Chimpanzees of Gombe* (Boston: Houghton Mifflin, 1990); Rachels, James, *Created from Animals: The Moral Implications of Darwinism* (New York: Oxford University Press, 1990); Savage-Rumbaugh, Sue, and Roger Lewin, *Kanzi: The Ape at the Brink of the Human Mind* (London: Doubleday, 1994); Singer, Peter, *Animal Liberation* (New York: New York Review of Books, 1990).

KARIN KARCHER

H

HILDROP, JOHN

John Hildrop (?–1756) was an English cleric and author of the benchmark work *Free Thoughts upon the Brute-Creation* (1742), in which he critiqued the work of a French Jesuit, Father Bougeant, for his view that animals have no reason, understanding, moral status, or immortal soul. Originally in the form of two letters to a lady, his work directly confronts the major elements of traditional Aristotelian/Thomistic thought and is one of the earliest and most sophisticated zoophile books ever published. He also critiqued John Locke's materialist view of animal rationality in Locke's *Essay Concerning Human Understanding* (1690): "Why does [Locke] take so much pains to persuade himself and us, that Rationality in Brutes must proceed from a quite different cause, from what it does in ourselves? What is he afraid of?" (15). Hildrop was rector of Wath in Yorkshire and chaplain to the earl of Ailesbury and Elgin. After Thomas Tryon, who introduced the word "rights" in the non-human context, his work may be classed as the earliest premodern zoophile treatise.

Selected Bibliography. Hildrop, John, *Free Thoughts upon the Brute-Creation, or, An Examination of Father Bougeant's "Philosophical Amusement, &c. on the Language of Beasts"* (London: R. Minors, 1742).

ANDREW LINZEY

HINDUISM. *See* RELIGION AND ANIMALS.

HOUGHTON, DOUGLAS

Labour politician, peer, and social reformer, member of Parliament for Sowerby (1949–1974), minister for social services (1964–1967), and chairman of the Parliamentary Labour Party (1967–1974), Douglas Houghton (1898–1996) devoted the last twenty-five years of his life to animal advocacy. He was chairman of the Committee for the Reform of Animal Experimentation and a vice president of the Royal Society for the Prevention of Cruelty to Animals* (RSPCA). He inaugurated the "Putting Animals into Politics" campaign (1976), which was significant in galvanizing political support for reforming measures, and was president of Animal Welfare Year (1976–1977). Houghton was personally instrumental in achieving a range of legislative changes including reform of the 1876 Cruelty to Animals Act governing animal experiments, which became (not uncontentiously) the Animals (Scientific Procedures) Act of 1986.

Selected Bibliography. Houghton, Douglas, Animals and the Law: Moral and Political Issues, in David A. Paterson and Richard D. Ryder (Eds.), *Animals' Rights: A Symposium* (London: Centaur Press, 1979), 209–215; Houghton, Douglas, Thoughts for the Future, in David A. Paterson (Ed.), *Humane Education: A Symposium* (London: Humane Education Council, 1981), 133–137; Houghton, Douglas, and Lord Platt, *Houghton/Platt Memorandum on Animal Experimentation to the House Secretary* (London: Committee for the Reform of Animal Experimentation, 1976).

ANDREW LINZEY

HUMANE EDUCATION MOVEMENT

Humane education is about kindness and respect. Most clearly identified with George Angell,* the founder of the Massachusetts Society for the Prevention of Cruelty to Animals (MSPCA), it is based on the assumption that if children learn to care for and respect animals, they will develop an empathetic (*see* EMPATHY FOR ANIMALS) or "feeling" personality that will guide them in their relations with people as well.

The general theme of being kind to animals was present in the very earliest publications printed for children. In the late 1700s and early 1800s a number of stories and books for children talked about the mistreatment of animals. The stories often had a strong moral theme that emphasized empathizing with the animals, and the evildoers came to a bad end because of their treatment of animals. This type of story culminated with the publication of *Black Beauty* by Anna Sewell* in 1877.

Early animal-protection work included elements of humane education. In the 1850s M. DeSally published "Method of Teaching Kindness to Animals" in the *Bulletin Annuel de la Société Protective des Animaux*. It was difficult for education to receive a high level of attention when an enormous amount of rescue and law-enforcement work was required. George Angell, who had a background as a teacher, placed a major emphasis of the early work of the MSPCA on promoting humane education. He understood that to teach children kindness would be the best way to prevent cruelty* to animals and people.

When Angell began to formalize the understanding of humane education in the 1870s, he found a fertile ground in the American educational system at the time. *McGuffey's Newly Revised Eclectic Reader*, published in 1843, included many stories on animals and nature. In that same era, the "common-school" philosophy of Horace Mann emphasized the important role that public education could play in providing students from many different backgrounds a common sense of culture and morals. Most valuable at the time was the concept that schools could play a significant role in helping to solve major social problems.

In 1882 Angell began to organize "Bands of Mercy" in schools across the country. These clubs encouraged children to learn about animals and to do things to help animals. Angell founded the American Humane Education Society (AHES) in 1889 "to carry Humane Education in all possible ways, into American schools and homes." AHES also promoted Bands of Mercy across the country. Twenty states, recognizing the importance of humane education for society in general, passed laws requiring its practice in the schools by 1922. Edwin Kirby Whitehead published the first humane education textbook in 1909, *Dumb Animals and How to Treat Them*, and Flora Helm followed with a *Manual of Moral and Humane Education*.

At the same time, the humane movement suffered the pains of evolution in a changing society. Many of the earliest humane societies, including the American Society for the Prevention of Cruelty to Animals* (ASPCA) and MSPCA, had been inspired by the need to protect the many horses used for transportation and work in America's cities and towns. When carriage and cart horses disappeared from streets and roads, the humane movement came to grips with new roles and challenges.

In the 1960s, America shook off the effects of the Great Depression and two world wars. People once again began to question their relationships with one another and the environment. New educational philosophies emerged. Earth Day and the developing environmental movement gave rise to environmental education, and humane educators were poised to move forward with new opportunities. New efforts have included curriculum development, teacher training, and teaching materials for classroom use. Most humane societies offer humane education programs, recognizing that the only certain way to prevent cruelty to animals is to help children learn the meaning of kindness. (*See also* EDUCATION AND THE USE OF ANIMALS.)

Selected Bibliography. Angell, George T., *Autobiographical Sketches and Personal Recollections* (Boston: Franklin Press: Rand, Avery and Co., 1884); Bank, Julie, and Stephen Zawistowski, The Evolution of Humane Education, *ASPCA Animal Watch*,

Humane education can involve observing animals in their natural habitats. Photo courtesy of the American Society for the Prevention of Cruelty to Animals.

Fall 1994; Good, H. G., *A History of American Education* (New York: Macmillan, 1956); Spring, Joel, *The American School, 1642–1985* (New York: Longman, 1986); Steele, Zulma, *Angel in a Top Hat* (New York: Harper and Brothers, 1942); Wells, Ellen B., and Anne Grimshaw, *The Annotated Black Beauty*, by Anna Sewell (London: J. A. Allen, 1989).

STEPHEN L. ZAWISTOWSKI

University-Level Humane Education

Woodrow Wilson, the only president of the United States who taught college and had a doctorate, noted that "it is easier to move a cemetery, than change a University curriculum." Of course, the curriculum does change, slowly and cautiously. The changes reflect not only new knowledge but new definitions of what is important to know. One area remarkably ignored is our relationship to animals.

An increasing proportion of people believe that companion,* laboratory, and farm animals should receive the best possible health care, including the latest advances in science and technology (*see* COMPANION ANIMALS AND PETS; FARM-ANIMAL WELFARE; LABORATORY ANIMAL USE). One approach is to develop a focused course of study for students involved in a variety of fields of inquiry addressing not only animal welfare,* but also issues related to the conservation of endangered animals (*see* EN-DANGERED SPECIES) and their environments. Such a curriculum has been developed at Purdue University. Like any curriculum, it reflects the strengths of the faculty and concerns of the present student body.

In 1982, Purdue University developed the Center for Applied Ethology and Human-Animal Interaction at the School of Veterinary Medicine to promote interdisciplinary activities in the university by serving as a focal point for the exchange of ideas and development of new information related to human-animal interactions and to disseminate information in an unbiased manner to students, scientists, consumers, and agricultural groups. The primary objectives of the program are to educate undergraduate students about the social, ethical, biological, behavioral, and economic aspects of animal care and use, provide students with a scientific and philosophic basis for care and use, and train students to resolve conflicts concerning the humane use of animals and to become leaders in policy development and implementation. There is ever-growing concern for and interest in our environment, the well-being* of animals, and the quality of our interactions with animals. This course of study provides the knowledge and skills to communicate and act on these issues. It also stimulates research to improve human and animal well-being.

Selected Bibliography. Beck, A. M., Animals and Society, in A. Goldberg and L. F. M. van Zutphen (Eds.), *World Congress on Alternatives and Animal Use in Life Sciences: Education, Research, Testing* (New York: Mary Ann Liebert, 1995), 59–64; Beck, A. M., and A. H. Katcher, *Between Pets and People: The Importance of Animal Companionship*, rev. ed. (West Lafayette, IN: Purdue University Press, 1996); Glickman, N. W., L. T. Glickman, M. E. Torrence, and A. M. Beck, Animal Welfare and Societal Concerns: An Interdisciplinary Curriculum, *Journal of Veterinary Medical Education* 18(2) (1991): 60–63; Pritchard, W. R. (Ed.), *Future Direction for Veterinary Medicine* (Durham, NC: Pew National Veterinary Education Program, 1989).

ALAN M. BECK

HUMANE SLAUGHTER ACT

The first humane slaughter bill ever presented in Congress was introduced by Senator Hubert Humphrey in 1955. Most European democracies had enacted humane slaughter legislation in the previous three decades. The U.S. Department of Agriculture (USDA) opposed the Humphrey bill and its companion House bill, saying that American enterprise could provide better humane slaughter than legislation could. The meat packers had managed to put off action for many years by claiming that they were studying the matter. They continued to pursue the "study" gambit in their vigorous effort to defeat mandatory humane slaughter legislation.

A number of members of the House Agriculture Committee joined Subcommittee Chairman W. R. Poage, a cattleman from Texas, in a visit to Chicago slaughterhouses to witness large-scale slaughtering practices first-hand. At that time in all the big slaughterhouses, cattle were stunned by swinging a heavy pole axe at their heads, sometimes as many as 13 times before they collapsed. Slaughtermen resorted to early morning alcoholic drinks to make their work endurable, but their aim with the heavy sledge-hammer was even worse as a result. The big meat packers' decades of so-called study had failed to come up with anything less cruel.

The U.S. House of Representatives voted overwhelmingly in favor of Poage's mandatory humane slaughter bill. The industry's lobbyists focused on the Senate Agriculture Committee, where they succeeded in having all of the effective protection for animals deleted. The American public was shocked. Editorials in leading newspapers expressed outrage. Senator Humphrey and 17 cosponsors introduced a bill restoring the mandatory language as passed by the House of Representatives. Over the passionate objections of the chairman of the Senate Agriculture Committee, Majority Leader Lyndon Johnson called up the bill for a vote.

Senator Humphrey led the seven-hour-long debate on the Senate floor. By a 43–40 vote, the Senate reversed the committee. President Dwight Eisenhower signed the bill into law, effective June 30, 1960.

The act covers 80% of U.S. plants by making it compulsory for all packing companies selling meat to the federal government to use humane methods. A last-minute attempt to undermine the legislation was made by the biggest buyer of meat for the U.S. government, the Military Subsistence Agency, purchaser of all meat for the armed forces, which tried to limit it to contracts exceeding $2,500. But Representatives W. R. Poage and Martha Griffiths joined Senator Humphrey in strongly opposing the exemption as illegal. As enacted, the bill provides that

> cattle, calves, horses, mules, sheep, swine, and other livestock, all animals are rendered insensible to pain by a single blow or gunshot or an electrical, chemical or other means that is rapid and effective, before being shackled, hoisted, thrown, cast, or cut;
> . . . or by slaughtering in accordance with the ritual requirements of the Jewish faith or any other religious faith that prescribes a method of slaughter whereby the animal suffers loss of consciousness by anemia of the brain caused by the simultaneous and instantaneous severance of the carotid arteries with a sharp instrument and *handling in connection with such slaughtering*.

In 1978, the Federal Meat Inspection Act was amended by a bill sponsored by Senator Robert Dole and Representative George E. Brown. Federal meat inspectors have the authority to prevent inhumane practices by withholding inspection until any cruel methods are corrected. Profits in the meat industry depend on speed in putting animals through "the line." Thus the threat of stopping the line provides a powerful incentive to avoid cruelty. The amended law also prohibits importation of meat from inhumanely slaughtered animals. USDA personnel inspect foreign plants to assure adherence to sanitary standards and, from 1978 on, have included humane standards. The Humane Slaughter Act does not cover small meat-packing plants that are not subject to federal inspection, nor does it require humane preslaughter handling for kosher-killed animals.

CHRISTINE STEVENS

HUMANE SOCIETIES. *See* AMERICAN SOCIETY FOR THE PREVENTION OF CRUELTY TO ANIMALS; HUMANE SOCIETY OF THE UNITED STATES; ROYAL SOCIETY FOR THE PREVENTION OF CRUELTY TO ANIMALS.

HUMANE SOCIETY OF THE UNITED STATES (HSUS)

The Humane Society of the United States (HSUS) was founded in 1954 with the mission to promote the humane treatment of animals and to foster respect, understanding, and compassion for all creatures. Since then, the HSUS has grown into the largest animal-protection organization in the world, with a full-time staff of over 200 and an active constituency of more than 3.5 million people. A nonprofit, charitable organization, the HSUS is funded by membership dues, contributions, and gifts.

From its beginning, the HSUS has sought to broaden traditional humane concerns to include a wide range of animal and environmental issues. A partial list of the major program concerns and accomplishments over the years includes the following:

1954–1964: Passage of the Humane Slaughter Act*; establishment of a system of regional offices to meet the needs of local societies and constituents; providing evidence leading to an embargo of monkeys shipped to the United States from India

1964–1974: Prominent role in the passage of the Laboratory Animal Welfare Act,* the Endangered Species Act, the Wild Free-roaming Horse and Burro Act, and the Marine Mammal Protection Act; establishment of the National Association for Humane and Environmental Education

1974–1984: Strong efforts to upgrade standards of animal shelters* and zoos*; establishment of a disaster-relief program for animals; promotion of national opposition to milk-fed veal (*see* VEAL CALVES); efforts toward a moratorium on commercial whaling; major campaigns investigating and publicizing the cruelties suffered by puppy-mill dogs*; major efforts against organized dogfighting

1984–present: Launching of the "Be a P.A.L.—Prevent a Litter" campaign to promote the importance of spaying and neutering; fighting the slaughter of dolphins through a consumer boycott of tuna caught in ways harmful to dolphins; launching of a nationwide antifur campaign; establishment of the "Beautiful Choice" program to promote cruelty-free products; promotion of alternatives* to the use of animals in research (*see* LABORATORY ANIMAL USE) and education*; facilitation of networking between animal-protection groups and groups working against child abuse and family and social violence; strong efforts for felony-level penalties for cruelty to animals

This diversity of interests is reflected in the structure of the HSUS. Specific program sections coordinate efforts in the areas of animal research issues, companion animals,* farm animals and bioethics, state and federal legislation, wildlife and habitat protection, investigations, and training. These activities are facilitated by nine regional offices throughout the United States. The national and global outreach of the HSUS is further supported by a family of organizations that come under the HSUS umbrella. These include the following:

Humane Society International (HSI) is the HSUS abroad, working on animal issues that cross many borders, including the trade in wild birds, the decimation of elephant populations, endangered-species* issues, marine mammal concerns and practices, and conditions affecting companion and farm animals worldwide.

The National Association for Humane and Environmental Education (NAHEE) has served as a resource for educators since 1973. Its publications include the *KIND News* monthly newspaper for elementary-school students and the *Student Network News* for secondary-school students.

The Center for Respect of Life and Environment (CRLE), founded in 1968, focuses on higher education, religion, the professions, and the arts in promoting a humane and sustainable future.

Earthkind is the global environmental arm of the HSUS and is committed to fostering humane, sustainable development and protecting biodiversity. It works with a sister organization in England and international offices including sites in Russia, Romania, Brazil, and Sri Lanka.

The International Center for Earth Concerns (ICEC) is dedicated to the development and implementation of nature-conservation projects that enhance and protect wild places, animals, and the environment.

The HSUS Wildlife Land Trust creates and maintains sanctuaries in which recreational and commercial hunting* and trapping* will never take place. Wild animals are protected by preserving their natural habitats and providing them sanctuary within those habitats.

RANDALL LOCKWOOD

HUMANISM. *See* ANTHROPOCENTRISM.

HUNTING

For 99% of human history, hunting and gathering have been the principal subsistence pattern. In the sweep of history, it is only in recent years that the purpose and effect of hunting have dramatically changed (though there is renewed debate about the role of aboriginal hunting in the loss of megafauna in the Pleistocene and other periods of human history). More specifically, humans altered the terms of the hunt in the second half of the 19th century. The change was precipitated not only by a worldview that differed starkly from that of Native peoples (*see* NATIVE PEOPLES AND ANIMALS), but also by the development of technologies and national and international economic markets, bound together by more efficient means of transport, including the first transcontinental railroad. In particular, the development of the repeating firearm had a profound impact on the nature of hunting. Hunting developed as a tool of commerce.

At the same time, the movement for hunting as sport emigrated from Europe and gained a foothold in the United States. The combined rise of market and sport hunting in the second half of the 19th century ushered in a period of unprecedented wildlife destruction. Commercial hunters slaughtered bison, elk, swans, egrets, and other wildlife by the tens of millions. Markets developed for buffalo hides and tongues. The millinery trade developed markets in bird feathers to adorn women's hats.

The hunters' destruction of wildlife provoked a backlash among people concerned about wildlife and among those concerned about long-term hunting opportunities. In the 1930s a new model developed, principally credited to the father of game management, Aldo Leopold. This was the science of game management, which imposed rules and regulations governing the sport of hunting. It marked the triumph of sport hunting over market hunting and created wildlife as a public resource to be managed for sustainable utilization. The states, goaded by the conservation lobby, imposed limits on the kill that all but banned commerce in hunted wildlife products.

Conservationists developed a game-management infrastructure in every state, imposing an agricultural model on the killing of wildlife. Wildlife was considered to be a crop to be harvested on an annual basis. The kill was not to exceed the capacity of the population to restore itself through reproduction. The newly formed state fish and game agencies and the federal Fish and Wildlife Service oversaw wildlife and created wildlife policy. During the same period, Congress created other funding sources to build the game-management infrastructure. In the mid-1930s, Congress passed the Pittman-Robertson Act, an excise tax on the sale of guns and ammunition with revenue to go to the states for game-management purposes. Congress also approved the Duck Stamp Act to set aside money to acquire wetlands so that waterfowl populations could be sustained at huntable levels.

The game-management model has dominated wildlife policy making since the 1930s. Game managers have been successful at directly limiting the killing of animals so as not to endanger species survival, but the toll on the lives of individual animals has been immense. Every year, sport hunters kill in excess of 200 million animals. According to the U.S. Fish and Wildlife Service, there are approximately 15 million Americans who hunt, about 7% of adult Americans. Most of them use firearms, but there is a growing primitive-weapons constituency who use bow and arrows and muzzleloaders to enhance the element of chase and sport.

There are thousands of rod and gun clubs across the United States and hundreds of organizations that work to promote and defend hunting. The largest among them are the National Rifle Association and the Safari Club International, which promotes worldwide hunting of rare and exotic wildlife. Hunters often justify their sport as a means of controlling wildlife populations, but disinterested biologists recognize that hunting is not necessary to control most animal populations. Some rural sociologists predict a steady decline of hunters as a percentage of the population well into the next century. The rise of an animal rights* ethic, the decline in hunter participation among young people, and the difficulties in accessing huntable lands provide support for that prediction.

Selected Bibliography. Amory, Cleveland, *Man Kind? Our Incredible War on Wildlife* (New York: Harper and Row, 1974); Cartmill, Matt, *A View to a Death in the Morning: Hunting and Nature through History* (Cambridge, MA: Harvard University Press, 1993); Kerasote, Ted, *Bloodties: Nature, Culture, and the Hunt* (New York: Random House, 1993); Mitchell, John G., *The Hunt* (New York: Alfred A. Knopf, 1980); Reisner, Marc, *Game Wars: The Undercover Pursuit of Wildlife Poachers* (New York: Viking, 1991); Swan, James, *In Defense of Hunting* (San Francisco: Harper, 1995).

WAYNE PACELLE

History of Ideas Surrounding Hunting

Although prehistoric people needed to hunt to survive, hunting has had little economic significance throughout most of the history of Western civilization. Its importance in Western thought derives chiefly from its symbolic meaning. That meaning has much to do with how we define hunting and distinguish it from butchery. Hunting is not simply a matter of killing animals. To count as quarry (a "kill"), the hunter's victim must be a wild animal. For the hunter, this means that it must be hostile: unfriendly to human beings, intolerant of their presence, and not submissive to their authority. The hunt is thus by definition an armed confrontation between the human domain and the wilderness, between culture and nature. The meanings that hunting has taken on in the history of Western thought reflect the varying values ascribed to culture and nature in this pretended confrontation.

Throughout Western history, the hunter has been seen as an ambiguous figure, sometimes a fighter against wilderness and sometimes a half-animal participant in it. The meaning of hunting accordingly varies with the meanings ascribed to the wilderness. For the Greeks and Romans, forests were generally threatening and scary places. In early Christian thought (*see* RELIGION AND ANIMALS, Christianity), the wilderness was a sort of natural symbol of hell, and the wild animals living there in rebellion against man's dominion were seen as typifying demons and sinners in rebellion against God. But this image was undermined by the counterimage of the hermit saint in the wilderness, attended by friendly wild animals that the saint's holiness had restored to the docility of Eden.

Other medieval changes in the symbolic meaning of wild places and creatures reflect changes in the social status of hunting. From the 10th century on, Europe's forests dwindled as improved techniques of agriculture fostered a surge in human population growth. Hunting gradually became the exclusive privilege of the aristocracy, who put the remaining forest patches off limits as hunting preserves and ruthlessly punished any peasants caught taking game. The deer, who are the symbolic inhabitants of the wilderness and give it its English name (etymologically a *wild-deer-ness*), became the main objects of the aristocratic hunt and took on an air of nobility in both folk ballads and high culture.

It was not until the early 1500s that the chase began to be viewed as cruel and to be invoked as a symbol of injustice and tyranny. Erasmus condemned the hunt in 1511 as a bestial amusement. Thomas More denounced it in *Utopia* (1516) as "the lowest and vilest form of butchery . . . [which] seeks nothing but pleasure from a poor little beast's slaughter and dismemberment." Similar revulsion toward hunting is evident in the essays of Montaigne and in the plays of Shakespeare. Antihunting sentiment also crops up in 16th-century hunting manuals, which from 1561 on contain rhymed complaints by the game animals denouncing the senseless cruelty of Man the Hunter.

The rise of antihunting sentiments in the 1500s reflected rising doubts about the importance of the boundary between people and animals. In 1580, Montaigne denied the existence of that boundary and concluded that "it is [only] by foolish pride and stubbornness that we set ourselves before the other animals and sequester ourselves from their condition and society." The erosion of the animal-human boundary in Western thought was accelerated by the scientific revolution of the 1600s and the associated mechanization of the Western world picture. Animal suffering* came to be more widely regarded as a serious evil, and hunting was increasingly attacked as immoral.

The romantic movement of the late 1700s brought about a radical transformation in Western images of wilderness. In romantic thought, nature ceased to be a system of laws and norms and became a place, a holy solitude in which one could escape man's polluting presence and commune with the Infinite. Romantic art and literature picture the hunter sometimes as a poet with a gun participating in the harmony of nature (e.g., James Fenimore Cooper's Natty Bumppo), but more often as a despoiler of nature and animal innocence (e.g., Samuel Taylor Coleridge's Ancient Mariner).

Western hunting has always been a characteristically male activity, often regarded as valuable training for the military elite and praised as a prototype of the just war. In the context of 19th-century European imperialism, this tradition gave birth to a third stereotype of the huntsman: the colonial White Hunter who dons a pith helmet and leads an army of servile natives on safari to assert his dominion over the conquered territory's land, animals, and people. At the height of Europe's empires in the late 1800s and early 1900s, a love of hunting commonly went hand in hand with imperialist politics, and anti-imperialism was often associated with antihunting sentiment. This link between hunting and the political right has persisted into our own time.

During the 20th century, the romantic idea of the sanctity of nature and the Nietzschean and Freudian picture of man as a sick animal have interacted to yield a vision of the wilderness as a place of timeless order and sanity, in opposition to the polluted and unstable domain of civilization and technology. However, hunters tend to regard the hunt as a healing participation in the natural order—what the hunting philosopher José Ortega y Gassett described as "a vacation from the human condition"—whereas opponents of hunting see it as an armed assault on the harmony of nature.

Both these attitudes are grounded in the romantic image of "nature" as a place with no people in it. If we reject that concept of nature and adopt instead a more scientific (and pre-romantic) conception of human beings and their works as part of nature, the distinction between wild and domestic animals evaporates (*see* DOMESTICATION). Hunting thereby loses its rationale and appears to us, as it did to More, as nothing but a species of butchery practiced for amusement. However, doing away with the opposition between the human and natural domains poses problems as well for the philosophy of animal rights.* The rights view generally assumes that the moral order and nature are separate realms and that what wild animals do to each other is a matter of moral indifference. But if the boundaries between people and animals and between culture and nature are imaginary, it is not clear why we should have a duty to prevent a wolf from eating a baby but not from eating a rabbit.

Selected Bibliography. Anderson, J. K., *Hunting in the Ancient World* (Berkeley: University of California Press, 1985); Cartmill, M., *A View to a Death in the Morning: Hunting and Nature through History* (Cambridge, MA: Harvard University Press, 1993); MacKenzie, J. M., *The Empire of Nature: Hunting, Conservation, and British Imperialism* (Manchester: Manchester University Press, 1988); Ortega y Gassett, J., *Meditations on Hunting* (New York: Scribner's, 1972); Thiebaux, M., *The Stag of Love: The Chase in Medieval Literature* (Ithaca, NY: Cornell University Press, 1974).

MATT CARTMILL

Environmental Ethics and Hunting

J. Baird Callicott's 1980 article "Animal Liberation: A Triangular Affair" stated two widely shared concerns about animal rights* views and environmental ethics.* One of these was that animal rights views are incompatible with sound environmental management because they would rule out all hunting as immoral, even when overpopulated herds threaten to degrade their habitat. However, it is at least possible for an animal rights view to endorse hunting, at least in the kinds of situations where environmentalists feel compelled to endorse hunting as a way of preventing habitat destruction. For instance, if it is true that the overpopulation that damages their habitat would also reduce the average welfare of individuals in the herd, then a utilitarian (see UTILITARIANISM) view like Peter Singer's could endorse hunting. Utilitarians evaluate policies in terms of their total impact on all affected individuals.

In his book *The Case for Animal Rights*, Tom Regan expressly opposes all hunting. He reasons that the defenders of hunting use utilitarian arguments and says that this fails to respect the rights of individual animals. If an individual has moral rights, then, on Regan's analysis, it is wrong to harm him or her simply because the total benefit to others will outweigh the harm to the individual. However, Regan never considers the application of his own principles to hunting scenarios. Regan defends the use of two nonutilitarian principles for deciding whom to harm when harm is inevitable. Of his two principles, the one applicable to hunting where overpopulation* threatens to degrade habitat is his "miniride principle," which directs one to harm the few rather than the many when the harms involved are all roughly comparable. Regan acknowledges that where it applies, this principle implies the same thing as utilitarianism, but for different reasons: it is the minimizing of rights violations that is at issue, not the magnitudes of benefit and harm in the total package. Where overpopulation threatens to reduce the future carrying capacity of the range, the miniride principle would seem to endorse hunting. If it is true that allowing the overpopulated herd to stabilize on its own would result in a lower sustainable population in the end, then a carefully regulated hunt could minimize the number of deaths.

Hunters give various reasons for hunting, including tradition and sport, but to the extent that hunting maximizes average well-being within the target population and/or minimizes the total death rate, hunting is not strictly incompatible with some animal rights philosophies. To the extent that the hunting environmentalists feel compelled to endorse these same things, animal rights philosophies are not strictly incompatible with the goals of environmentalists.

Selected Bibliography. Callicott, J. Baird, Animal Liberation: A Triangular Affair, *Environmental Ethics* 2 (1980): 311–338; Leopold, Aldo, The Land Ethic, in *A Sand County Almanac* (New York: Oxford University Press, 1949), 201–226; Regan, Tom,

The Case for Animal Rights (Berkeley: University of California Press, 1983); Varner, Gary, Can Animal Rights Activists Be Environmentalists? in Don E. Marietta, Jr., and Lester Embree (Eds.), *Environmental Philosophy and Environmental Activism* (Lanham, MD: Rowman and Littlefield, 1995), 169–201.

GARY VARNER

Hunting in the United States

Hunting traditions in the United States of America stem, in part, from a reaction to the rigid and elitist forms of northern Europe. In the United States, access to natural resources, including wildlife, is connected to the concept of private property ownership by way of the Magna Carta, the Charter of the Forest, and English common law. Until the beginning of the modern era (about 1815), hunting was more or less unregulated in the United States. Anyone with the desire, guile, and ability was able to take wildlife in more or less unrestricted numbers and of unrestricted kinds. However, as constitutional law became more sophisticated, restrictions came to be applied to hunting. In general, the regulation of hunting is reserved to the states in the United States through the police power of the state. However, the interstate commerce clause of the U.S. Constitution and the ability to enter into international treaties have been used by the federal government to exert increasing control over hunting, especially on publicly owned lands. Legal decisions since before the Civil War have progressively changed the view of wildlife from wild nature, which allowed anyone to reduce wild animals to private property at will, to commonly held resources owned by the states, to commonly held resources of which the state is a recent custodian. The issue between the federal government and the states over the control of hunting in particular and wildlife in general remains a question of legal tension.

As this view of wildlife has changed, so has the view of hunting. Hunting has progressed from an activity without restrictions or culturally important implications to one of extreme cultural importance to a minority in society. In the United States, hunting has emerged as an activity primarily carried out by white males who have been initiated into hunting by their fathers or other, older male members of their immediate family. As such, hunting constitutes a culturally important activity psychologically centered on issues of the family.

Recent studies indicate a small but steady decline in the number of white males taking part in hunting. Hunting by females and nonwhite males has increased, but this is a very small percentage of the total hunting population. No single factor can be identified as a primary cause for the declining participation, but most appear to be issues of changing family values and recreational activity. Influence of the animal rights movement* and other cultural pressures do not account for a statistically significant amount of the decline. Pressure to change some of the more egregious activities, such as the use of wild animals as living targets or hunting purely for the sake of securing a trophy without the consumption of the meat, will likely result in significant changes in the types of hunting permitted by the states and the federal government and by hunters themselves.

Selected Bibliography. Baker, Rob, and Ellen Draper (Eds.), The Hunter, *Parabola: The Magazine of Myth and Tradition* 16(2) (1991): special issue; Duda, Mark D., Steven J. Bissell, and Kira C. Young, Factors Related to Hunting and Fishing Participation in the United States, *Transactions of the 61st North American Wildlife and Natural Resource Conference* (Washington, DC: Wildlife Management Institute, 1996), 324–337; Hendee, John C., A Multiple-Satisfaction Approach to Game Management, *Wildlife Society Bulletin* 2(3) (1974): 104–113; Leopold, Aldo, *Game Management* (Madison: University of Wisconsin Press, 1986).

STEVEN J. BISSELL

Fair Chase

Fair chase is but one component of the more general concept of ethical hunting or angling and specifically concerns the way a hunter interacts with the quarry. This concept addresses the balance between the hunter and the hunted, a balance that allows the hunter to occasionally succeed while animals generally avoid being taken. Many states have laws and regulations pertaining to fair chase. However, many of these are intended to restrict hunting behavior that gives one hunter an unfair advantage over another, not to restrict the hunter's advantage over game. Thus fair chase is still sometimes interpreted as a matter of humans' duties of fairness to other humans and only indirectly or secondarily, if at all, of fairness to the animal chased. Current issues in fair chase include the appropriateness of put-and-take hunting and angling, baiting (*see* BEAR BAITING), electronic trail monitoring, group hunts and game drives, the use of dogs,* tournament hunting and fishing, and road hunting.

The regulatory process may be used to define standards of conduct for hunting and fishing, but regulations ensure only the minimum of ethical behavior. A sportsperson committed to the ideal of fair chase goes beyond the regulations and exhibits a voluntary respect for and decency toward animals. Such commitment includes limitations of behaviors and gadgetry that compensate for poor hunting skills and that minimize the animal's reasonable and natural chance to escape. It also prohibits hunting and game-management practices that cause the quarry to behave unnaturally, to the hunter's advantage (an example of this would be hunting near a feeder, near a mineral block, or over planted food plots). Above all, implicit in fair chase is the attitude of respect for the animal. What constitutes fair chase is always contextual, and both hunters and the nonhunting public provide the relevant context.

Selected Bibliography. *Fair Chase* (Boone and Crockett Club), Spring and Summer 1996 issues; Kerasote, Ted, *Bloodties* (New York: Random House, 1993); Ortega y Gasset, José, *Meditations on Hunting* (New York: Charles Scribner's Sons, 1972); Posewitz, Jim, *Beyond Fair Chase* (Helena, MT: Falcon Press, 1994); Wisconsin Natural Resources Board, *Report of Ethics and Fair Chase Committee*, 1996.

ANN S. CAUSEY

I

IACUCs. *See* INSTITUTIONAL ANIMAL CARE AND USE COMMITTEES.

INDIVIDUALITY. *See* ANIMAL INDIVIDUALITY.

INDUCED MOLTING. *See* CHICKENS.

INSTITUTIONAL ANIMAL CARE AND USE COMMITTEES (IACUCs)

Since 1985, with extensive revision of the Laboratory Animal Welfare Act* and the adoption of new policies by the National Institutes of Health, most institutions in the United States that conduct animal research have relied on an institutional animal care and use committee (IACUC) to determine whether research meets generally accepted ethical standards for the use of animals. Before 1985, such committees were generally called animal care committees, and while they had some oversight of the care and housing of laboratory animals, they did not review the actual research procedures. Now, however, any organization that receives federal funds must follow Public Health Services (PHS) policies on animal research. Institutions engaged in interstate commerce in covered species of animals (mammals, with the exception of mice,* rats, and animals used in agricultural practice) fall under U.S. Department of Agriculture (USDA) regulations, particularly the Laboratory Animal Welfare Act. Both sets of regulations require an IACUC to ensure that the institution follows all applicable regulations, and that any proposal to use animals in research has been reviewed.

An IACUC must include (*a*) a veterinarian,* (*b*) someone who does not use animals for research (typically referred to as "the nonscientist"), and (*c*) someone who does not work for the institution (*see* INSTITUTIONAL ANIMAL CARE AND USE COMMITTEES [IACUCs], Nonaffiliated Members). The two main duties of an IACUC are to review all proposals or protocols for use of covered species of animals and to ensure compliance with all government regulations. Practices vary widely depending on the size of the institution, the amount and range of animal research, and policies set up by the individual IACUC.

The whole system of IACUCs is based on the starting point that animal research is justified as long as it is carried out as well as possible, given the research goals. The questions IACUCs consider are almost never of the form "Should we be doing research on animals?" but rather, "Given that Dr. Smith is investigating *x*, has she shown that the study requires the use of this many animals of this species, and that she has designed the procedure to use appropriate care of the animals, including anesthetics and analgesics?" Granted that starting point, there are still at least two other ethical issues raised by the practice of using IACUCs to regulate research: the scope of an IACUC's authority, and the assumption that self-regulation is the best way to bring institutions into compliance with appropriate standards for ethical research.

With regard to scope, many animals are not covered by the relevant regulations, most notably, rats and mice are not currently covered by USDA regulations, and farm animals used for "production"-oriented research also fall into an ambiguous category. No cold-blooded species is covered by USDA regulations, and no invertebrate is covered by PHS policy. Moreover, many IACUCs have adopted the policy that "issues of scientific merit" fall outside the scope of their decision-making process. This has the effect of restricting, sometimes in significant ways, the nature of the deliberation process when trying to decide whether a particular proposal should be approved. Few attempts have been made to evaluate or ground these scope restrictions in a well-formulated ethical theory.

The second ethical issue focuses on the fact that IACUCs are a way in which research institutions regulate themselves. Some other countries, for example, Sweden (*see* ANIMAL ETHICS COMMITTEES [SWEDEN]), have adopted systems of outside regulation. Arguments that have been advanced in favor of outside regulation include a higher probability of impartial and consistent standards that might also better reflect the standards of the general public. Arguments in favor of institution-based systems such as IACUCs include increased flexibility and the fact that outside review, while feasible in localized areas with a small amount of research, would not be practical in the United States. A broader perspective on the inside/outside issue might ask whether the review process should be carried out primarily by those inside the research community, or primarily by ordinary citizens who do not themselves carry out research. In most review systems today, including the U.S. system of IACUCs, the majority of decision makers (on a typical IACUC, the proportion may be six or eight to one) are people who themselves are or have been engaged in animal research.

LILLY-MARLENE RUSSOW

Regulatory Requirements

The two major U.S. regulatory systems governing laboratory animal use,* the Laboratory Animal Welfare Act* and the Public Health Service Policy on Humane Care and Use of Laboratory Animals, require IACUCs. Both systems have similar requirements on IACUC membership, duties, and authority.

Committees must have at least three members. At least one doctor of veterinary medicine must serve on the committee. In addition, at least one person on the committee must have no other affiliation with the research institution (*see* INSTITUTIONAL ANIMAL CARE AND USE COMMITTEES [IACUCS], Nonaffiliated Members). According to the Laboratory Animal Welfare Act, this person should "provide representation for general community interests in the proper care and treatment of animals." Before a research project involving animals can go forward, it must be reviewed by the IACUC.

There are advantages and disadvantages of IACUCs. The federal government has often adopted a system of institutional committee oversight to address ethical issues in research. Institutional committees were first adopted in the 1970s as a means of monitoring research involving human subjects. Institutional committees are also used to address problems involving scientific misconduct and financial conflicts of interest affecting researchers.

Committee oversight systems reduce government expenses by assigning most of the monitoring responsibilities to research institutions, rather than to government officials. Researchers also are more likely to respect and cooperate with a committee of their colleagues than with a group of government "outsiders."

Although committees must comply with certain general rules, they have a great deal of flexibility and freedom to tailor the rules to their specific institution's situation. The committee's mixed membership is intended to allow diverse values to shape ethical decision making. The hope is that this approach will produce reasonable positions on a variety of controversial bioethical issues.

Yet the committee system has its critics as well. Institutions bear the financial and other burdens of administering the oversight system; faculty and staff must put aside their other duties to serve on the committees. Because the federal rules are somewhat general, different individual committees can reach different decisions on proposed research. Animal advocates also question whether the inclusion of one public member can prevent the scientific viewpoint from dominating in IACUC deliberations. They argue that committees would be more effective if one member were assigned to represent the interests of animals against proresearch interests. Thus far, however, these advocates have not persuaded Congress to revise the rules governing IACUCs.

Committees also face challenges in developing an effective approach to working with the scientists whose projects they evaluate, and in establishing meaningful programs for training on humane approaches to animal care and experimentation. They must also develop a defensible approach to recruiting and selecting new committee members, particularly the persons chosen from outside the institution.

Many of the issues facing IACUCs reflect general uncertainty over the appropriate use of animals in science. Persons favoring the elimination of or drastic reduction in laboratory animal use are unlikely to see IACUCs as providing meaningful oversight of animal research. On the other hand, persons who believe that scientists should have complete control over their experiments are likely to label IACUC activities an unjustified invasion of scientific freedom.

Congress and other government officials designed the IACUC system to implement a third ethical perspective. This view is that animal research is ethical if it is conducted to advance important social goals and if harm to laboratory animals is reduced to the minimum necessary to achieve these goals. IACUCs will continue to operate within this ethical framework unless advocates of another view successfully persuade Congress to alter the current regulatory approach.

Selected Bibliography. Animal Welfare Act, United States Code, vol. 7, sections 2131–2159 (1994); Dresser, Rebecca, Developing Standards in Animal Research Review, *Journal of the American Veterinary Medical Association* 194(9) (1989): 1184–1191; Office for Protection from Research Risks, *Institutional Animal Care and Use Committee Guidebook*, NIH Publication no. 92–3415 (Washington, DC: U.S. Department of Health and Human Services, 1992); *Public Health Service Policy on Humane Care and Use of Laboratory Animals* (Washington, DC: U.S. Department of Health and Human Services, 1986; available from Office for Protection from Research Risks, National Institutes of Health, 6100 Executive Blvd., MSC 7507, Rockville, MD 20892–7507); Rowan, Andrew, Ethical Review and the Animal Care and Use Committee, *Hastings Center Report* [Special Supplement] 20(3) (1990): 19–24.

REBECCA DRESSER

Nonaffiliated Members

Laws stipulate that institutional animal care and use committees (IACUCs) should include a person or persons who are not affiliated with the research facility to represent the concerns of the community about animal care and use. These members are referred to as nonaffiliated members (NAMs). NAMs review research proposals submitted to the IACUC and participate in meetings of the committee. Questions about the proposals can be raised, and the researcher has the opportunity to answer these questions. Although some committees require unanimous approval for passage of a proposal, most committees require a simple majority vote. Thus in most research facilities, a NAM cannot block a proposal.

Only anecdotal information is available concerning the views of individuals being selected as NAMs. Nonetheless, Barbara Orlans states that individuals who are selected are typically not known within their communities as animal advocates. In fact, people with possible biases (e.g., practicing scientists or staff of provivisectionist organizations) have reportedly sat on these committees. Levin and Stephens have proposed that NAMs should be community members known for their advocacy of animal protection. They propose that these people should be neither mouthpieces for the facility nor spies for local activists. Rather, they should be advocates for the research animals operating "within an imperfect oversight mechanism."

Some feel uncomfortable if the NAM is or was a practicing scientist, for they believe that such a person cannot be an advocate for the animals. However, this issue should be resolvable if NAMs are chosen after careful deliberation. As we learn more about the effectiveness of NAMs in the past (e.g., backgrounds and records), we will be able to make recommendations for the future.

Selected Bibliography. Levin, L. H., and M. L. Stephens, Appointing Animal Protectionists to Institutional Animal Care and Use Committees, *Animal Welfare Information Center Newsletter* 5 (4) (1994/1995): 1–10; Orlans, F. B., *In the Name of Science: Issues in Responsible Animal Experimentation* (New York: Oxford University Press, 1993); Orlans, F. B., R. C. Simmonds, and W. J. Dodds (Eds.), Effective Animal Care and Use Committees [Special issue], *Laboratory Animal Science*, January 1987; U.S. Congress, *Health Research Extension Act of 1985*, Public Law 99–158, November 20, 1985; U.S. Congress, Text of "Improved Standards for Laboratory Animals Act," *Congressional Record* 131(175) (1985): H12335–H12336.

MARJORIE BEKOFF

INTELLIGENCE OF ANIMALS. *See* ANIMAL COGNITION.

ISLAM. *See* RELIGION AND ANIMALS.

J

JAINISM. *See* RELIGION AND ANIMALS.

JUDAISM. *See* RELIGION AND ANIMALS.

K

KANT, IMMANUEL

Immanuel Kant (1724–1804) was born in the town of Königsberg, a small city on the eastern frontier of what would later become Germany. Königsberg was also the birthplace, in the last quarter of the 17th century, of the Pietist movement. The Pietists were a Christian sect holding strict, moralistic beliefs similar to those of the American Puritans. At the age of eight, Kant began his formal education at a school emphasizing Pietist teachings and virtues. Although Kant retained throughout his life the highest regard for moral virtue, particularly that of doing one's duty, his eight years of Pietist education led to his lifelong suspicion and dislike of religious enthusiasms in general and evangelical religion in particular.

Kantian ethics stresses that the origin of moral values lies in rational (reasoning) consistency. "What if everybody did that?" is the common moral idea forming the starting point for Kant's analysis of moral values. Kant concluded that in all we do, we should show respect for all rational beings. However, he did not believe that any nonhuman animals were rational beings; thus he believed that the well-being* of animals was not, by itself, a morally significant matter. He believed that we should be kind and fair to animals only because this would reinforce being kind and fair to humans. Kant's position on the moral insignificance of animals is developed in his *Lectures on Ethics*, which is primarily a student record of Kant's 1780–1781 course on "practical philosophy."

Selected Bibliography. Copleston, Frederick, Kant's Life and Character, in *A History of Philosophy*, vol. 6, pt. 1 (Garden City, NY: Doubleday, 1960); Durant, Will, Kant Himself, in *The Story of Philosophy* (New York: Simon and Schuster, 1926);

Greene, Theodore Meyer (Ed.), Introduction, in *Kant Selections* (New York: Charles Scribner's Sons, 1929); Korner, S., Some Notes on Kant's Life and Personality, in *Kant* (Baltimore: Penguin Books, 1955).

STEVE F. SAPONTZIS

KROGH PRINCIPLE

The Krogh principle is one of the guiding principles of animal investigations. In a lecture delivered in 1929, Danish physiologist August Krogh (1874–1949) said, "For a large number of problems there will be some animal of choice, or a few such animals, on which it can be most conveniently studied" (quoted in Krebs 1975, 221). While there is no nonhuman animal upon which all problems can be conveniently studied, for most problems there exists a convenient animal model (*see* ANIMAL MODELS, Real-World Analogies).

Animal researchers have generally adopted the Krogh principle. They seek out species whose members have, for any problem of interest, anatomical structures of useful size or arrangement, or physiological and biochemical processes that make it easy to conduct their experiment. This principle is primarily applicable in the context of basic research. It is less clear how it is to be applied in the context of applied research, especially where the aim is to make predictions about humans.

Selected Bibliography. Bernard, C., *An Introduction to the Study of Experimental Medicine* (1865; Paris: Henry Schuman, 1949); Gold, L., T. Slone, N. Manley, and L. Bernstein, Target Organs in Chronic Bioassays of 533 Chemical Carcinogens, *Environmental Health Perspectives* (1991): 233–246; Krebs, H., The August Krogh Principle, *Journal of Experimental Zoology* 194 (1975): 221–226; LaFollette, H., and N. Shanks, *Brute Science: Dilemmas of Animal Experimentation* (London: Routledge, 1996); Lave, L. B., F. K. Ennever, H. S. Rosenkranz, and G. S. Omenn, Information Value of the Rodent Bioassay, *Nature* 336 (1988): 631–633; Nishimura, H., and K. Shiota, Summary of Comparative Embryology and Teratology, in J. Wilson and F. Fraser (Eds.), *Handbook of Teratology*, vol. 3 (New York: Plenum Press, 1978), 119–154.

HUGH LaFOLLETTE AND NIALL SHANKS

L

LABORATORY ANIMAL USE

The use of laboratory animals creates an ethical dilemma for humans, offering the possibility of improvements for human health but also causing possible pain* and discomfort to animals. One result of this dilemma has been a growing resolve to find alternatives* in research, teaching, and testing. At the same time, finding alternatives has become an established goal of many organizations and has been enacted as a requirement in some legislation. This idea that it is desirable to develop new methods reflects changing attitudes toward animals. More and more people, both inside and outside the scientific community, have come to believe that it is worth considerable effort and cost to reduce discomfort of laboratory animals. This shift in consciousness was accelerated by publicity concerning the Draize test in rabbits (see ACTIVISM FOR ANIMALS), which was used to test new cosmetic and pharmaceutical products for eye and skin irritancy. Henry Spira's concerted campaign begun in 1979 against the Draize test unleashed a growing movement against causing animals discomfort. In 1981 the cosmetics industry awarded a one-million-dollar fund to Johns Hopkins School of Hygiene and Public Health to establish the Center for Alternatives to Animal Testing (CAAT).

Since then, the concept of alternatives has grown. It has become more respectable among scientists to endorse alternatives and to work to validate alternative testing methods. The Laboratory Animal Welfare Act* regulations specify that scientists must explore alternatives "to procedures that cause more than momentary pain or distress to an animal." Pain and suffering,* and their alleviation through anesthesia and analgesia, are the issues of primary concern. In the United States, CAAT has led industrial and governmental efforts to validate alternatives for toxicity testing.

Laboratory-housed animals are used in two main settings: company testing of products to assure safety and university research and testing. Although teaching involves fewer animals than research or testing, it is the use that could most easily be replaced with other methods. Veterinary schools are leading the way in creating computer software, soft-tissue models, and interactive videodiscs. Rather than performing dissections, veterinary students use anatomical specimens that have been preserved by plastination and systems developed for teaching psychomotor skills. These methods can be adapted at relatively modest cost for use in other educational settings (*see* EDUCATION AND THE USE OF ANIMALS).

The most energetic efforts to develop alternatives are in toxicity testing. The formation of the European Community has resulted in international legislation and funding for animal welfare and alternative methods: the European Centre for the Validation of Alternative Methods is spearheading an international coordinated effort to end the use of animal testing for cosmetic products in Europe. (*See also* ANTIVIVISECTIONISM; CATS; CHIMPANZEES; DOGS; INSTITUTIONAL ANIMAL CARE AND USE COMMITTEES; MICE.)

Selected Bibliography. Hart, L. A., The Animal Subjects Protocol Process: Applying the Three Rs, *Lab Animal* 24(5) (1995): 40–43; Rowan, A. N., G. M. Loew, and J. C. Weer, *The Animal Research Controversy: Protest, Process, and Public Policy— An Analysis of Strategic Issues* (Grafton, MA: Center for Animals and Public Policy, Tufts University School of Veterinary Medicine, 1995); Russell, W. M. S., and R. L. Burch, *The Principles of Humane Experimental Technique* (London: Methuen, 1959); U.S. Government. Title 9 Code of Federal Regulations (9 CFR), Part 3, Animal Welfare, Standards, Final Rule, *Federal Register: Rules and Regulations* 56(32) (February 15, 1991): 6426–6505; Zasloff, R. L., Alternatives in Veterinary Medical Education, *In Vitro Toxicology* 7(2) (1994): 185.

LYNETTE A. HART

Evolution and Animal Experimentation

Tests on animal subjects are designed to uncover the causal mechanisms that produce and direct the course of a disease or condition in animals. Researchers claim that these results can then be extended by analogy to humans. However, one result of evolution is that characteristics found in members of one species may be absent in members of another; for example, rats lack gall bladders. Furthermore, because organisms are intact systems composed of mutually interacting parts or subsystems, the interactions of one organism's subsystems may differ from the relationships in an organism of another species.

Evolution leads us to expect that different species will achieve many of

Laboratory Animal Use: Computer software accompanied by anatomical specimens preserved with plastination is replacing the formalin specimens and dissection formerly used in anatomy instruction. Here, Rick Hayes of the University of California at Davis Computer Assisted Learning Facility introduces an instructional program on the heart to Joe Epperson, a high-school teacher who will use the program for his students. Photo by Lynette A. Hart.

Associate Professor Sue Stover uses bones of racehorses who have died naturally for research and teaching. Photo by Lynette A. Hart.

the same biological functions. Moreover, biological organisms are usually "built" from similar parts. However, these organisms have faced different evolutionary pressures. In short, even if members of two species achieve similar biological functions, we cannot conclude that they have similar underlying causal mechanisms. Similar functions may be achieved by different causal routes.

Even a seemingly small change in an organism will almost certainly be associated with a variety of other changes that may be biomedically significant. Evolutionary theory tells us that animal models* cannot simply be assumed to be relevant, directly or otherwise, to human biomedical phenomena. Such relevance must be established empirically, and this will involve tests on humans as well as animals.

Selected Bibliography. Amdur, M., J. Doull, and C. Klaassen (Eds.), *Casarett and Doull's Toxicology*, 5th ed. (New York: McGraw-Hill, 1996), 226–281; American Medical Association (AMA), *The Use of Animals in Biomedical Research: The Challenge and Response*, rev. ed. (Chicago: American Medical Association, 1992); Futuyma, D. J., *Evolutionary Biology* (Sunderland, MA: Sinauer Associates, 1986); LaFollette, H., and N. Shanks, *Brute Science: Dilemmas of Animal Experimentation* (London: Routledge, 1996); Nesse, R., and G. Williams, *Why We Get Sick: The New Science of Darwinian Medicine* (New York: Times Books, 1994); Nomura, T., M. Katsuki, M. Yokoyama, and Y. Tajima, Future Perspectives in the Development of New Animal Models, in *Animal Models: Assessing the Scope of Their Use in Biomedical Research* (New York: Alan R. Liss, 1987), 337–353.

HUGH LaFOLLETTE AND NIALL SHANKS

Number and Species of Laboratory Animals Used Worldwide

Worldwide, the total number of animals used annually in biomedical research, testing, and education is well in excess of 41 million. Some estimates are as high as 100 million animals per year. The table on pages 216–217 lists figures from 17 countries for which data are available. The United States is probably the largest user, followed by Japan and France, in that order. The figure of 41 million is an underestimate because many countries that use laboratory animals do not count them.

Collection of official statistics on use of laboratory animals first started in the United Kingdom in 1960. The United Kingdom has reliable, detailed data on numbers and species of animals used and the purposes for which they are used—data not matched by any other country. The United Kingdom has the strongest law of all countries; it was enacted in 1876 when animal experimentation in Europe was rapidly gaining in popularity. Almost a century passed before any other countries started to collect any comparable data. For instance, the United States passed its first law governing laboratory animals in 1966 (*see* LABORATORY ANIMAL WELFARE ACT OF 1966) and started collecting data on numbers used in 1973. Spain, Italy, and Portugal started collecting data in 1991.

Two countries with the best statistical records (the United Kingdom and the Netherlands) have reported approximately a 50% decline in the numbers of animals used since the mid-1970s. In the United Kingdom, numbers peaked in the 1970s to over 5 million per year, but declined to under 3 million by 1994. In the United States, the official statistics are not complete enough to make it possible to assess trends.

It is probably true that the most commonly used species in all countries are rodents such as rats and mice.* In the United Kingdom, the data are as follows: rats, mice, and other rodents, bred specially for the purpose, comprise 83% of all animals used; fish,* birds, amphibians,* and reptiles* account for 12%; small mammals other than rodents, mostly rabbits and ferrets, 3%; sheep, cows, pigs,* and other large mammals, 1.3%; dogs* and cats,* 0.4% (in the United Kingdom, unlike the United States, no strays or unwanted pets can be used); and monkeys such as marmosets and macaques, 0.2%. In the United Kingdom, the great apes (chimpanzees,* orangutans, and gorillas) have not been used since 1980. In the United States in 1996, approximately 2,000 chimpanzees were being used for research.

The Netherlands has the most complete data on the degree (duration and severity) of animal pain* or distress* resulting from animal experimentation. Researchers have to report the numbers of animals used in one of three categories: minor, moderate, or severe animal pain or distress.

Number of Laboratory Animals Used in Research by Country (in Thousands)

United States (1995)[a]	13,955,000
Japan (1991)[b]	12,236,000
France (1991)	3,646,000
United Kingdom (1994)	2,842,000
Germany (1993)	2,080,000
Canada (1993)	2,042,000
Netherlands (1994)	771,000
Switzerland (1994)	724,000
Italy (1991)	683,000
Australia, Victoria, and New South Wales (1991)	565,000
Spain (1991)	559,000
Sweden (1994)	352,000
Denmark (1991)	304,000
New Zealand (1993)	292,000

Portugal (1991)	87,000
Greece (1991)	25,000
Ireland (1991)	25,000
Total	**41,188,000**

Note: Numbers of animals are given to the nearest thousand. The year of count follows the country name in parentheses. Because of different criteria for counting (in the United Kingdom, for instance, procedures are counted rather than number of animals), the figures may not be directly comparable from country to country.

[a] The United States Department of Agriculture counts only about 10% of all animals used in experimentation. The most used species—rats, mice, and birds—are not protected under the relevant legislation and are therefore not counted. In 1995, the number of animals officially counted was 1,395,000. For this table, this figure has been multiplied by 10 to achieve approximate comparability with data from other countries.

[b] Number of animals sold (not necessarily used).

Sources: Official national statistics, except for the United States (U.S. Department of Agriculture; see note *a*) and Japan (Japanese Society of Laboratory Animals; see note *b*).

Selected Bibliography. Facts and Figures on Animal Research in Great Britain [Brochure], Research Defence Society, London, October 1995; Orlans, F. B., Data on Animal Experimentation in the U.S.: What They Do and Do Not Show," *Perspectives in Biology and Medicine* 37(2) (1994): 217–231; Statistics 1995, *Animal Experimentation in the Netherlands* (Rijswijk, Netherlands: Veterinary Public Health Inspectorate in the Netherlands, 1995); Stephens, M. L., Chimpanzees in Laboratories: Distribution and Types of Research, from the proceedings of the first conference organized by People against Chimpanzee Experiments entitled "Poor Model Man: Experimenting on Chimpanzees," *ATLA* 23(5) (1995): 579–683; Straughan, D. W., First European Commission Report on Statistics of Animal Use, *ATLA* 22 (1994): 289–292.

F. BARBARA ORLANS

Housing and Handling of Nonhuman Primates

Nonhuman primates in research laboratories have traditionally been kept and handled in ways that suit the convenience of the investigator rather than the needs* of the animal subject. The animals are used as scientific tools with little consideration of the fact that they are sentient beings (*see* SENTIENTISM) experiencing boredom (*see* ANIMAL BOREDOM), frustration, anxiety, fear,* pain,* discomfort, and well-being* in ways similar to those of human primates. Typically, each animal is given an identification number rather than a name, as a conceptualized safeguard for "scientific objectivity."

Primates are social animals who are biologically adapted to live in a complex, ever-challenging environment, but they were commonly housed in an extremely boring environment, the barren single cage. Deprived of social companionship and basic stimuli for the expression of species-characteristic

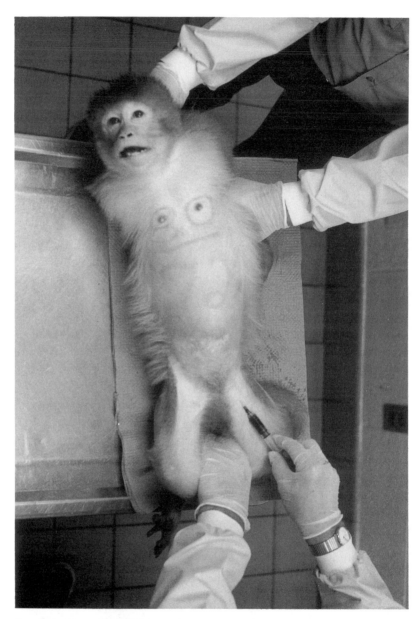

Research data collected from a distressed monkey are "distressed" and hence of little scientific value. *Source:* H. Davis and D. Balfour (Eds.), *The Inevitable Bond: Examining Scientist-Animal Interactions,* 1992. Reprinted with the permission of Cambridge University Press.

behaviors, the singly caged individual often developed symptoms of behavioral and mental disease, such as self-aggression, withdrawal, and passivity. Handling procedures traditionally implied that individuals were forcefully subdued, thereby experiencing extreme anxiety, fear, and discomfort. Typical reactions to the immobilization distress are struggling, fear-induced diarrhea, screaming, alarm vocalization, and increased stress-hormone (cortisol) secretion.

Animal technicians and animal caretakers have long recognized that the conventional housing and handling techniques of nonhuman primates are not adequate because they disregard basic requirements for the subjects' well-being. These techniques, however, not only raise ethical but also methodological concerns that are being gradually acknowledged by a growing number of scientists and veterinarians. Public pressure finally led in the time period 1985–1991 to the comprehensive amendment of federal animal welfare* regulations that prompted the development of more humane housing and handling techniques for nonhuman primates assigned to research.

In the wild, primates live in cohesive troops. In many cases, housing them in groups rather than in single cages may therefore be the ideal way to account for their social needs. The risk of aggression, however, is significant when new groups are formed from singly caged animals. Pair housing offers a safe and practical alternative to group housing. Successful pair formations of previously singly caged individuals have been documented in recent years for numerous species. Even rhesus monkeys, who are commonly believed to be particularly intolerant and hence unsuitable for social housing, can be transferred without special risk from solitary housing to permanent pair-housing arrangements if the two partners are first given the chance to get to know each other during a brief noncontact familiarization period.

Individuals afflicted with behavioral pathologies tend to abandon their peculiar habits once they are transferred to a compatible pair-housing situation. Paired animals spend approximately the same amount of time interacting with each other as do wild animals. This suggests that being transferred from single housing to permanent pair housing improves the animals' well-being by providing them with an appropriate environment for their social needs. The presence of another member of the same species also serves as a buffer against fear-inducing situations (e.g., being restrained in a "chair" during a physiological experiment) that the singly caged subject is lacking.

A companion is undoubtedly the best remedy against boredom. Species-appropriate distraction, however, can also be provided by enriching the complexity of the animals' living space (*see* ENRICHMENT FOR ANIMALS). The installation of perches, shelves, or swings no longer restricts the animals to an unnatural, permanent terrestrial lifestyle but opens up the vertical dimension and allows the animals to exhibit arboreal activities and natural, that is, vertical, flight responses. Unlike toys, elevated structures retain their

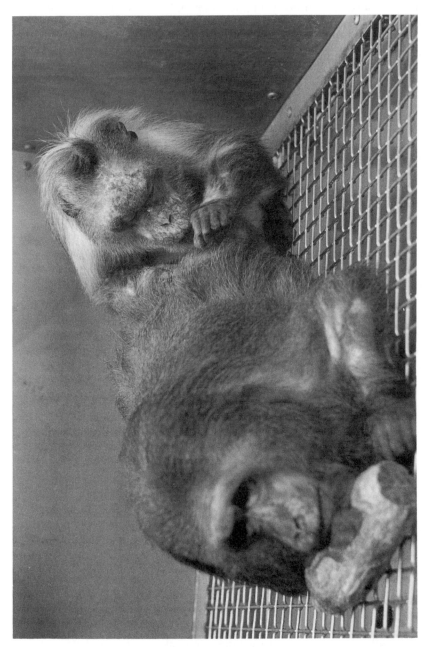

Pair housing allows nonhuman primates to express their social disposition. Here, two stumptailed macaques spend approximately one-fourth of the day interacting with each other in species-typical ways. *Source: Animal Technology* 45 (1994), with permission of the editor.

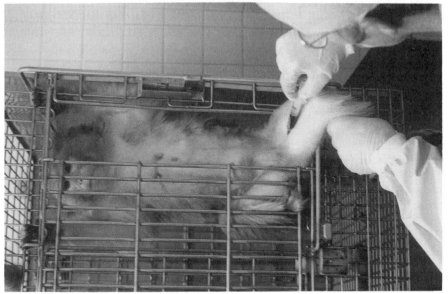

Nonhuman primates are intelligent and can learn quickly to cooperate during common handling procedures such as insulin injection or blood collection. Photos by Viktor Reinhardt.

stimulatory value over time because they trigger behaviors that would be crucial for the animals' survival in the wild.

With patience, gentle firmness, and positive reinforcement, primates can easily be trained to cooperate during capture; blood collection; systemic, oral, and topical drug application; urine collection, and veterinary examination. The training challenges the animals' intelligence, offers them—and the research personnel—some distraction, eliminates distress responses, and avoids possibly dangerous defensive reactions triggered by fear. Scientific data collected from such an animal are distinguished by a high degree of reliability because they are not biased by distress responses. (*See also* CHIMPANZEES; GREAT APE PROJECT.)

Selected Bibliography. Nowak, M. A., and A. J. Petto (Eds.), *Through the Looking Glass: Issues of Psychological Well-being in Captive Nonhuman Primates* (Washington, DC: American Psychological Association, 1991); Reinhardt, V., Social Enrichment for Laboratory Primates: A Critical Review, *Laboratory Primate Newsletter* 29 (1990): 7–11; Reinhardt, V., C. Liss, and C. Stevens, Restraint Methods of Laboratory Nonhuman Primates: A Critical Review, *Animal Welfare* 4 (1995): 221–238; Segal, E. F. (Ed.), *Housing, Care, and Psychological Wellbeing of Captive and Laboratory Primates* (Park Ridge, NJ: Noyes Publications, 1989); U.S. Department of Agriculture, Animal Welfare, Standards, Final Rule, *Federal Register* 56 (1991): 6426–6505.

VIKTOR REINHARDT

Sacrifice

Different language is used for killing different categories of animals. Companion animals* are "euthanized" (*see* EUTHANASIA), farm animals are "slaughtered" (*see* TRANSPORTATION AND SLAUGHTER), and research animals are "sacrificed." Unlike the first two terms, however, use of the term "sacrifice" has been particularly controversial.

Spokespersons from the scientific community have called upon its members not to use the term "sacrifice" because it is unnecessary, too regularly used, and meaningless and because it has religious and unscientific connotations. In recent years there has been a serious effort to delete the term from biological journals and grant proposals as part of a trend in this century to remove subjectivity and personalization from science. Some individuals critical of animal experimentation have also challenged its use because it makes it easier for researchers to kill animals and glorifies a practice that, in their opinion, should be seriously questioned if not stopped.

Despite official efforts to ban the term, "sacrifice" can still be overheard in the laboratory conversations of scientists and technicians as well as in the presentations of scientific papers at professional meetings. Direct observation of scientists and technicians has led sociologists to conclude that sacrifice is not used in the religious sense, but rather in a broader sacred sense within the scientific community. According to sociologists, sacrifice means more than simply killing laboratory specimens; it is part of a sequence of procedures that transform the everyday meaning of animals into "tools" having a clear and valuable place in laboratories. Although sociologists agree that this transformation enables researchers to use animals in experiments, they disagree about the processes that create this transformation.

On the one hand, Michael Lynch argues that the transformation entails a single social process where the "naturalistic" animal found in nature is redefined as an "analytic" object signifying data and having only research value. The animal's death* has meaning only to the extent that it assists research. On the other hand, Arnold Arluke maintains that the transformation involves two opposing social processes. Like Lynch, Arluke argues that laboratory animal sacrifice involves the stripping away of the everyday or nonscientific identity of animals so that they can be regarded as instruments or data. Arluke also contends that sacrifice involves a process of identification with lab animals. Some researchers, especially those who have routine contact with nonhuman primates or domestic animals (see DOMESTICATION), attribute human qualities to them (see ANTHROPOMORPHISM). For these researchers, the animal's death has personal meaning. The concept of sacrifice embraces both of these tendencies by acknowledging the simultaneous distancing from and identification with laboratory animals that occur in research settings.

Rather than getting rid of the term "sacrifice," the metaphor can be institutionalized by creating and openly acknowledging group rituals commemorating the death of laboratory animals (see LABORATORY ANIMAL USE, Memorial Services for Animal Research Subjects). Rituals link individuals and culture by pulling together, in a personally meaningful way, the paradoxes of existence into something sensible and the fragmentation of reality into something whole.

Selected Bibliography. Arluke, Arnold, Sacrificial Symbolism in Animal Experimentation: Object or Pet? *Anthrozoös* 2 (1988): 98–117; Douglas, Mary, *Natural Symbols* (New York: Pantheon Books, 1970); Hubert, H., and M. Mauss, *Sacrifice: Its Nature and Function* (Chicago: University of Chicago Press, 1964); Lynch, Michael, Sacrifice and the Transformation of the Animal Body into a Scientific Object: Laboratory Culture and Ritual Practice in the Neurosciences, *Social Studies of Science* 18 (1988): 265–289.

ARNOLD ARLUKE

Memorial Services for Animal Research Subjects

For several years, the University of Guelph has been the site of an event that appears to be unique in North America: a memorial service to acknowledge animals used in research and teaching. This event brings together students, staff, and faculty for a simple yet dignified ceremony recognizing the role of animals in our community. Although the original idea for the service

Memorial Services for Animal Research: Members of the University of Guelph community gather annually to acknowledge animals used in research and teaching. Photo by Trina Koster, courtesy of the Office of Research, University of Guelph, Ontario, Canada.

arose from elaborate Buddhist-based (*see* RELIGION AND ANIMALS, Buddhism) rituals held at Japanese primate research facilities, a more secular approach was chosen. The idea was not to offer thanks nor to plead forgiveness, but rather simply to acknowledge the role of animals in research and teaching, and that without them, this work would be greatly altered.

Temple Grandin advocates simple rituals of acknowledgment to prevent the development of mechanistic attitudes (*see* DESCARTES, RENÉ) toward animals. It is relatively common for slaughter-plant workers, for example, to perform their jobs in an automatic, almost unconscious fashion, without emotion and without apparent concern for the animals they are killing or butchering. Similarly, Bernard Rollin describes the process of intellectual compartmentalization in the context of animal-based research. Compartmentalization occurs when one's theoretical (intellectual, professional) and commonsense (personal, emotional) responses are quite different, even mutually exclusive. For example, a researcher may conduct painful research on animals in the course of his or her work without a thought to the subjective experiences of the subjects, yet show sincere and intense concern for the pain* experienced by the family dog following surgery. Like mechanization, compartmentalization distances the individual from the task he or she must perform and from his or her intuitive or emotional responses to it. In light of such tendencies, rituals can promote and maintain greater respect for animals.

Following the first memorial service in the spring of 1993, a stone marker was erected at the center of the University of Guelph campus. The plaque reads: "In recognition of the animals used by the University of Guelph community in support of excellence in teaching and research." This marker serves as a very tangible focal point for our community's recognition of animals' roles in our work. Those who have initiated activities such as the memorial gatherings believe that they can only lead to greater awareness of the issues surrounding the use of animals in research and teaching, and ultimately to better animal care.

Selected Bibliography. Asquith, P., The Monkey Memorial Service of Japanese Primatologists, *Royal Anthropological Institute News* 54 (1983): 3–4; Grandin, T., Behavior of Slaughter Plant and Auction Employees toward the Animals, *Anthrozoös* 1(4) (1988): 205–213; Rollin, B., *The Unheeded Cry: Animal Consciousness, Animal Pain, and Science* (New York: Oxford University Press, 1989); Strauss, S., Mind and Matter: Dead Animals Merit a Prayer and a Wake from Researchers, *Globe and Mail* (Toronto, Ontario), March 30, 1991; Taylor, A., and H. Davis, Acknowledging Animals: A Memorial Service for Teaching and Research Animals, *Anthrozoös* 6(4) (1993): 221–225.

ALLISON A. TAYLOR AND HANK DAVIS

LABORATORY ANIMAL WELFARE ACT OF 1966

In the mid-1960s the scientific community and animal welfarists squared off over proposed federal legislation regulating the sale and transportation of animals bound for research laboratories. Public furor over the subject spurred the 89th Congress to pass the Laboratory Animal Welfare Act of 1966 (LAWA). Although originally put forth as a simple pet-protection act, LAWA became the foundation of U.S. animal welfare* law.

In 1965, only two federal laws existed addressing the protection of animals in the United States (Humane Slaughter Act,* 1958, and Wild Horses Act, 1959). About this same time, Americans became increasingly concerned over an increase in pet thefts. Animal welfare organizations like the Animal Welfare Institute (AWI) and the Humane Society of the United States (HSUS)* attributed this increase to the growth in biomedical research and the corresponding demand for test subjects. One particular Pennsylvania "dognapping" incident and a graphic pictorial in *Life* magazine showing the condition of dogs* kept at the home of a New York State dog dealer appear to have been the reason behind the LAWA's origination. The volume of mail received by Congress and the White House urging passage of protective legislation for animals surpassed the total correspondence addressing the issues of civil rights and the Vietnam War.

During the congressional sessions, lawmakers considered over thirty different bills. Representative Joseph Resnick (Democrat of New York) offered the first of these after an unsuccessful attempt to retrieve a pet (*see* COMPANION ANIMALS AND PETS) taken from a Pennsylvania family and sold to a New York hospital. In order to stop the trafficking in stolen pets, most of the bills provided for the licensing of animal dealers and the research institutions that dealt with them. Others established humane treatment standards and required that animal facilities be open for inspection by the U.S. Department of Agriculture (USDA). The biomedical research community strenuously opposed the latter bills, believing that such conditions would lead to interference with experiment protocols. A self-policing system was proposed and debated. The debate moved beyond the elimination of pet theft to the appropriateness of certain research. In hearings before the House Agriculture Subcommittee on Feeds and Livestock on March 7–8, 1966, over 150 people and organizations offered testimony. Legislation that incorporated elements of licensing, humane standards, and USDA inspections carried the day. HR 13881, presented by subcommittee chairman W. R. Poage (Democrat of Texas), passed the House in April 1966 by a vote of 352–10. The Senate accepted it 85–0 that June. Reading from remarks prepared by Supreme Court Justice Abe Fortas, a longtime supporter of animal welfare, President Lyndon Johnson signed the bill into law in a modest White House ceremony on August 24, 1966.

Although the original LAWA did not reach beyond the laboratory door, subsequent amendments in 1970, 1976, 1985, and 1990 extended the range of care provided to animals. Among them were provisions on the administration of pain-relieving drugs, minimum-size requirements for holding cages, and the establishment of institutional review boards (*see* INSTITUTIONAL ANIMAL CARE AND USE COMMITTEES [IACUCs]) to minimize or prevent duplication of experiments and examine their protocols. The Laboratory Animal Welfare Act of 1966 is significant on several historical levels. The language and tactics displayed by both supporters and opponents of the legislation represent an example of the activism* displayed

President Lyndon Johnson congratulates Supreme Court Justice Abe Fortas after signing the Laboratory Animal Welfare Act of 1966. *Source*: 2915-34a, 8/24/66, photographer unknown, Lyndon Baines Johnson Presidential Library Archives, Austin, Texas, with permission.

by the varied social and political movements of this troubled era in American history. It is notable further as an example in which both animal welfarists and scientists displayed concern for the welfare of animals. Moreover, it is an expression of how the role of nonhuman animals in society is perceived by the human public.

Selected Bibliography. Finsen, Lawrence, and Susan Finsen, Historical Roots, in *The Animal Rights Movement in America: From Compassion to Respect* (New York: Twayne, 1994); Francione, Gary, The Federal Animal Welfare Act, in *Animals, Property, and the Law* (Philadelphia: Temple University Press, 1995); Lederer, Susan, The Controversy over Animal Experimentation in America, 1880–1914, in Nicolaas Rupke (Ed.), *Vivisection in Historical Perspective* (London: Croom Helm, 1987); Rowan, Andrew, *Of Mice, Models, and Men* (Albany: State University of New York Press, 1984); Stevens, Christine, Laboratory Animal Welfare, in Animal Welfare Institute, *Animals and Their Legal Rights: A Survey of American Laws from 1641 to 1990*, 4th ed. (Washington, DC: Animal Welfare Institute, 1990).

LARRY D. TERRY

Law (Federal) Governing Animal Research

During the 1960s, vivid press coverage both of kidnapping of family pets that were then sold for research and also of the conditions under which dog dealers who sold animals to research facilities kept these animals aroused public fear of having their pets kidnapped and sold for research. Congress reacted to these concerns by passing the Laboratory Animal Welfare Act of 1966, which mainly licensed and regulated animal suppliers and did little to assure the well-being of animals used in research. By the 1970s, however, more substantive concerns about animal research had surfaced in society.

Growing public suspicions and misgivings about animal research were solidified in the early 1980s when a number of serious examples of animal abuse in research facilities were revealed, including instances at the University of Pennsylvania Head Injury Laboratory and the laboratory of Edward Taub (*see* SILVER SPRING MONKEYS), both of which situations involved abuse, improper care, and neglect of nonhuman primates. By the mid-1980s, public confidence in the research community's ability to regulate itself in the area of animal care and use was sufficiently eroded to demand federal legislation.

In 1976, a group of Colorado citizens consisting of two laboratory animal veterinarians,* a humane advocate and attorney, and a philosopher began proposing legislation that would enforce self-regulation by local animal care and use committees. These committees would review research projects before they began in order to make sure that everything possible was being done to assure that animal pain,* distress,* and suffering* were minimized. The committees would also assure that facilities were adequate, and that systems of care assured proper animal husbandry.

In 1985, the key concepts proposed by the Colorado group were passed by Congress as components of two pieces of legislation, despite vigorous opposition from certain portions of the research community. The first piece of legislation was passed as an amendment to the Laboratory Animal Welfare Act and was entitled the Improved Standards for Laboratory Animals Act. The second piece of legislation, complementing the first, was the Health Research Extension Act. The major provisos of the Laboratory Animal Welfare Act amendment were as follows:

1. Establishment of an institutional animal care and use committee* (IACUC) whose members must include a veterinarian and a person not affiliated with the research facility.

2. A directive to the U.S. Department of Agriculture (USDA), which enforces the law, to establish standards for exercise for dogs.

3. Establishment of standards for a physical environment for primates that enhances their psychological well-being.*

4. Establishment of standards of adequate veterinary care, including use of anesthetics, analgesics (painkillers), and tranquilizers.

5. Prohibition of the use of paralytics (drugs that cause paralysis) without anesthetics for surgical procedures.

6. The investigator must provide proof of having considered alternatives to painful procedures.

7. Multiple surgery is prohibited except for "scientific necessity."

8. The IACUC must inspect facilities at least semiannually, review protocols, and file an inspection report detailing violations and deficiencies.

9. The USDA was mandated to establish an animal welfare* information service at the National Agricultural Library to provide information aimed at eliminating duplicative animal research, reducing or replacing animal use, minimizing animal pain and suffering, and training animal users.

10. Each research institution must train animal users in the items enumerated in (9) and in any other ways of minimizing animal suffering.

11. The USDA should effect a working relationship with the National Institutes of Health (NIH).

The Health Research Extension Act turned NIH guidelines for proper care and use of animals into law. (NIH had long promoted reasonable guidelines for animal care but had had no mechanism for enforcing them.) Violations could result in seizure of all federal money to an institution. Between the two laws, virtually all vertebrate animals used in research in the United States, with the exception of farm animals used in agricultural research and rats and mice* used in private-industry research, are now legally covered. Many IACUCs apply the same standards to agricultural researchers vis-à-vis pain and suffering as they do to animals used in biomedical research.

Researchers are becoming increasingly sophisticated about animal pain, suffering, and distress and how to control them in the face of federal law that assumes the existence of animal pain, thought (*see* ANIMAL COGNITION), and feeling.* Many researchers now admit that attention to pain and distress results in better data. Researchers are also gradually becoming aware of the ethical issues in animal research. Consequently, researchers are increasingly looking into housing systems that better take into account animals' psychological and biological needs.*

Selected Bibliography. Newcomer, Christian, Laws, Regulations, and Policies Pertaining to the Welfare of Laboratory Animals, in B. E. Rollin and M. L. Kesel (Eds.), *The Experimental Animal in Biomedical Research*, vol. 1 (Boca Raton, FL: CRC Press, 1990); Rollin, Bernard E., *Animal Rights and Human Morality*, rev. ed. (Buffalo, NY: Prometheus Books, 1992); Rollin, Bernard E., Laws Relevant to Animal Research in the United States, in A. A. Tuffery (Ed.), *Laboratory Animals: An Introduction for Experimenters* (London: John Wiley, 1995); Rollin, Bernard E., *The Unheeded Cry: Animal Consciousness, Animal Pain, and Science* (Oxford: Oxford University Press, 1989); Russow, Lilly-Marlene, NIH Guidelines and Animal Welfare, in James M. Humber and Robert F. Almeder (Eds.), *Biomedical Ethic Review: 1990* (Clifton, NJ: Humana Press, 1991): 229–252.

BERNARD E. ROLLIN

LAW AND ANIMALS

Most Western legal systems, including that of the United States, include two primary normative entities: *persons* and *property*. Persons are both natural entities (human beings) and nonnatural entities (such as corporations) that are regarded as having rights and duties within the system. It is generally recognized that property is that which exists only as a means to the ends of persons and that property cannot have rights or duties. Animals are regarded as property. Indeed, the domestication* and ownership of animals is closely related to the idea of property or money. For example, the word "cattle" comes from the same etymological root as "capital," and in many European languages, "cattle" was originally synonymous with "chattel" and "capital."

There are two primary types of defense offered to support the status of animals as property. The first is the religious justification supposedly in Genesis (1:20–28) in which man is given "dominion over the fish of the sea, and over the birds of the air, and over the cattle, and over all the earth, and over every creeping thing that creeps upon the earth." The second type of justification for the status of animals as property is based on the notion that nonhumans possess some inherent "defect" that makes them inferior to humans. Although the two justifications are related, there are differences. The religious justification does not necessarily depend on any particular measurable differences between humans and animals and can rest solely on the notion of divine ordering. The view that animals are inherently "inferior" to humans rests on supposedly scientific observations, such as the inability of animals to use language* or to think rationally.

Although animals are regarded as property, the law, reflecting moral thought, has long recognized that animals who feel pain* are different from other sorts of property. This recognition has led to the development of restrictions on the use that humans may make of their animal property. These restrictions require that humans treat animals "humanely" and that they not make animals suffer "unnecessarily."

Legal welfarism is the version of animal welfare* theory embodied in current law. Legal welfarism requires that we balance human and animal interests to determine whether particular conduct is "humane," or whether particular suffering* is "necessary." The problem is that human beings are rightholders as a general matter and are holders of the right to own property in particular. The standard of "humane" treatment or "unnecessary" suffering is not determined by reference to some ideal moral notion; that is, we do not look to the allegedly cruel act and then "balance" in order to determine its legality or the legality of the activity of which the cruelty is a part. If the act is *causally* necessary to a legally sanctioned activity (it is customarily regarded as part of the activity), then the act is regarded as *morally* and *legally* "necessary."

Courts have long held that animal-protection laws do not prohibit "cruelty" as that term is used in ordinary language; rather, cruelty "must refer to something done for no legitimate purpose" (*Lewis v. Femor*, 18 Q. B., U.S. 532, 534 [1887]). This explains why the law does not prohibit the farmer from castrating or dehorning animals without anesthesia; these actions, although extremely painful, facilitate the socially approved use of the animal as food. If, however, a farmer allows animals to starve to death for no good reason, then the law will punish that conduct because the farmer has inflicted pain and death* outside a socially recognized practice, and the conduct results in the completely uncalled-for wasting of animal resources.

Legal welfarism accounts for why, despite a widespread moral norm against "unnecessary" suffering, animals are exploited for virtually every conceivable purpose, including entertainment. As long as we are willing to tolerate the use of animals for entertainment (or for food or for science or for clothing) as a general matter, and as long as animals are regarded as property, the law has no ready way of interpreting a regulation on the use of animal property as anything more than not allowing conduct that goes beyond what is required to allow the activity. That is, virtually the only conduct that is proscribed is the infliction of *gratuitous* or uncalled-for suffering and death because that would result in a "waste" of animal property and an overall lessening of social wealth. As a result, courts have generally deferred to the customary activity of animal exploiters as establishing standards for the efficient use of animal property. (*See also* LABORATORY ANIMAL WELFARE ACT OF 1966, Law [Federal] Governing Animal Research.)

Selected Bibliography. Cohn, Priscilla N., Animals as Property and the Law, in Roberta Kevelson (Ed.), *Law and Semiotics*, vol. 1 (New York: Plenum Press, 1987); Francione, Gary L., Animal Rights and Animal Welfare, *Rutgers Law Review* 48 (1996): 397–469; Francione, Gary L., Animals, Property, and Legal Welfarism: "Unnecessary" Suffering and the "Humane" Treatment of Animals, *Rutgers Law Review* 46 (1994): 721–770; Francione, Gary L., *Animals, Property, and the Law* (Philadelphia: Temple University Press, 1995); Francione, Gary L., *Rain without Thunder: The Ideology of the Animal Rights Movement* (Philadelphia: Temple University Press, 1996).

GARY L. FRANCIONE

LEARNED HELPLESSNESS IN ANIMALS

Learned helplessness arises from experiencing aversive stimuli under conditions an organism cannot control. The phenomenon is called "learned helplessness" because animals learn that there is nothing they can do to prevent or to terminate the aversive stimuli. Learned helplessness first addressed theoretical questions about the nature of learning and later became a model for studies of stress.* The study of learned helplessness is defended by some researchers because it concerns new mechanisms of learning and establishes a set of conditions like those experienced by many humans that result in depression and posttraumatic stress syndrome (PTSS), psychological disorders we must understand if we are to cure these people.

Three impairments constitute the learned helplessness syndrome: (1) a reduction in behaviors to cope with any aversive challenge, (2) an impairment in attention and ability to learn, and (3) overt emotional passivity combined with chronic stress reactions. These effects are not merely the result of experiencing the aversive events themselves but rather of the uncontrollability and unpredictability of the events.

Learned helplessness is now known to be general across a wide range of species and a wide range of conditions, including the natural experiences of being attacked and injured by members of the same species. Martin Seligman generalized the theory that lack of control over one's experiences impairs later normal functioning and proposed that learned helplessness could provide a model of reactive depression in humans.

The controversy about the ethics of performing studies of learned helplessness continues. It is maintained by differences in basic beliefs about whether or not the gains in understanding and relief provided to unhealthy animals and humans by virtue of this research offset the degree of exploitation of the research subjects (*see* UTILITARIANISM).

Selected Bibliography. Maier, S. F., and M. E. P. Seligman, Learned Helplessness: Theory and Evidence, *Journal of Experimental Psychology: General* 105 (1976): 3–46; Overmier, J. B., and V. M. LoLordo, Learned Helplessness, in W. O'Donohue (Ed.), *Learning and Behavior Therapy* (Needham, MA: Allyn & Bacon, 1997), 352–372; Overmier, J. B., and J. Patterson, Animal Models of Psychopathology, in P.

Soubrie, P. Simon, and D. Widlocher (Eds.), *Animal Models of Psychiatric Disorders*, vol. 1, *Selected Models of Anxiety, Depression, and Psychosis* (Basel, Switzerland: Karger, 1988), 1–35; Peterson, C., S. F. Maier, and M. E. P. Seligman, *Learned Helplessness: A Theory for the Age of Personal Control* (New York: Oxford University Press, 1993); Seligman, M. E. P., *Helplessness: On Depression, Development, and Death* (San Francisco: Freeman, 1975).

J. BRUCE OVERMIER

LEFFINGWELL, ALBERT T.

Albert T. Leffingwell (1845–1916) was an author, medical doctor, and the most significant medical critic of vivisection in the United States between 1880 and 1915, producing numerous articles, books, and pamphlets on the subject. Independent wealth permitted Leffingwell to travel and write extensively on a variety of subjects. Leffingwell was not an antivivisectionist (*see* ANTIVIVISECTIONISM) but a regulationist, with special concern for the link between animal experimentation and unethical experiments with human subjects. He founded the American Society for the Regulation of Vivisection. Leffingwell served as president of the American Humane Association and was the author of *The Vivisection Question* (1901), *The Vivisection Controversy* (1908), *American Meat* (1910), and *An Ethical Problem* (1915). Leffingwell's essay "Vivisection in America" was included as an appendix to the American edition of *Animals' Rights* (1894).

Selected Bibliography. Albert Tracy Leffingwell, in Irving A. Watson (Ed.), *Physicians and Surgeons of America* (Concord, NH: Republican Press, 1896); Albert Tracy Leffingwell, *New York Times*, September 2, 1916; Leffingwell, Albert Tracy, *American Meat* (New York: Theo Schulte, 1910); Leffingwell, Albert Tracy, *An Ethical Problem; or, Sidelights upon Scientific Experimentation on Man and Animals* (New York: C. P. Farrell; London: G. Bell and Sons, 1915); Leffingwell, Albert Tracy, *The Vivisection Controversy: Essays and Criticisms* (London: London and Provincial Anti-Vivisection Society, 1908); Leffingwell, Albert Tracy, *The Vivisection Question* (1st ed., New Haven: Tuttle, Morehouse and Taylor, 1901; 2nd ed., Chicago: Vivisection Reform Society, 1907).

BERNARD UNTI

LEGAL WELFARISM. See LAW AND ANIMALS.

LEWIS, C. S.

An English theologian and writer, fellow of Magdalen College, Oxford, and subsequently professor of medieval and Renaissance literature at Cambridge, C. S. Lewis (1898–1963) held that the infliction of pain on animals was an evil and that carnivorousness was a result of the Satanic corruption of nature (see *The Problem of Pain*). Lewis was a convinced antivivisectionist (*see* ANTIVIVISECTIONISM) and wrote a major tract on the subject in 1947 for the New England Anti-Vivisection Society. His writings have laid the foundation for a more compassionate theological view of animals. Of special interest are his fictional works, in which he envisages a paradisal world where humans are freed from predation* and live in peace with animals (see, especially, *Perelandra*). Lewis is prophetic in warning of the dangers of modern technological power over nature: "What we call Man's power over Nature turns out to be a power exercised by some men over other men with Nature as its instrument" (*The Abolition of Man*, 35).

Selected Bibliography. Joad, C. E. M., and C. S. Lewis, The Pains of Animals, *The Month* 3(2) (February 1950); 95–102, reprinted in *God in the Dock* (Grand Rapids, MI: Eerdmans, 1970); Lewis, C. S., *The Abolition of Man* (Oxford: Oxford University Press, 1943); Lewis, C. S., *Perelandra* (New York: Macmillan, 1944); Lewis, C. S., *The Problem of Pain* (London: Geoffrey Bles, 1940).

ANDREW LINZEY

LIBERATION ETHICS. *See* ANIMAL LIBERATION ETHICS.

LIND-AF-HAGEBY, EMILIA AUGUSTA LOUISE

An indefatigable animal advocate and campaigner whose activism dominated the British scene during the first half of the 20th century, Emilia Augusta Louise Lind-af-Hageby (1878–1963) stood at the center of one of the most contentious episodes in the history of antivivisectionism,* the Brown Dog Incident. In 1901, Lind-af-Hageby and her friend Leisa Schartau enrolled at the London School of Medicine for Women to seek medical degrees in order to fight vivisection. The two recorded their experiences in diaries and later exposed the fact that a brown terrier dog had, in contravention of the Cruelty to Animals Act, been vivisected, revived, and used in another procedure. Lind-af-Hageby and her codefendent Stephen Coleridge lost the court case stemming from the publication of her work *The Shambles of Science*, but their efforts galvanized a coalition of antivivisectionists, trade unionists, and suffragettes who confronted medical students in the streets of Battersea, where a statue commemorating the dog's death was raised in 1906. Later, in collaboration with Nina, duchess of Hamilton and Brandon, Lind-af-Hageby operated the Animal Defence and Anti-Vivisection Society, founded in 1906, and pursued an active career in antivivisection, slaughterhouse reform, and related causes. Lind-af-Hageby became a naturalized British subject in 1912. During World War I, her group maintained three veterinary hospitals for sick and wounded horses. In 1954, she purchased Hamilton's Ferne Estate in Dorset, setting it up as an animal sanctuary in memory of her longtime collaborator.

Selected Bibliography. Lansbury, Coral, *The Old Brown Dog: Women, Workers, and Vivisection in Edwardian England* (Madison: University of Wisconsin Press, 1985); Lind-af-Hageby, Louise, and Leisa K. Schartau, *The Shambles of Science: Extracts from the Diary of Two Students of Physiology* (London: Ernest Bell, 1903); *Proceedings of the International Anti-Vivisection and Animal Protection Congress, Philadelphia, 1926* (Philadelphia: American Anti-Vivisection Society, 1926); Vyvyan, John, *The Dark Face of Science* (London: Michael Joseph, 1971).

BERNARD UNTI

M

MANNING, HENRY EDWARD

Henry Edward Manning (1808–1892) was appointed Roman Catholic archbishop of Westminster in 1865. Together with Lord Shaftesbury,* Frances Power Cobbe,* and George Hoggan he founded the world's first antivivisection (*see* ANTIVIVISECTIONISM) society, the Victoria Street Society for the Protection of Animals from Vivisection, served as one of its vice presidents, and spoke at its first general meeting in 1876. Manning argued that we owe a "sevenfold obligation" of mercy to the Creator and therefore to animals as God's creatures. Science should be free to pursue its work, but only within moral limits. The deliberate infliction of suffering* on animals exceeded one of these limits and was judged incompatible with the primary obligation of mercy. Manning was also very active in the Society for the Prevention of Cruelty to Children.

Selected Bibliography. Coleridge, Stephen, *Great Testimony* (London and New York: John Lane, 1918); Kramer, Molly Beer, and Andrew Linzey, Vivisection, in Paul Barry Clarke and Andrew Linzey (Eds.), *Dictionary of Ethics, Theology, and Society* (London and New York: Routledge, 1996), 870–874; Manning, Henry Edward, Speech to the Victoria Street Society for the Protection of Animals from Vivisection, March 9, 1887, in *Speeches against Vivisection* (London: National Anti-Vivisection Society and the Catholic Study Circle for the Welfare of Animals, 1977), extract in Andrew Linzey and Tom Regan (Eds.), *Animals and Christianity: A Book of Readings* (London: SPCK; New York: Crossroad, 1988), 165–166; Roamer, Stanley, *Cardinal Manning as Presented in His Own Letters and Notes* (London: Eliot Stock, 1896); Stevenson, Lloyd G., Religious Elements in the Background of the British Anti-vivisection Movement, *Yale Journal of Biology and Medicine*, November 1956, 125–157.

ANDREW LINZEY

MARGINAL CASES

The argument from marginal cases (AMC) is used to dispute the claims of those who argue that only humans have special moral rights. The AMC is supported by pointing out the inconsistencies between the ways we treat animals and the ways we treat humans. The AMC attacks the commonly held view that special forms of moral respect (the right to life, for example) are appropriate for all members of the human species and only members of the human species.

If asked to provide reasons for the view that all and only humans have special moral rights, people will usually mention some trait or ability that humans have (intelligence, language, or the like). Such traits or abilities are thus regarded as "morally important properties." The AMC claims that whatever morally important property only humans have will not be had by all humans, and whatever morally important property all humans have will not be had by only humans.

To see how the AMC works in action, imagine someone arguing that only humans have a right to life because of their comparatively high intelligence (*see* ANIMAL COGNITION, Intelligence). The AMC replies that if high intelligence is the reason why it is in most circumstances seriously wrong to kill human beings, then it cannot be seriously wrong to kill all human beings because there are many humans (the "marginal cases") who, because of a mental handicap, do not possess intelligence greater than that of many non-humans.

The argument from marginal cases has been important in the work of prominent animal rights* philosophers, particularly in writings of Tom Regan. It has been attacked a great deal too, and the critics have been both foes and friends of "animal rights." The philosopher R. G. Frey points out one important weakness of the argument from marginal cases. The AMC by itself cannot show that animals have greater moral worth than humans have typically believed. Rather, it shows only that there is an inconsistency in the way we treat animals and some humans and leaves it up to us whether the inconsistency will be fixed by treating the humans worse or the animals better.

Other critics such as Arthur Caplan have also explored the idea that "marginal" members of our own species have special ties of affection to "normal" human beings, and that the special moral status of the humans labeled marginal comes from their connection to normal humans. A seriously mentally handicapped infant may be deeply loved by a "normal" adult, and any harm to the infant thus becomes a harm to the adult. This criticism of the AMC, of course, only works regarding mentally handicapped humans who enjoy such important relationships and so does not show that all humans have a kind of moral importance that all animals lack. Caplan's criticism also suggests that it would be wrong to kill animals who were deeply loved by normal humans.

A reason for rejecting the AMC that often comes up in discussion of the argument is the claim that all human beings, no matter how handicapped, have souls, and that no animal, no matter how smart, does. This argument, to those who support the AMC, seems to make the ethical defense of using animals for food, clothing, or experimentation a religious doctrine, one that cannot be supported outside of particular faiths, and that therefore seems an inappropriate basis for making laws and policies in a nonreligious state. To appreciate the force of this point, recall that some religious traditions have thought that some humans, for example, people of color or women, are not the spiritual equals of other humans.

The AMC is also open to criticism by those who are in favor of extending and deepening the moral seriousness with which we regard animals. Steven F. Sapontzis has argued that the AMC distorts the significance of the non-humans by suggesting that we see them as "impaired" versions of human beings. He has maintained that the moral standing* of animals is not based on their resemblances to handicapped humans, but rather on the fact that many of them behave in morally admirable ways we should respect, such as being loyal, or caring, or courageous.

Despite all of these criticisms, the argument from marginal cases has a simplicity, directness, and power that makes it hard to ignore. A very thorough and thoughtful book-length discussion is Evelyn Pluhar's work *Beyond Prejudice: The Moral Significance of Human and Nonhuman Animals*.

Selected Bibliography. Caplan, Arthur, Is Xenografting Morally Wrong? *Transplantation Proceedings* 24 (1992): 722–727; Frey, R. G., *Interests and Rights: The Case against Animals* (Oxford: Clarendon Press, 1980); Frey, R. G., Vivisection, Morals, and Medicine, *Journal of Medical Ethics* 9 (1983) 95–104; Nelson, James Lindemann, Animals, Handicapped Children, and the Tragedy of Marginal Cases, *Journal of Medical Ethics* 14 (1988): 191–194; Pluhar, Evelyn B., *Beyond Prejudice: The Moral Significance of Human and Nonhuman Animals* (Durham, NC: Duke University Press, 1995); Regan, Tom, An Analysis and Defense of One Argument Concerning Animal Rights, *Inquiry* 22 (1979): 189–220; Sapontzis, Steven F., Are Animals Moral Beings? *American Philosophical Quarterly* 17 (1980): 45–52; Sapontzis, Steven F., Speciesism, *Between the Species* 4 (1988): 97–99.

JAMES LINDEMANN NELSON

Categorical and Biconditional Versions

The argument from marginal cases (AMC) has been one of the most powerful weapons in the contemporary debate about nonhuman animal rights.* There are two basic versions of the AMC. The *categorical* version claims that marginal humans have moral rights and concludes that nonhumans who are relevantly similar to these humans also have moral rights. The *biconditional* version maintains that the moral status of relevantly similar "marginal" humans and nonhumans is equivalent: the nonhumans have moral rights *if and only if*—hence the name of this version of the AMC—the humans have such rights.

Several objections have been made to both versions of the AMC. Some people are concerned that the argument is unfair to "marginal" humans. Many mentally disadvantaged humans are capable of going to school, learning trades, and speaking. These abilities are not possessed by any nonhuman animals, so far as we know. Defenders of the AMC can fully agree that many mentally disadvantaged humans are more capable than nonhuman animals. Nevertheless, quite a few severely damaged, sentient (*see* SENTIENTISM) humans are far less capable than many nonhuman animals. Empirical evidence supports the contention that some humans and some nonhumans are roughly comparable in terms of their intellects, emotional capacities, and other capabilities. While some humans outstrip some nonhumans on this score, the reverse also appears to hold.

Another rather more serious charge of unfairness has been made against the AMC. Humans who become mentally incapacitated are unfortunate because they have been deprived of their personhood. Humans who are born with severe mental limitations are also unfortunate, one might argue, because they do not possess the potential for becoming normal members of their species. In contrast, the nonhumans used in laboratories and farms are likely to be normal members of their species. Thus there is a morally relevant difference between "marginal" humans and mentally and emotionally comparable nonhumans. Fairness dictates that we not add yet another huge burden to the unfortunate humans' life. The normal nonhuman, then, rather than the "marginal" human, should be sacrificed to benefit persons.

AMC supporters could respond as follows. The objection assumes that "marginal" humans are already morally significant. Only a morally significant being can be treated fairly or unfairly. But what makes them morally significant, in the context of the objection? It cannot be the "misfortune" itself, since this would make the objection circular. If it is the fact that they are capable of preferring pleasure to pain,* this also holds for many nonhumans. Thus the latter would be morally significant also. In the case of two obviously morally significant beings, for example, two human persons who are alike apart from the fact that one of them is missing a leg and the other has two, we would not consider it justified to steal from the human with two legs rather than the human with one leg because the latter is already more burdened than the former. A choice that would be fair to both individuals is the refusal to sacrifice either.

Another approach to criticizing the AMC is to deny moral status to both "marginal" humans and sentient nonhumans, but deny that unacceptable consequences would follow in practice. A. V. Townsend, for example, has argued that many humans, incapable of personhood in the strict sense, do not have rights, as is the case for similarly limited sentient nonhumans. Thus he rejects the categorical version of the argument while accepting its biconditional form. But he does claim that persons must treat these humans as if they have rights. Otherwise, when distinctions among humans are blurred, genuine rights holders are threatened; this allegedly does not hold for the case of nonhumans. Peter Carruthers has made essentially the same argument.

Animal rights supporters can counter that this is a textbook example of the "slippery slope" fallacy: without further evidence, it is assumed that treating marginal humans as we now treat nonhuman animals would lead to denial of persons' rights. Indeed, history and anthropology offer several examples of societies whose members had no difficulty in distinguishing between "marginal" and typical humans. After all, humans excel in their discriminatory powers, even when the characteristics chosen as the basis of that discrimination are morally irrelevant (e.g., race or gender).

According to the final, very serious objection made by Alan Holland, the AMC is at best a useless addition to the case constructed for nonhuman animal rights and at worst an unexploded bomb that could take out many humans as well as nonhumans. The biconditional version of the AMC claims the moral equivalence of marginal humans and sentient nonhumans. There is nothing in the argument to stop a person from rejecting the moral significance of both groups.

Although this last objection is strong, we nevertheless cannot conclude that the AMC is rhetorically or psychologically superfluous. Both opponents and supporters of nonhuman animal rights should confront the following questions: If it were to be wrong to "harvest" the organs of a severely retarded human to save the life of a normal human adult, is it also wrong to sacrifice a baboon or pig for the same purpose (assuming that transspecies transplants become medically feasible; see XENOGRAFT)? In general, is it wrong to treat sentient "nonpersons" as resources for persons? Both versions of the AMC challenge all parties to the debate to do some very fundamental moral thinking.

Selected Bibliography. Carruthers, Peter, *The Animals Issues: Moral Theory in Practice* (Cambridge: Cambridge University Press, 1992); Frey, R. G., The Significance of Agency and Marginal Cases, *Philosophica* 39(1) (1987): 39–46; Holland, Alan, On Behalf of a Moderate Speciesism, *Journal of Applied Philosophy* 1(2) (1984): 281–291; Narveson, Jan, On the Case for Animal Rights, *Monist* 70(1) (1987): 31–49;

Nelson, James, Xenograft and Partial Affections, *Between the Species* 2(2) (1986): 70–80; Pluhar, Evelyn, *Beyond Prejudice: The Moral Significance of Human and Nonhuman Animals* (Durham, NC: Duke University Press, 1995); Regan, Tom, An Examination and Defense of One Argument Concerning Animal Rights, in *All That Dwell Therein* (Berkeley: University of California Press, 1982), 113–147; Rollin, Bernard, *Animal Rights and Human Morality*, rev. ed. (Buffalo: Prometheus Books, 1992); Sauvage-Rumbaugh, Sue, and Roger Lewin, *Kanzi: The Ape at the Brink of the Human Mind* (New York: John Wiley and Sons, 1994); Singer, Peter, *Animal Liberation* (New York: New York Review of Books, 1990); Townsend, Peter, Radical Vegetarians, *Australasian Journal of Philosophy* 57(1) (1979): 85–93.

EVELYN PLUHAR

MARTIN, RICHARD

Richard Martin (1754–1834) was a leader in the establishment of the first animal-protection society and the passage of the first British legislation to protect animals. His lifelong practice of assisting those in need led to his nickname, "Humanity Dick." Building on the earlier effort of Thomas Erskine* to legislate against wanton cruelty to animals, and influenced by the gentleman-farmer John Lawrence, Martin secured passage of the Ill-Treatment of Cattle Act (1822). After this success, Martin regularly secured the passage of anticruelty legislation, the final initiative being his Bill to Prevent the Cruel and Improper Treatment of Dogs in 1826. Martin's enthusiastic participation in blood sports almost certainly helped to shape the Royal* Society for the Prevention of Cruelty to Animals'* long-standing neutrality (abandoned in 1976) on hunting* for sport.

Selected Bibliography. Lynam, Shevawn, *Humanity Dick: A Biography of Richard Martin* (London: Hamilton, 1975); Pain, Wellesley, *Richard Martin* (London: Leonard Parsons, 1925).

BERNARD UNTI

MARY ELLEN

Mary Ellen (about 1864–1956) was adopted as a young child in the 1870s with no identification other than a reference from the family doctor of the family that adopted her. She subsequently came to the attention of a social worker, Etta Wheeler, as a terrible case of child abuse. This small child was confined during hot weather, provided little clothing in cold weather, and beaten daily. Police and other institutions Wheeler approached about Mary Ellen were sympathetic, but while there were laws to protect children, these laws did not provide an effective means to remove them from their home. Etta Wheeler approached Henry Bergh,* founder and president of the American Society for the Prevention of Cruelty to Animals (ASPCA).* Bergh was influential and known for his sense of justice. He decided to investigate.

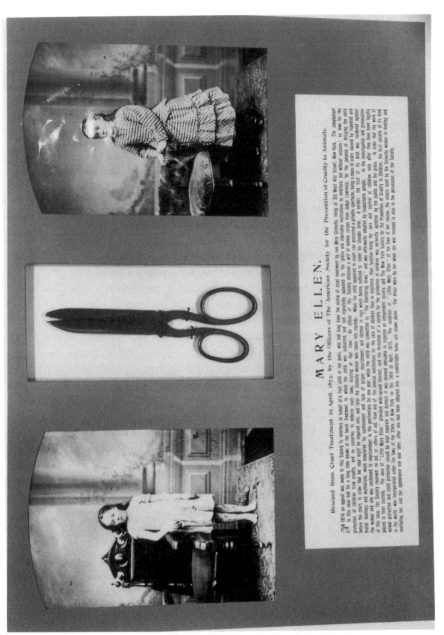

Portraits of Mary Ellen, rescued from cruel treatment by officers of the American Society for the Prevention of Cruelty to Animals. Photo courtesy of the American Society for the Prevention of Cruelty to Animals.

When one of his men, posing as a census worker, observed Mary Ellen and reported on the condition of the child, Bergh and his attorney Elbridge Gerry went into action. Judge Abraham R. Lawrence of the New York State Supreme Court issued a special warrant provided by section 65 of the habeas corpus act, and Mary Ellen was forcibly removed from the home. The court case that followed aroused a great public outcry regarding the treatment of children. Mary Ellen was brought into the court wearing ragged clothes and wrapped in an old blanket. She was thin and fragile, was bruised, and was cut across the face where her adopted mother had struck her with a pair of scissors. Two charges were brought against her adopted mother, who was found guilty of assault and battery and sentenced to one year of hard labor in the city penitentiary.

During the trial Henry Bergh made it clear that he was acting as a private citizen and was in no way functioning in his official capacity as president of the ASPCA. However, a rumor circulated that Bergh rescued Mary Ellen because if nothing else, she should be provided with the same protection as an animal. This may largely be due to the reporting of the case by Jacob Riis, who clearly saw the Mary Ellen case as a watershed in the establishment of children's rights. To clearly separate the protection of children from the protection of animals, Bergh, Gerry, and James Wright formed the New York Society for the Prevention of Cruelty to Children on December 15, 1874. Gerry's use of the writ of habeas corpus had been hailed in the press as brilliant, and it provided the first effective means to intervene in the rescue of children in abusive situations.

Selected Bibliography. Lazoritz, Stephen, and Eric A. Shelman, Before Mary Ellen, *Child Abuse and Neglect* 20 (1996): 235–237; Loeper, John J., *Crusade for Kindness: Henry Bergh and the ASPCA* (New York: Atheneum, 1991); Steele, Zulma, *Angel in a Top Hat* (New York: Harper and Brothers, 1942); Stevens, Peter, and Marian Eide, The First Chapter of Childrens' Rights, *American Heritage*, July/August 1990, 84–91; Watkins, Sallie A., The Mary Ellen Myth: Correcting Child Welfare History, *Social Work* 35 (1990): 500–503.

STEPHEN L. ZAWISTOWSKI

METAMORPHOSIS. *See* ANIMAL PRESENCE.

MICE

The mouse is the most typical laboratory mammal, and mice account for a large majority of all mammals used in research in the United States and Europe. Despite their tiny size, mice show remarkable genetic similarities to humans and can be used to study human genetic diseases. With their small body size, adaptability, and high reproductive rate, they are relatively economical and easy to maintain.

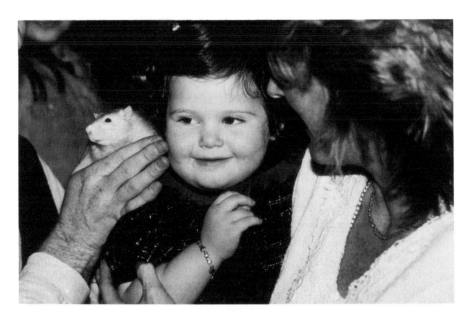

Although rats and mice in the past were viewed as pests or laboratory animals, they are increasingly favored as companions. Photo by Joan Borinstein.

Research studies of the mouse genome and the human genome are complementary. Through selective breeding and genetic manipulation, many thousands of distinct strains of mice now exist, some of which serve as models for specific human diseases. The mouse is the only mammal available in so many different genetic strains. For example, two different mutations (genetic variations) have resulted in mice with impaired immune status, making it possible to greatly advance studies of immune function, tumor growth, and various human genetic diseases.

Mice are also used in the development, preparation, and safety testing of vaccines. Mice played a central role in the development of whooping cough and yellow fever vaccines. Experimental vaccines are under development for human hepatitis A, sickle-cell anemia, and malaria. Approaches are being explored with vaccines for treating cancer and producing contraception.

In the 1960s, a pregnancy test required killing a rabbit that had been injected with the woman's urine and examining the rabbit's ovary for follicular growth. Today, a simple test kit allows a woman to conduct a pregnancy test herself. One essential ingredient in this and other biological test kits is a monoclonal antibody, a protein formed in the blood that specifically responds to a certain hormone or protein. Often, monoclonal antibodies are produced in mice by an injection of an antigen, a substance that induces the formation of a specific antibody. To boost the production of antibodies, an irritating adjuvant is also injected into the mouse. Subsequently, antibody-rich fluid builds up in the peritoneal cavity within the abdomen of the mouse; the fluid is aspirated with a large-gauge needle. Antibody production is one of the more controversial uses of animals, and one that is sharply regulated or prohibited in several European countries, where methods of antibody production not requiring animals are primarily used.

Mice can also be patented. The Harvard mouse that carries an oncogene for breast cancer was patented in the United States in 1988. This mouse strain is used in cancer research to screen carcinogens and anticancer drugs. Whether patents such as this should be issued to other mouse strains, or even simply to specific DNA chains identified in the human or mouse genome, is hotly debated, particularly within the European Community. Whether mice, or any animal for that matter, carrying unique genes should be patentable is a question not yet resolved.

Selected Bibliography. Hart, L. A., and A. Mitchell, The (Almost) All-Purpose Laboratory Mouse, in N. E. Johnston (Ed.), *Animals in Science Conference: Perspectives on Their Use, Care, and Welfare* (Melbourne, Australia: Monash University, 1995), 184–195; Herzog, H. A., The Moral Status of Mice, *ILAR News* 31(1) (1989): 4–7; Orlans, F. B., *In the Name of Science: Issues in Responsible Animal Experimentation* (New York: Oxford University Press, 1993); Rowan, A. N., *Of Mice, Models, and Men: A Critical Evaluation of Animal Research* (Albany: State University of New York Press, 1984); Smith, J. A., and K. M. Boyd (Eds.), *Lives in the Balance: The Ethics of Using Animals in Biomedical Research* (Oxford: Oxford University Press, 1991).

LYNETTE A. HART

MISOTHERY

The term "misothery" is derived from Greek *misein*, to hate, and *therion*, beast or animal, and literally means hatred and contempt for animals. Since animals are so representative of nature in general, misothery can mean hatred and contempt for nature, especially its animal-like aspects. One writer, for example, has described nature as "red in tooth and claw," that is, bloodthirsty like a predatory animal. In another version of the same idea, we say, "It is a dog-eat-dog world." These are misotherous ideas, for they see animals and nature as vicious, cruel, and base.

"Misothery" was constructed because of its similarity to the word "misogyny," a reasonably common word for an attitude of hatred and contempt toward women. The similarity of the two words reflects the similarity of the two bodies of attitudes and ideas. In both cases, the ideas reduce the power, status, and dignity of others. Misogyny reduces female power, status, and dignity and thus aids and abets the supremacy of males under patriarchy. Misothery reduces the power, status, and dignity of animals and nature and thus aids and abets the supremacy of human beings under dominionism.*

Just as agrarian society invented beliefs to reduce women, it also invented beliefs or ideologies about animals that reduced them in the scheme of life. Among these are the ideas that animals are too base and insensitive to feel physical pain* or emotional suffering.*

Selected Bibliography. Fisher, Elizabeth, *Woman's Creation* (Garden City, NY: Anchor Press/Doubleday, 1979); Gray, Elizabeth Dodson, *Green Paradise Lost* (Wellesley, MA: Roundtable Press, 1981); Nash, Roderick, *Wilderness and the American Mind*, 3rd ed. (New Haven: Yale University Press, 1982); Serpell, James, *In the Company of Animals* (London: Basil Blackwell, 1986); Tuan, Yi-Fu, *Dominance and Affection* (New Haven: Yale University Press, 1984).

JIM MASON

MOORE, JOHN HOWARD

John Howard Moore (1862–1916) was a Chicago schoolteacher whose work *The Universal Kinship* (1906) was lauded by Henry Salt* as one of the most important humanitarian titles of its era. This work alone distinguishes Moore as perhaps the first organic American intellectual in the realm of animal rights.* It was through Moore's intercession that his brother-in-law, Clarence Darrow, became a supporting member of the Humanitarian League. Moore contributed articles and essays to numerous humane and vegetarian (*see* VEGETARIANISM) publications. He was also the author of *Better World Philosophy* (1899), *The New Ethics* (1907), and *Savage Survivals* (1916). Moore's work was marked by the conviction that the science of evolution provided an affirmation of the humane ethic.

Selected Bibliography. Magel, Charles (Ed.), *The Universal Kinship* by J. Howard Moore, with appendices including letters from Moore to Henry S. Salt, a eulogy by Clarence Darrow, a bibliographic essay and an introduction by Charles Magel (Fontwell: Centaur Press, 1992); Moore, J. Howard, *Better World Philosophy: A Sociological Synthesis* (Chicago: The Ward Waugh Company, 1899); Moore, J. Howard, *The New Ethics* (London: E. Bell, York House, 1907); Moore, J. Howard, *Savage Survivals* (Chicago: Charles H. Kerr, 1916); Nash, Roderick, *The Rights of Nature* (Madison: University of Wisconsin Press, 1989); Salt, Henry S., *Company I Have Kept* (London: George Allen and Unwin, 1930); Salt, Henry S., *Seventy Years among Savages* (London: George Allen and Unwin; New York: T. Seltzer, 1921).

BERNARD UNTI

MORAL AGENCY AND ANIMALS

Moral agents are those who can recognize what is morally right or wrong and attempt to do what is right and avoid what is wrong. We commonly believe that only human beings have this ability and that not even all humans do. Very young children are commonly believed not to know the difference between right and wrong. For example, a two-year-old may take a shiny paperweight when visiting a neighbor's house, but we do not call this stealing, because the child does not yet understand the idea of respecting another person's property. We say the same thing when a dog* takes someone's slipper and uses it for a chew toy. The difference is that we expect the child to come to understand moral concepts as he matures, while we believe the dog incapable of such understanding throughout its life.

This difference has been cited by many philosophers as indicating a morally very important difference between humans and other animals. These philosophers consider reciprocity a fundamental consideration in morality, particularly in matters of moral rights and obligations. They claim that a person can be morally obligated to respect the rights only of those who respect his rights. It follows from this idea that if certain kinds of beings are not able to recognize and respect the rights of others, those others cannot be obligated to recognize and respect their rights. Particularly, if animals cannot recognize and respect human rights, then humans cannot be obligated to recognize and respect animal rights.*

There are two responses to this challenge to animal rights based on moral agency. First, the assumption that moral obligations rest on reciprocity can be challenged. In the case of very young children, their inability to recognize and respect the rights of others does not detract from our moral obligations to them. In fact, we acknowledge strong moral obligations to protect and care for young children, even though they are not moral agents. We acknowledge similar obligations to other humans incapable of moral agency, such as those who are severely retarded or brain damaged and elderly people suffering severe dementia. Consequently, even if animals cannot be moral agents, it is still possible that we have moral obligations to protect their lives and to care for them, since moral obligations are not all based on reciprocity.

A second line of response involves narrowing the gap between humans and animals based on moral agency. This is done by arguing that the difference here is a matter of degree (see CONTINUITY). We commonly attribute moral virtues (see VIRTUE ETHICS) to animals. For example, we refer to loyal dogs, courageous lions, and industrious beavers. There are many accounts of members of herds of a wide variety of animals standing guard while other members of the herd graze. There are also many observations of adults of a wide variety of animals being devoted parents, even putting their lives in danger to protect their young. There are even observations of porpoises and other wild animals saving the lives of humans who were in distress. These observations tend to indicate that even if animals cannot recognize and respect such human concepts as property* rights, there are a great many moral values they can and do recognize.

Many people discount these examples of moral virtue among animals by contending that they are just matters of instinct. These people believe that in order to be a moral agent, a being must be able to recognize and act on a general moral principle, like "Do unto others as you would have them do unto you." Again, our ordinary moral practice does not support restricting what counts as moral agency in this way. For example, many humans are good parents because they love their children. They do not protect and care for their children because they recognize some general moral principle; rather, they recognize that their children need protection and care and out of love—sometimes called maternal and paternal "instinct"—they provide that protection and care.

When an individual recognizes that another is in need and acts to help, then he or she has recognized what is morally important in the situation and has responded in a morally good way. Many animals appear to be capable of such recognition and response. Consequently, even though animals are not capable of recognizing and respecting some of the moral values that mature humans can recognize and respect, they are not entirely lacking in moral agency.

Selected Bibliography. Clark, Stephen R. L., *The Nature of the Beast* (Oxford: Oxford University Press, 1982); Cohen, Carl, The Case for the Use of Animals in Biomedical Research, *New England Journal of Medicine* 315 (1986): 865–870; Darwin, Charles, *The Descent of Man and Selection in Relation to Sex* (London: John Murray, 1871); Kant, Immanuel, First Section: Transition from the Common Rational Knowledge of Morals to the Philosophical, in *Foundations of the Metaphysics of Morals*, trans. Lewis White Beck (Indianapolis: Liberal Arts Press, 1959); Sapontzis, S. F., Being Rational and Acting Morally, *Morals, Reason, and Animals* (Philadelphia: Temple University Press, 1987).

STEVE F. SAPONTZIS

MORAL STANDING OF ANIMALS

Intelligence and adaptation in animals are often difficult to understand without acknowledging that animals exhibit understanding, intention, thought, imaginativeness, and various forms of communication (*see* ANIMAL COGNITION; ANIMAL COMMUNICATION). Many actions suggest adaptive and creative forms of judgment. If one attributes these capacities to animals, then they are credited with capacities similar to human capacities, which suggests that animals merit at least some of the moral protections humans enjoy.

Prior to the work of Charles Darwin,* many biologists and philosophers argued that despite the anatomical similarities between humans and apes, humans are distinguished by the possession of *reason*, *speech*, and *moral sensibility*. Darwin thought, by contrast, that animals have various powers of deliberation and decision making, excellent memories, a strong suggestion of imagination in their movements and sounds while dreaming, and the like. He wrote about the intelligence, sympathy, pride, and love of animals. Darwin also criticized the hypothesis that major cognitive differences separate apes and humans. The ultimate importance of his theory is that it is not only complex biological structures and functions that are shared in the evolutionary struggle, but cognitive abilities as well.

Problems about whether animals have higher-level cognitive capacities are connected to questions of moral (and legal) standing. The term "standing" refers to "one's place in the community in the estimation of others; one's relative position in social, commercial, or moral relations; one's repute, grade, or rank" (*Black's Law Dictionary*). In a weak sense, "standing" refers to a status, grade, or rank of moral importance. In a strong sense, "standing" means to have rights or the functional equivalent of rights.

People assign a more significant standing to an animal by granting that it is similar to an intact adult human being. Its standing is still further enhanced by attributing something like personhood or autonomy* to it. A category such as "person" or "autonomous agent" (*see* AUTONOMY OF ANIMALS) raises the animal to a position similar to that occupied by those who have rights (*see* GREAT APE PROJECT). A widely shared view today is that if animals have capacities for understanding, intending, and suffering,* these morally significant properties themselves confer some moral standing.

Much recent discussion about standing has centered on the criteria for being a person, under the assumption that all and only persons have the relevant distinctive properties. Several philosophers have produced arguments along the following lines: One is a person if and only if one possesses certain cognitive properties; the possession of these properties gives an entity moral standing. A list of the conditions for being a person and thus acquiring moral standing includes (1) self-consciousness (of oneself as existing over time), (2) capacity to engage in purposive sequences of actions, (3) capacity to appreciate reasons for acting, (4) capacity to communicate with other persons using a language, (5) capacity to make moral judgments, and (6) rationality. Many believe that more than one of these conditions is required to be a person. As long as high-level cognitive criteria are required, animals cannot qualify for significant moral standing. But if less demanding cognitive capacities are employed, animals might acquire a significant range of moral protections.

Cognitive criteria can help us in our examination of the moral standing of animals. Perhaps a noncognitive property is sufficient to confer some measure of moral standing. At least two kinds of properties need to be considered: (1) properties of sensation (or perception), for example, feeling pain,* and (2) properties of emotion, for example, love and pride.

Selected Bibliography. Cavalieri, Paola, and Peter Singer (Eds.), *The Great Ape Project* (London: Fourth Estate, 1993); Frey, Raymond, Moral Standing, the Value of Lives, and Speciesism, *Between the Species* 4(3) (Summer 1988): 191–201;

Midgley, Mary, Persons and Non-Persons, in Peter Singer (Ed.), *In Defense of Animals* (Oxford: Basil Blackwell, 1985); Rachels, James, *Created from Animals: The Moral Implications of Darwinism* (New York: Oxford University Press, 1990); Regan, Tom, *The Case for Animal Rights* (Berkeley: University of California Press, 1983).

TOM L. BEAUCHAMP

N

NATIVE AMERICANS. *See* NATIVE PEOPLES AND ANIMALS.

NATIVE PEOPLES AND ANIMALS

Contemporary Native Americans and tribes throughout the world, whether Walpiri of central Australia, Bimin-Kuskusmin of Papua New Guinea, the Tasaday of the southern Philippines, or the Qollahuaya of Mt. Kaata in Bolivia, all practice elaborate rituals that worship and are devoted to other animal species. While there is no doubt that our human ancestors hunted (*see* HUNTING) and frequently sacrificed animals, there is also a growing body of evidence suggesting the widespread habit of vegetarianism.*

Today, few totally vegetarian communities exist, but there are some. In India, for example, where cows are traditionally revered, there are over 100 million vegetarians. Among them are nearly 1 million desert Bishnoi, a subsect of Hindus (*see* RELIGION AND ANIMALS, Hinduism) who live in the Thar Desert of Rajasthan State. They worship a medieval saint who claimed that all plants and animals are sacred and must be respected. Further to the south, the Todas of the Niligiri massif in the Indian state of Tamil Nadu have been vegetarian for at least 1,200 years. They worship animal life, particularly the buffalo, around which they have formulated an extensive set of rituals and beliefs that utterly encompass their way of life.

Throughout much of India, nearly 10 million Jains (*see* RELIGION AND ANIMALS, Jainism) are also strict vegetarians. The Jains will not partake of any profession that exploits animals. Theirs can be said to have been the first communal ecological ethic. The Jains ethically and scientifically heralded the beginnings of animal liberation.*

These three Indian communities, Bishnoi, Toda, and Jain, are traditional societies with a visionary reverence for animals that has utterly defined the sphere of their respective professional, personal, and social lives. Other traditions have followed these examples to varying degrees: Quakers, Zoroastrians, Seventh-Day Adventists, countless Buddhists of various sects and paths (see RELIGION AND ANIMALS, Buddhism), and many others.

Selected Bibliography. Spencer, Colin, *The Heretic's Feast: A History Of Vegetarianism* (London: Fourth Estate, 1993); Tobias, Michael, The Anthropology of Conscience, *Society and Animals* 4(1) (1996): 65–73; Tobias, Michael, *Life Force: The World of Jainism* (Fremont, CA: Asian Humanities Press, 1991); Tobias, Michael, *A Naked Man* (Fremont, CA: Jain Publishing, 1994); Tobias, Michael (Ed.), *Mountain People* (Norman: University of Oklahoma Press, 1986).

MICHAEL TOBIAS

Native Americans' Early Uses of Animals

Many, if not all, pre-Columbian Native American nations used animals in the production of medical treatments and education. The common view of Native medicine has been shamanistic, but although ritual did, and still does, play an important role in Native American medicine, there was extensive use of practical therapy. The more practical therapies included the use of plants and animal parts to treat specific medical conditions.

Most Native American nations, with the notable exception of the Aztecs, did not engage in internal surgical practices. Furthermore, in many Native American nations post-mortem (after-death) examinations were not conducted on the dead for religious reasons. Most of the information Native Americans had about internal anatomy came from their dissection of animals during the butchering process. It has been documented that from the analogy with animals, Native Americans knew the function of internal organs and knew that the brain was the organ of thought.

Besides providing anatomy lessons, animals were utilized in observational "research." By noting particular animal behaviors, especially the interaction between animals and plants, Native Americans gained information about the nutritional and medicinal properties of many plant substances. For example, the bear in Ojibwa culture is a medicinal animal, believed to be given the secrets of the Mide (medicine) by Kitshi Manido (Great Spirit). Because of this belief, the Ojibwa would carefully observe the bear in its environment. These examples demonstrate that besides using animals for food and clothing, early Native Americans also used animals to gain information about themselves and their environment.

Although a number of Native American herbal remedies for medical discomforts have been adapted by medical organizations, the use of animal products in medical treatment has not received the same attention. Animal products were used in a number of medical remedies in many Native American nations. Moose and bear fat were used by the Ojibwa to treat skin wounds and to ensure healthy skin in the extreme temperatures. Deer tendons were used as suture material by numerous tribes. The Yukon treated scurvy by ingestion of the animal adrenal glands. Fish oil, because of its high iodine content, was used to treat goiters in Eskimo and Aleut nations. Some South American nations treated epilepsy through "shock treatment" with electric eels. A type of injection device was used by some Native American nations well before the invention of the syringe in 1904. Such devices were constructed from the bladder of a deer or duck connected to a reed or quill of the porcupine. These syringes were used to clean wounds or to inject herbal medicine into the wound.

The examples listed here demonstrate that Native Americans' unique relationship with animals included their use in research and medicine. By documenting both the physiological and behavioral properties of animals, we can learn more about animals, including ourselves (*see* ANIMAL MODELS).

Selected Bibliography. Altman, J., *Organic Foundations of Animal Behavior* (New York: Holt, Rinehart and Winston, 1966); Aronson, L. R., Levels of Integration and Organization: A Revaluation of the Evolutionary Scale, in G. Greenberg and E. Tobach (Eds.), *Behavioral Evolution and Integrative Levels* (Hillsdale, NJ: Lawrence Erlbaum Associates, 1984); Hershman, M. J., and K. M. Campion, American Indian Medicine, *Journal of the Royal Society of Medicine* 78, (1985): 432–434; Hoffman, W. J., *The Midewiwin or Grand Medicine Society of the Ojibwa*, Seventh Annual Report of the Bureau of American Ethnology (Washington, DC: Government Printing Office, 1885–1886), 149–300; Major, R. C., Aboriginal American Medicine North of Mexico, *Annals of Medical History* 10(6) (1938): 534–49; Vogel, V. J., *American Indian Medicine* (Norman: University of Oklahoma Press, 1970).

LISA M. SAVAGE

Native Americans' Relationships with Animals: All Our Relations

The relationships between animals and Native Americans are as varied as are the more than four hundred different tribal nations that existed in pre-Columbian North America. Native people were and in many cases still remain deeply tied to the particular ecosystems in their regions of the continent. Some based their lives on agriculture, some on the ocean and salmon fishing, others on the hunting of hooved animals. However, certain generalizations about the relationships between Native Americans and animals can be made. One of the most important generalizations is that animals are not seen by the American Indian as dumb beasts whose lives are ruled only by instinct, but as individuals—thinking, feeling beings with families, beings worthy of respect. They are the "animal people."

In the truest sense of the word, animals are seen as relatives to human beings. Many Native traditions, such as those of the Cherokee or the Lakota, tell how certain animals were direct ancestors. The idea of clan often comes from a tradition of direct descendants from one animal or another—a frog, an eagle, a bear. If a person belongs to the Bear Clan, it may be that the clan's origin is in the form of a bear who married a human woman and produced offspring. The border between the worlds of the animal people and the human beings is easily crossed. A human being may go and live among the animals and become a bear or a deer as easily as an animal may take on human shape and live among human beings. Sometimes these "animal people" have great power and are to be feared. Through the Midwest and West tales are still common of the Deer Woman who comes to gatherings to lure off young men and harm them. Beneath her long dress, she has hooves, not feet. Such beliefs are extremely widespread and are reinforced by stories and ceremonies.

Animals often appear in traditional stories as teachers. Humans can learn many things from the animal people. Traditional stories tell us how flute songs came from the birds, how medicine plants were shown to the humans by the bears, and how humans were taught how to work together and to care for their children by watching the behavior of the wolves as they hunted and cared for their cubs.

At times, Native American people find it necessary to hunt the animals to ensure their own survival. However, even hunting* is seen as being done in cooperation with the animals. Although the animals's body is killed, its spirit survives and may punish a disrespectful or greedy hunter. It is only through the animals' consent that they allow themselves to be hunted. Further, the hunting of animals that are pregnant or caring for young ones whose survival depends upon the mother is usually forbidden. Many of the "new ideas" about game laws, closed seasons, and limited harvesting of game animals appear to have their roots in Native American traditions that have existed for thousands of years.

Animals are frequently kept as pets or companions (*see* COMPANION ANIMALS AND PETS). In the Northeast among the Iroquois, orphaned beavers were often suckled by Native women and adopted into the family. Dogs* were kept as pets and used for hunting throughout the continent. According to the traditional stories of the Abenaki, the dog was not domesticated, but chose to live with the human beings because it liked them. To this day, the dog in a Native American household is often viewed not as a possession but as a family member. The fact that in some Native American cultures dogs were sometimes eaten or sacrificed, as in the Seneca White Dog Sacrifice so that the dog's spirit could take a message to the Great Spirit, did not diminish the respect for the dog or its place in the household.

In the traditions of the many different Native peoples of North America, animals are almost universally seen as equals to humans on the circle of life.

The word "circle" is especially appropriate, for all living things, animals and humans alike, are viewed as part of a great circle. No part of that circle is more important than another, but all parts of that circle are affected when one part is broken. In the eyes of the Native American, animals are all our relations.

Selected Bibliography. Brown, J. E., *Animals of the Soul: Sacred Animals of the Oglala Sioux* (Rockport, MA: Element, 1992); Caduto, M., and J. Bruchac, *Keepers of the Animals* (Golden, CO: Fulcrum Publishing, 1991); Cornell, G., Native American Contributions to the Formation of the Modern Conservation Ethic (Ph.D. dissertation, Michigan State University, 1982); Hughes, J. D., *American Indian Ecology* (El Paso: Texas Western Press, 1983); Vecsey, C., and R. W. Venables, *American Indian Environments: Ecological Issues in Native American History* (Syracuse, NY: Syracuse University Press, 1980).

JOSEPH BRUCHAC

NEEDS OF ANIMALS

A need can be defined as a requirement that is a consequence of the biology of the animal to obtain a particular resource or respond to a particular environmental or bodily stimulus. Animals have a range of functional systems controlling body temperature, nutritional state, and social interactions. Together, these functional systems allow the individual to control its interactions with its environment and hence to keep each aspect of its state within a tolerable range. When an animal acts to return to this tolerable range, we say that it has a need.

Some needs are for particular resources, such as water or heat. However, the means of obtaining a particular objective have also become important to the individual animal. For example, various species will work, in the sense of carrying out actions that result in food presentation, even in the presence of food. Hence pigs* need to root in soil or some similar ground, hens need to dust-bathe, and animals of these species need to build a nest before giving birth or laying eggs. Needs therefore range from those that can be satisfied in a simple way, for example, by ingesting water, to complex ones involving a variety of sensory input or sufficient contact with other members of the species.

Some reports and laws refer separately to physiological needs and behavioral or ethological needs. However, while the recognition of a need may depend on an effect on the physiology of an animal or the urgent and energetic attempts of an animal to show a particular behavior, the need is in the brain of the individual. Hence the need itself is not physiological or behavioral but may be satisfied only when some physiological imbalance is prevented or corrected, or when some particular behavior is shown.

Some needs are associated with feelings (*see* FEELINGS OF ANIMALS), and these feelings are likely to change when the need is satisfied. If the existence of a feeling increases the chances that the individual will carry out some adaptive action and hence be more likely to survive, the capacity to have such a feeling is likely to have evolved by natural selection. Further, if the state of an individual in certain conditions is desirable from an evolutionary viewpoint, there should be a propensity for that individual to have good feelings. On the other hand, if a state is one that should be quickly altered, it should be associated with unpleasant feelings that prompt avoidance or some other action. Feelings are part of a mechanism to achieve an end, just as adrenal responses or temperature regulatory behavior are mechanisms to achieve an end.

Research on needs is of two kinds. Preference tests (*see* PREFERENCE AND MOTIVATION TESTING) in which the strength of positive preference is quantified give information about what is important to the subject animal. Studies in which a need is not satisfied and the extent of poor welfare is assessed using indicators of abnormal behavior, negative preference, physiology, immunosuppression, disease, injury, and so on, also indicate the importance of the resource concerned in terms of biological effects.

Selected Bibliography. Broom, D. M., Animal Welfare Defined in Terms of Attempts to Cope with the Environment, *Acta Agricultural Scandinavica, Section A, Animal Science*, Supplement 27 (1996): 22–28; Broom, D. M., and K. G. Johnson, *Stress and Animal Welfare* (London: Chapman and Hall, 1993); Hughes, B. O., and I. J. H. Duncan, The Notion of Ethological "Need," Models of Motivation, and Animal Welfare, *Animal Behaviour* 36 (1988): 1696–1707; Toates, F., and P. Jensen, Ethological and Psychological Models of Motivation: Towards a Synthesis, in J.-A. Meyer and S. Wilson (Eds.), *From Animals to Animats* (Cambridge, MA: MIT Press, 1991), 194–205; Vestergaard, K., The Regulation of Dustbathing and Other Behaviour Patterns in the Laying Hen: A Lorenzian Approach, in R. Moss (Ed.), The Laying Hen and Its Environment, *Current Topics in Veterinary Medicine and Animal Science* 8 (1980): 101–113.

DONALD M. BROOM

NEW WELFARISM. *See* ANIMAL RIGHTS.

NIETZSCHE, FRIEDRICH WILHELM

Friedrich Wilhelm Nietzsche (1844–1900) was a German philosopher influenced by Arthur Schopenhauer,* especially by his rejection of Immanuel Kant's* view that duties to animals are only indirect duties to humankind. Nietzsche particularly developed Schopenhauer's critique that the Kantian view leads to a lack of compassion for animals. Nietzsche held that the "deeper minds of all ages have had pity for animals" and that the "sight of blind suffering is the spring of the deepest emotion." From this it follows that pity for animals is a virtue (*see* VIRTUE ETHICS), if not an imperative.

Nietzsche argued that "Nature" is an order in need of higher transformation personified by the artist, philosopher, and saint: "Finally, Nature needs the saint," for "in him the ego has melted away, and the suffering of his life is, practically, no longer felt as an individual, but as the spring of the deepest sympathy and intimacy with all living creatures." Moreover, nature needs to attain the "high state of man" so "that she may be delivered from herself" ("Schopenhauer as Educator," 149–155). In short, the suffering* of animals and nature await moral transformation by an enlightened humanity.

Selected Bibliography. Nietzsche, Friedrich, Pity for Animals [Extract], in Paul Clarke and Andrew Linzey (Eds.), *Political Theory and Animal Rights* (London: Pluto Press, 1990), 148–152; Nietzsche, Friedrich Wilhelm, Schopenhauer as Educator (1874), translated by Adrian Collins, in *Thoughts out of Season* (Edinburgh: T. N. Foulis, 1909), pt. 2, 149–155.

ANDREW LINZEY

NORWEGIAN INVENTORY OF AUDIOVISUALS (NORINA)

There is an increasing demand for information on alternatives* to the use of animals in teaching. Since 1991 information has been collected on audiovisual aids that may be used as animal alternatives or supplements in the biomedical sciences at all levels from primary schools to university. This information is available as an English-language database known as Norwegian Inventory of Audiovisuals (NORINA). NORINA contains information on over 3,500 audiovisuals and their suppliers. Each audiovisual has been designated a category describing the type of animal alternative (e.g., computer program or video film) and one designating appropriate area(s) of use (e.g., anatomy, dissection, or physiology). NORINA is a nonprofit venture funded by external support from animal welfare* organizations. Personal copies may be purchased for IBM Windows or Macintosh computers where the database program Filemaker Pro is already installed. NORINA is currently in use in 15 countries worldwide. (*See also* "Resources on Animal Welfare and Humane Education.") NORINA's web site addresses are http://oslovet.veths.no and http://www.bio.mq.eu.au/NORINA. Further information about NORINA is available by e-mail from Karina Smith (karina .smith@veths.no) or by telefax (+47 22 96 45 35).

ADRIAN SMITH AND KARINA SMITH

Record number   37
Program/Contact person's name:

The Rat Stack

Type of record: Computer Program

Category   Anatomy (animal) & Dissection (animal)

Computer type: Macintosh

Version: 1989   Updated: 04-06-96

Price: £80.00 for site licence (when ordering more than one version add £5.00 extra per version)

Program source/Contact person's address:

Dr. David Dewhurst, Sheffield Bioscience Programs, 11 Robinson Drive, The Park, Pannal Ash, Harrogate HG2 9DJ, England

Telephone: Int-44-423-520495

Telefax: Not available yet

Details:

Highly interactive Hypercard stack teaching the functional anatomy of the rat.   It is suitable for independent learning, tutorial revision and could be used as a realistic alternative to animal dissection.   Scanned photographs and diagrams at different stages of dissection are presented in a manner comparable to removing layers to reveal the layer beneath.   Each image becomes annotated as the cursor is moved across the image with the mouse and clicking the mouse button when pointing to an organ or structure of interest leads the user further through the dissection in that area.   In some cases detailed anatomical and physiological information is available.   Test mode for questions with scoring of correct or nearly correct answers.   Users may ask the program to highlight a structure on an image or ask for additional textual information about a structure or about the rat in general.   This information is stored in a database and may be altered by a person with a password to the 'input text' part of the program.   Additional on-line questions may be added using this feature.   Text in the scrolling box is part of a Hyper Text facility and any word followed by an asterisk may be clicked on to see the definition from the database associated with that word.

Comments & references:

The hard disk version contains more images than the floppy disk version and it must run from a hard disk drive or a high density floppy drive.   To upgrade from the floppy version to the hard-disc version, send the floppy master disc back together with £5.00, and a hard-disc version will be sent.   Suitable for High School, College and Undergraduate students of Biology.   Extra software required: Hypercard.   A map of the dissection is available (divided into seven sections) showing an icon of each image and highlighting in black areas of the dissection the user has already visited.   Clicking the mouse button on a heading shows the images available in that section.   Clicking on any icon or double-clicking a heading will take the user straight to that part of the dissection.   Screens are annotated when in 'browse' or 'browse & test' modes.   In test mode questions are generated from the image on-screen and a tally of correct (and nearly correct) answers is kept while the program is in use.   Presently uses scanned graphics, intention to use interactive video.

Program author:
Megan Quentin-Baxter and David Dewhurst

Norwegian Inventory of Audiovisuals: An example of a printout from NORINA. Courtesy of Adrian Smith.

O

OBJECTIFICATION OF ANIMALS

In 1995, the Summit for Animals, an informal collection of national and grass-roots animal protection organizations, passed a resolution stating, in part, "We resolve to use language that enhances the social and moral status of animals from objects or things to individuals with needs and interests of their own." Collectively called the "linguistic turn," a current view in several academic fields holds that language plays an important role in the way we see, think about, and, ultimately, treat entities in both the cultural and natural world.

Numerous areas that need change have been identified. The most important and perhaps the most difficult to bring about is the use of the term "animal," which has come to mean "as distinguished from human." In this use, the conflicting terms "human" and "animal" deny that human beings are part of the animal kingdom. More critically, this usage reinforces the notion that animal is inferior to human.

Other linguistic habits support the lower status of animals. In many settings, such as the farm and the research laboratory, animals are not named. Further, they are referred to as "it" rather than "he" or "she" and "which" rather than "who." These uses decrease the value of animals by depriving an animal of his or her individuality (see ANIMAL INDIVIDUALITY), including his or her identity as a member of a particular gender. This practice is also seen in language used by hunters and "wildlife" managers when they refer to "the deer" as a species rather than a group of individuals.

In farm and laboratory settings, language operates to deprive animals other than humans of even this identity as members of a particular species. Rather than "the rat" or "the monkey," investigators typically refer to animals in the lab as "the animal." A final decrease in value occurs when they are referred to as less than even this already-weakened notion of animal. On the farm, that individual cow is "beef" or "meat on the hoof," while in the laboratory that individual rat is an "organism," a generic living being, or a "preparation," a living physiological or behavioral process.

In the scientific laboratory setting, additional practices support the devaluing of animals. Many scientists use the term "anthropomorphism"* as a criticism of both scientific and popular accounts that use psychological terms to describe animals other than humans. For example, terms like "intended," "anticipated," and "felt" and attributions like "play," "grief," and "deceit" to animals other than humans are avoided because their use is necessarily committing the error of anthropomorphism. This prohibition against terms implying consciousness in animals other than humans is a continuation of the ideas of the philosopher René Descartes,* in whose view such animals were mechanical beings, without psychology, without minds. Consistent with this view, pain,* suffering,* and death* accompanying either the conditions of an experiment or the conditions under which animals in the laboratory are kept are typically not described as such. For example, an animal is said to be "food deprived" rather than "hungry" or subjected to "aversive stimulation" rather than "experiencing pain." The death of an animal is obscured by various terms such as "collected," "harvested," "sacrificed," or "anesthetized and then exsanguinated."

Selected Bibliography. Birke L., and J. Smith, Animals in Experimental Reports: The Rhetoric of Science, *Society and Animals* 3 (1995): 23–42; Dunayer, J., Sexist Words, Speciesist Roots, in C. Adams and J. Donovan (Eds.), *Animals and Women: Feminist Theoretical Explorations* (Durham, NC: Duke University Press, 1995), 11–32; Phillips, M. T., Proper Names and the Social Construction of Biography: The Negative Case of Laboratory Animals, *Qualitative Sociology* 17(2) (1994): 119–143; Shapiro, K., The Death of an Animal: Ontological Vulnerability and Harm, *Between the Species* 5(4) (1989): 183–195.

KENNETH J. SHAPIRO

OVERPOPULATION

As of 1997 the human population on earth was approximately 5.7 billion. With a global increase of over 2% annually and an average fertility rate (the number of children per woman) averaging 3.5 worldwide, scientists generally agree that *Homo sapiens sapiens* will reach 12 billion by early in the 21st century. The aggressive and widespread use and the commercial development of previously wild land by 12 billion humans does not bode well for the fast-diminishing wildlife on the planet. Our NPP rate (net primary production, the amount of arable land that has been overtaken for whatever purpose by human beings) is approaching 50% of the entire land area of the planet.

Human overpopulation is at the root of the many causes of poverty or greed that motivate destruction of habitat and wildlife. Vast human numbers also generate an enormous appetite for goods that result from the slaughter of hundreds of species of animals, from cows to emus to alligators, for fast-food hamburgers or the 12 billion leather shoes that will be sold annually by the year 2000 just to keep pace with the fast-growing human population.

At the Rio Summit in 1992, climate change was the focus. At the 1994 Cairo Summit, population was the issue. But in neither instance was wildlife or the rights of habitat focused on. Human overpopulation, however, has an impact on nature. Animal rights* are incompatible with a human population of 12 billion. Given the inevitability of several billion more people on the planet, regardless of whatever new family-planning successes are likely to come about, it is now clear that the legal, political, and cultural advocacy of animal rights can reverse negative human impact on the biosphere.

Selected Bibliography. Abernethy, Virginia D., *Population Politics: The Choices That Shape Our Future* (New York: Insight Books, 1993); Tobias, M., The Dynamics of Environmental Despair and Optimism, *Population and Environment*, September 1996; Tobias, M., *World War III: Population and the Biosphere at the End of the Millennium* (Santa Fe, NM: Bear and Co., 1994); Wilson, E. O., *The Diversity of Life* (Cambridge, MA: Belknap Press of Harvard University Press, 1992); *World Resources: A Report by the World Resources Institute, with the United Nations Environment Program and the United Nations Development Program* (New York: Oxford University Press, 1993), 119.

MICHAEL TOBIAS

OXFORD GROUP

Oxford Group is a title used by Richard Ryder to describe those intellectuals associated with the city of Oxford, England, who ignited the modern interest in the moral status of animals. The novelist Brigid Brophy* (1929–1995) had broken the long silence on this subject in 1965 with her article "The Rights of Animals" in the *Sunday Times* (October 10). In 1969 Ryder, a psychologist working in Oxford, published his first attacks upon animal abuse and was contacted by Brophy, who introduced him to three young Oxford University postgraduate philosophers, John Harris and, from Canada, Stanley and Roslind Godlovitch. In 1971 they and Ruth Harrison contributed to the book *Animals, Men, and Morals*, edited by Harris and the Godlovitches, the first serious work on animal rights* since Henry Salt's *Animals' Rights Considered in Relation to Social Progress*, first published in 1892.

Three other Oxford writers of distinction joined the Oxford Group a little later: Andrew Linzey (who published *Animal Rights: A Christian Assessment* in 1976), Stephen Clark (*The Moral Status of Animals*, 1977; 1984, paperback ed.) and Peter Singer, who, as an Australian postgraduate student and lecturer at University College, Oxford (1969–1973), met the group and reviewed *Animals, Men, and Morals* for the *New York Review of Books* under the title "Animal Liberation" (April 5, 1973). So successful was this review that Singer was invited to publish a book on the subject, which he proceeded to do in 1975 while in New York lecturing in philosophy at New York University. This book took the message across the Atlantic.

The members of the Oxford Group also published and circulated leaflets and organized protests against animal experimentation; Ryder initiated lawful demonstrations against otter hunting* and hare coursing, sometimes supported by the Godlovitches and John Harris. (Otter hunting was outlawed in England in 1976.) In February 1975 Ryder's *Victims of Science* created a major stir in Britain and helped focus attention on speciesism* and, in particular, on the abuse of animals in research.

Linzey and Ryder instigated the first International Conference on Animal Rights, which was held at Trinity College, Cambridge in 1977, from which followed *Animal Rights: A Symposium* (1979), and both participated in the modernization of the Royal Society for the Prevention of Cruelty to Animals (RSPCA).* Associates of the group included Patrick Corbett, Mary Midgley, Colin McGinn, Jon Wynne-Tyson, Michael Peters, and David Wood. By 1978 the Oxford Group had dispersed, although Linzey went on to become the leading Christian theologian of animal rights and in 1993 returned to Oxford as International Federation of Animal Welfare (IFAW) Fellow at Mansfield College. Tom Regan, the preeminent American-born philosopher of animal rights, also passed through Oxford in 1973, where he came into contact with the movement.

Selected Bibliography. Godlovitch, Stanley, Roslind Godlovitch, and John Harris (Eds.), *Animals, Men, and Morals: An Enquiry into the Maltreatment of Non-Humans* (London: Victor Gollancz, 1971); Paterson, David, and Richard Ryder (Eds.), *Animal Rights: A Symposium* (London: Centaur Press, 1979); Regan, Tom, and Peter Singer (Eds.), *Animal Rights and Human Obligations* (Englewood Cliffs, NJ: Prentice-Hall, 1976); Ryder Richard D., *Animal Revolution: Changing Attitudes Towards Speciesism* (Oxford: Basil Blackwell, 1989); Ryder, Richard D., Speciesism [Leaflet], privately printed in Oxford, 1970; Ryder, Richard D., *Victims of Science: The Use of Animals in Research* (London: Davis-Poynter, 1975); Singer, Peter, *Animal Liberation* (New York: New York Review of Books, 1990).

RICHARD D. RYDER

P

PAIN

Pain is an unpleasant sensation or range of unpleasant sensations that can protect animals from physical damage or threats of damage from external forces. It involves specialized receptors (nociceptors) in the skin and viscera (body organs) that when stimulated, result in impulses passing along afferent nerves to the central nervous system (CNS), specifically to the cerebral cortex, where the actual feeling or experience of pain is felt. Rapid motor responses before the sensation of pain is actually felt by the animal (within tenths of a second), such as withdrawal of a limb, are spinal reflexes to the painful stimulus and help in the protective aspect of this sensory function. There are descending pain pathways from the brain that can moderate or gate afferent (sensory) impulses to the CNS, thus reducing the magnitude of any perceived pain. In animals who are self-aware or self-conscious (see ANIMAL COGNITION), there may be further integration of the afferent pain nerve impulses that reach the CNS through neurons connecting to other areas of the CNS so that earlier experiences are reflected in the conscious responses an animal may make independent of the rapid reflex response.

Responses to a painful stimulus that last for more than a few seconds are likely to represent an animal's conscious awareness of persistent pain, for example, vocalization, licking at the affected site, or rolling (as in colic). There is growing evidence, contrary to what was once thought, that very young animals, and even human fetuses in the last trimester of pregnancy, may feel pain. This is because the descending pain inhibitory pathways do not develop for some time after birth in many species, and so pain in such young animals and older fetuses may possibly be more than that felt by older animals whose nervous system has fully developed. Pain may be an integral part of other aspects of suffering*—an animal in pain from a broken leg may be fearful of being moved or touched, as well as being distressed by its inability to move normally.

Selected Bibliography. DeGrazia, D., and A. Rowan, Pain, Suffering, and Anxiety in Animals and Humans, *Theoretical Medicine* 12 (1991): 193–211; Fitzgerald, M., Neurobiology of Foetal and Neonatal Pain, in Patrick Wall and Ronald Melzack (Eds.), *Textbook of Pain*, 3rd ed. (London: Churchill Livingstone, 1994), 153–163; Institute of Laboratory Animal Resources and National Research Council, Committee on Pain and Distress in Laboratory Animals, *Recognition and Alleviation of Pain and Distress in Laboratory Animals* (Washington, DC: National Academy Press, 1992); International Association for the Study of Pain, Guidelines on Painful Experiments: Report of the International Association for the Study of Pain Subcommittee on Taxonomy, *Pain* 6 (1979): 249–252; Melzack, R., and P. Wall, *The Challenge of Pain* (Harmondsworth, UK: Penguin Books, 1982); Morton, D. B., Recognition and Assessment of Adverse Effects in Animals, in N. E. Johnston (Ed.), *Proceedings of Animals in Science Conference: Perspectives on Their Use, Care, and Welfare* (Melbourne, Australia: Monash University, 1995), 131–148; Morton, D. B., and P. H. M. Griffiths, Guidelines on the Recognition of Pain, Distress, and Discomfort in Experimental Animals and an Hypothesis for Assessment, *Veterinary Record* 116 (1985): 431–436.

DAVID B. MORTON

Experimental Analysis of Pain

The nervous system, including pain neural mechanisms, is similar across vertebrates, as are the basic processes that allow events to become learned signals for pain and to evoke "fear"*-mediated defense reactions. These similarities suggest that the neural bases of pain and fear (or anxiety) and their behavioral expression are evolutionarily old traits. Therefore, what we learn in animal experiments can lead us to an understanding of the human condition.

Whether all organisms can experience pain is a complex definitional matter that has defied widespread agreement. Certainly most all animal organisms, from unicellular to vertebrates, respond to contact with tissue-damaging stimuli. While all organisms respond to such stimuli, some argue that to experience "pain," the organism must have at least a *nervous system* (as does the planarian), or that it must have a *central* nervous system (CNS, as does the octopus), or that it must have a *cerebrally* anchored nervous system (as do all modern vertebrates). It is perhaps best to argue that so long as the individuals of a species (1) appear to react strongly to a stimulus and (2) can learn to anticipate that stimulus to defend against it, we should consider that the organism experiences pain.

Scientists study pain, including learning based on pain, because it is important to organisms' survival in a harmful and threatening world. Many different methods of producing experimental pain have been developed, standardized, and used to explore the physiological and psychological responses to pain. These include bone "crush," formalin injection, tail pinch, and electrocutaneous stimulations (commonly called "shocks"), among others. These methods have been used because they activate different aspects of the pain systems. The first two methods always cause tissue damage, the third does so only if prolonged, and the last does not, unless it is quite prolonged at extremely high intensities.

Electrocutaneous stimulation ("shock") is the most commonly used method in both physiological and behavioral experiments because our knowledge of physics allows very precise measurement and control of the stimulation. Also, at commonly used values, it directly activates the nociceptive/pain neuronal signal fibers called fast A fibers (which signal potential tissue damage) without any risk of tissue damage. These cutaneous electrical shocks "fool" the nervous system into responding as if it were in imminent danger of tissue damage, with the consequent activation of appropriate physiological, behavioral, and emotional systems to respond to this threat. It is this ability to elicit these responses without any genuine danger of tissue damage that has lead to the widespread use of electrical shocks in the study of pain and its physiological, behavioral, and psychological consequences.

Electrical shocks mimic some natural sources of pain. For example, when a strange rat intrudes into a colony of other rats, it is attacked and bitten. Such attacked rats show a pattern of physiological and behavioral changes that is exactly duplicated by subjecting a rat to a series of relatively brief electric shocks over which the rat has no control, but without the tissue injury inherent in the natural event. However, because studying pain often requires producing pain in animals, ethical questions are raised.

Selected Bibliography. Bolles, R. C., and M. Fanselow, A Perceptual-Defensive-Recuperative Model of Fear and Pain, *Behavioral and Brain Sciences* 3 (1980): 291–323; Gibson, R. H., Electrical Stimulation of Pain and Touch, in D. R. Kenshalo (Ed.), *The Skin Senses* (Springfield, IL: Charles C. Thomas, 1968), 223–261; Kelley, D. D., Central Representations of Pain and Analgesia, in E. Kandel and J. Schwartz (Eds.), *Principles of Neural Science*, 2nd ed. (New York: Elsevier, 1985), 331–343; Kitchell, R. L., H. H. Erickson, E. Carstens, and L. E. Davis (Eds.), *Animal Pain: Perception and Alleviation* (Bethesda, MD: American Physiological Society, 1983); Liebeskind, J., and I. Paul, Psychological and Physiological Mechanisms of Pain, *Annual Review of Psychology* 28 (1977): 41–60.

J. BRUCE OVERMIER

Invertebrates and Pain

While most people assume that vertebrates (animals with backbones) perceive pain, the situation is not as clear for most invertebrates (animals without backbones). However, the common octopus, with its large central nervous system and complex behaviors, has been given the benefit of the doubt in Great Britain and is now protected under the Animals (Scientific Procedures) Act of 1986.

Some argue that insects do not perceive pain but that it is difficult to be certain. For example, some researchers argue that insects do not perceive pain although they might still avoid some aversive stimuli. Others are also uncertain about insect pain but believe that insects should be given the benefit of the doubt. The conclusion that insects do not perceive pain is based on several lines of reasoning.

First, although insects have complex nervous systems, they lack the well-developed central processing mechanisms found in mammals and other vertebrates (and the octopus) that appear to be necessary to feel (perceive) pain. Second, insects have apparently not been shown to have a nerve fiber system equivalent to the nociceptive (pain) fibers found in mammals. However, this does not mean that they do not have some nerve fibers that carry nociceptive signals. Third, the behavior of insects when faced with noxious or harmful stimuli can usually be explained as a startle or nociceptive protective reflex. In some cases (for example, locusts being eaten by fellow locusts), insects display no signs that the tissue damage that is occurring is aversive.

The conclusion that insects do not perceive pain appears to contradict the claim that pain confers important survival advantages. However, simple nociceptor neural reflex loops (producing the startle reflex) that involve no pain perception could confer sufficient evolutionary advantage in short-lived animals (like insects) that rely on a survival strategy involving the production of very large numbers of individuals. If insects and most other invertebrates do not perceive pain, this would be relevant for ethical systems that rely on sentience as an important criterion of moral considerability.

Selected Bibliography. DeGrazia, D., and A. Rowan, Pain, Suffering, and Anxiety in Animals and Humans, *Theoretical Medicine* 12 (1991): 193–211; Eisemann, C. H., W. K. Jorgensen, D. J. Merrit, M. J. Rice, B. W. Cribb, P. D. Webb, and M. P. Zalucki, Do Insects Feel Pain? A Biological View, *Experientia* 40 (1984): 164–167; Fiorito, G., Is There Pain in Invertebrates? *Behavioral Processes* 12 (1986): 383–386; Lummis, S.C.R., GABA Receptors in Insects, *Comparative Biochemistry and Physiology, C* 95 (1990): 1–8; Wells, M. J., *Octopus* (London: Chapman and Hall, 1978); Wigglesworth, V. B., Do Insects Feel Pain? *Antenna* 4 (1980): 8–9; Young, J. Z., *The Anatomy of the Nervous System of Octopus Vulgaris* (Oxford: Clarendon Press, 1971); Young, J. Z., The Organization of a Memory System, *Proceedings of the Royal Society, Series B* 163 (1965): 285–320.

ANDREW N. ROWAN

Invasiveness Scales

A major consideration in the justification of an animal experiment is how much pain or suffering* the animal experiences. Among the questions to be answered are the following: How sick or incapacitated is the animal as the result of the experimental procedure? What is the duration and severity of the pain or distress*? Will the normal health or mental state of the animal be interfered with? What is the sum total of harms that will befall the animal? National policies of several countries require that the degree of animal pain and distress be assessed as either minor, moderate, or severe. Classification systems are variously called "invasiveness scales," "severity bandings," or, colloquially, "pain scales."

Classifying pain and harm in animal experiments is of fairly recent origin. In 1979, Sweden was the first country to adopt an invasiveness scale as national policy. Since then, several other countries have followed suit, including Canada, the Netherlands, Switzerland, Germany, Finland, Australia, and the United Kingdom.

Despite years of effort by animal protectionists to get such a system adopted, an invasiveness scale is not required by national policy in the United States. However, a few American institutional animal care and use committees (IACUCs)* do use it voluntarily in their review procedures. In 1987, an invasiveness scale was officially proposed by the U.S. Department of Agriculture, but was dropped because of opposition from the biomedical community. Opponents charged that such ranking is unworkable because it is too difficult, and that classifying animal pain and suffering goes beyond congressional intent. However, in 1996, the idea was revived, and this policy reform is actively sought. The rationale for this reform is the belief that assessment of the degree of animal pain and suffering is essential to judging ethical acceptability.

Recognizing and evaluating animal pain involves assessment of many factors. A number of people have described species-specific signs of pain in rats, rabbits, guinea pigs, dogs,* cats,* and monkeys. The signs include changes in posture or appearance, vocalizing, temperament, depression, locomotion, and immobility, as well as clinical signs in cardiovascular, respiratory, nervous, and musculoskeletal systems. A report of a United Kingdom working group recommended that a scoring system be used that ranks various factors such as whether or not the animal is conscious (see ANIMAL COGNITION, Conscious Experience of Animals) throughout the procedure; the use of restraint (its duration and whether it is continuous or discontinuous); tissue sensitivity; organ risk; mortality; level of pain; distress; deprivation of normal physiological function or activity; and other factors. The higher the combined score, the greater the severity. All of these indicators translate into minor, moderate, or severe ranking and present a continuum, with no clear dividing line between categories. However, over time, people who have studied the animals' responses to experimental procedures become fairly consistent in their judgments.

Invasiveness scales can guide the application of the Three Rs (*see* ALTERNATIVES TO ANIMAL EXPERIMENTS) when an investigator and IACUC seek to modify a proposed protocol by reducing the invasiveness. Sometimes modifications can lower the severity level from severe to moderate or from moderate to minor or to zero when a nonanimal model exists.

An invasiveness scale provides a conceptual basis for a policy on the use of animals in education.* One proposal is to link the degree of permitted invasiveness of a project with the educational level of the student. At the primary- and secondary-school level, a vast array of projects are available to teach the principles of biology that either use invertebrate species or that use vertebrate animals in noninvasive ways. At this level, infliction of any animal pain is not permissible. Only as their educational level advances (at the college level) should students be permitted to conduct minimally invasive vertebrate studies. At a later point in training, usually at the graduate-school level, the goal of the experiment shifts from educational to the search for new, significant knowledge. According to this view, only when the purpose of the experiment is to seek new knowledge and the investigator is highly trained should moderate levels of animal pain or suffering be permitted. Even so, restrictions on the level of "permissible" pain and suffering are needed. The following list summarizes different categories of invasiveness of animal experiments used in the Netherlands.

Procedures Having Minor Effect on Animals

Simple blood sampling

Vaginal smear sampling

Force-feeding of innocuous substances

Taking of X-rays in unanesthetized animals

Killing without prior sedation

Terminal experiments under anesthesia

Procedures Having Moderate Effect on Animals

Frequent blood sampling

Insertion of indwelling cannulae or catheters

Immobilization or restraint (e.g., primate chairs, inhalation chamber)

Skin transplantation

Caesarian section

Recovery from anesthesia

Procedures Having Severe Effect on Animals

Total bleeding without anesthesia

Production of genetic defects, e.g., muscular dystrophy or haemophilia

Prolonged deprivation of food, water, or sleep
Carcinogenicity research with tumor induction
Induction of convulsions
LD50 tests

Selected Bibliography. Laboratory Animal Science Association, Report of the Working Party on "The Assessment and Control of the Severity of Scientific Procedures on Laboratory Animals," *Laboratory Animals* 24 (1990): 97–130; Morton, D. B., and P. H. M. Griffiths, Guidelines on the Recognition of Pain, Distress, and Discomfort in Experimental Animals, and an Hypothesis for Assessment, *Veterinary Record* 116 (April 20, 1985): 431–436; Orlans, F. B., *In the Name of Science: Issues in Responsible Animal Experimentation* (New York: Oxford University Press, 1993), 118–127; Orlans, F. B., Invasiveness Scales for Animal Pain and Distress, *Lab Animal* 25(6) (June 1996): 23–25.

F. BARBARA ORLANS

PAINISM

"Painism" is a term coined by Richard Ryder to describe the theory that moral value is based upon the individual's experience of pain,* that pain is the only evil, and that the main moral objective is to reduce the pain of others, particularly that of the maximum sufferer. Painism is not a speciesistic view (*see* SPECIESISM). Furthermore, painism applies as a universal morality and not one limited only to certain areas of conduct such as the treatment of nonhuman animals. The concept of painism has the advantage of concentrating attention upon pain (suffering*). Ryder defines pain broadly to include all negative experiences, all forms of suffering. He uses the words "painient" and "painism" to mean, respectively, having the capacity to feel pain (and those possessing this capacity) and the principle that morality should be based upon such a capacity.

Utilitarianism* is based upon the recognition of the importance of pain. However, Ryder rejects the trading off of pains and pleasures between individuals that is a central feature of utilitarianism. Painism concentrates on the conscious experience of individuals (*see* ANIMAL COGNITION, Conscious Experience of Animals). Ryder recognizes that his ethical theory is an attempt to bring together different aspects of utilitarianism (its emphasis upon pain) with the rights tradition (its emphasis upon the supreme importance of the individual). Pleasures are also to be taken into account, but extremes of pain outweigh extremes of pleasure. Ryder agrees with the philosopher Jeremy Bentham that the morally important question is "Can they suffer?" not "Can they reason?"

The theory of painism has emerged from what was sometimes previously termed "sentientism."* Andrew Linzey had used this term approvingly, while John Rodman had attacked it on the grounds that it established too narrow a moral circle. Ryder eventually rejected "sentientism" in favor of "painism" on three grounds: (1) that "sentientism" might be deemed to refer to any sort of feeling or sensation; (2) that "sentientism" and "sentient" were words not popularly understood, whereas "painism" and "painient" could be easily grasped and would thus be of greater use politically; and (3) that these words usefully fill some significant gaps in the English language.

Selected Bibliography. Ryder, Richard D., *Painism: Ethics, Animal Rights, and Environmentalism* (Cardiff: University of Wales, 1991); Ryder, Richard D., Painism: The Ethics of Animal Rights and the Environment, in *Animal Welfare and the Environment* (London: Gerald Duckworth, 1992), 196–210; Ryder, Richard D., *Animal Revolution: Changing Attitudes towards Speciesism* (Oxford: Basil Blackwell, 1989).

RICHARD D. RYDER

PESTICIDES. *See* GENETIC ENGINEERING.

PET THEFT

In the late 1940s and early 1950s, as government funding for biomedical research increased, the demand for animals to use in research also grew. Commercial breeders of dogs* and cats* were virtually nonexistent. To fill the demand for dogs and cats, the research community turned to the city pound or shelter* or to dealers who acquired animals from pounds or other sources and then resold them to research facilities. This practice became known as "pound seizure."

As controversial as pound seizure was, it was not as controversial as pet theft. Unscrupulous individuals stole dogs and cats from suburban neighborhoods or rural farms. They found a large supply of free-roaming, unidentified pets and a huge demand from animal dealers who asked few questions about the sources of the animals. The person who stole the animals became known as a "buncher": He traveled around an area gathering up "bunches" of animals and then sold them to a dealer, who then sold them to a research facility.

In the mid-1960s, the theft of pets for use in research became so prevalent that Congress was asked to pass a law to stop the practice and to regulate individuals who sold dogs and cats for research purposes. In 1966, Congress passed the Research and Experimentation—Dogs and Cats Act, later to be retitled the Laboratory Animal Welfare Act.* But the practice of stealing animals for research or other purposes did not end. Attempts by Congress in the late 1980s to revisit the issue brought a denial from the U.S. Department of Agriculture (USDA), the agency charged with enforcing the Laboratory Animal Welfare Act, that there was still a problem. During the hearing, representatives of the Humane Society of the United States (HSUS),* the American Humane Association (AHA), and the Massachusetts

Society for the Prevention of Cruelty to Animals (MSPCA) testified that hundreds of animals had been stolen directly from their owners' property and sold for research purposes. Another illegal practice made known to the House members at the hearing was theft by deception. In this common scam, an unscrupulous individual responds to a "free to good home" ad and assures the owner that the pet will be given a new home in the country. In reality, the animal is sold immediately to a research supplier. In 1990, the Laboratory Animal Welfare Act was amended in a way that sought to cut off the supply of stolen pets for research and experimentation, but the practice has only slightly decreased by increasing pet owners' awareness of the need to protect their dogs and cats from thieves and those claiming to give them a new home.

Auction sales and trade days have become another growing source for stolen animals. Begun in the 1800s as a place for people to trade their wares—trading a handmade quilt for a few chickens* or a plow for a cow and a pig—these events were often a primary social gathering in the rural southeastern and midwestern United States. Evolving in the mid-1900s to become more of a giant flea market, they have now become a major transfer point for stolen pets. Investigations conducted by many local and national animal-protection organizations have uncovered thousands of illegally obtained dogs and cats being bought, sold, traded, and transferred hundreds of miles from their homes through one or more of these events.

Attempts to regulate auction sales and trade days have generally failed. The USDA, which questions whether stolen animals even move through these events, has neither the personnel nor the authority to police each trade day or auction sale. Further increasing the problem is the lack of authority that local humane organizations or animal control personnel have in trying to investigate these events.

Because many of the dealers who participate in trade days or auction sales are federally licensed by the USDA, state and local law-enforcement authorities mistakenly believe that they also do not have the power to step in. The USDA, however, has determined that state and local governments do have the right to pass and enforce laws that are stronger than the federal Laboratory Animal Welfare Act; in other words, federally licensed or registered facilities are not exempt from complying with state and local laws simply because their facilities are also regulated under the Laboratory Animal Welfare Act.

Two bills to address the issue of trade in stolen animals were stalled in the 104th Congress and died without any action taken. Another bill (H.R. 594) on the issue was filed in the 105th Congress. Sponsored by Congressmen Charles T. Canady (Republican–Florida) and George E. Brown (Democrat–California), the bill would abolish Class B dealers and seek to accomplish what the original Laboratory Animal Welfare Act of 1966 sought to do: end the practice of selling stolen pets for research purposes.

Selected Bibliography. Animal Welfare Institute, *Animals and Their Legal Rights: A Survey of American Laws from 1641 to 1990*, 4th ed. (Washington, DC: Animal Welfare Institute, 1990), 74; Concentration Camp for Dogs, *Life*, February 4, 1966; Dunn, Michael, Assistant Secretary, U.S. Department of Agriculture, correspondence to the Humane Society of the United States, June 1996; U.S. House of Representatives, Committee on Agriculture, Subcommittee on Department Operations, Research, and Foreign Agriculture, *Congressional Committee Hearing Record*, September 28, 1988; U.S. Senate, Committee on Commerce, *Senate Committee Hearing Record*, March 25 and 28 and May 25, 1966.

MARTHA ARMSTRONG

PETS. *See* COMPANION ANIMALS AND PETS.

PIGS

Domestic pigs are canny and sensitive animals, with strong urges to forage, explore, and interact socially. These characteristics were inherited from their ancestor, the Euro-Asian wild boar (*Sus scrofa* L.). Historically, pigs were either herded in woods or housed in pens. In the Euro-American civilization, they were always regarded with some scorn, which was probably often connected with rough treatment. Their way of life has been altered during the last 50 years by intensive husbandry and selective breeding. Through selection for fast growth and high-yielding carcass characteristics, pigs became heavier and more muscular, whereas the relative weight of bones and the heart decreased. They are prone to overheating and heart failure in stressful situations (see STRESS) and to leg problems, especially if they have little exercise and/or when they are housed on slippery or rough slatted floors. Breeding for fast growth also boosted pigs' appetite. While growing pigs and lactating sows can be fed to satiation, gestating sows cannot, because they will get fat. Hence they must be kept in a permanent, even if only "subjective," state of hunger.

Most pigs today are housed in barren environments that conflict with their behavioral makeup. The most pressing problems are the following:

Absence of bedding. Straw, which in older housing systems provided dry floor comfort, an outlet for exploratory and foraging activities, and a source of dietary fiber, has disappeared from most piggeries (*see* PREFERENCE AND MOTIVATION TESTING).

Restriction of movement. Almost all pregnant sows in North America and many in Europe are confined in small crates. This, combined with hunger and absence of bedding, leads to continual chewing on bars or other repetitive stereotypic behaviors (*see* STEREOTYPIES IN ANIMALS) and causes constant stress, as revealed by elevated levels of corticosteroid ("stress") hormones. Oral stereotypies could be reduced by a high-fiber diet, but this is rarely done. In small piglets, lack of space in small pens suppresses social play, which may hamper normal development of their social skills.

Thermoregulation. For adult pigs, temperatures above 25° C (77° F) pose a challenge, as they cannot sweat. In nature, they cool themselves by rolling in mud (wallowing).

Body cleanliness. If space allows, pigs defecate and urinate in one location and never lie in a fouled place. They are forced to do so, however, when they are kept in groups of high spatial density or confined in crates.

Social behavior. When unfamiliar pigs meet, they perceive each other as intruders, and intense fighting invariably begins. Numerous, although superficial, injuries are inflicted by biting. As confined spaces prevent the losing individuals from fleeing, attacks last several days, with the losers becoming distressed (*see* DISTRESS IN ANIMALS).

Farrowing and nursing. Hormonal changes preceding parturition prompt the sow to seek a half-hidden place and build a nest. Almost all parturient and lactating sows are housed in unbedded farrowing crates. The prevented locomotion and nest-building efforts result in agitation, futile nest-building movements, and elevated levels of the stress hormone cortisol.

Surgery on small piglets. The majority of piglets are subjected to tooth trimming, tail trimming, and castration (males). No anesthesia is given. Tooth trimming and tail trimming (*see* DOCKING) are performed to prevent damage to sow's teats and to the littermates, and to prevent mutual tail biting.

Weaning. While the natural age of weaning is 4 months, piglets are most often weaned at 3 to 5 weeks. The method of weaning at 8 to 16 days, based on strict hygiene and mandatory antibiotics in food, is becoming more common. However, weaning before 3 weeks of age causes intense distress reactions and disturbed behavior among the piglets, such as suckling-related belly nosing and nibbling of agemates.

Human-swine interactions. Rough treatment, such as hitting, kicking, and using pain-inflicting devices, makes pigs fearful of humans. They are then difficult to handle, get easily excited, and produce less well in terms of growth and reproduction.

Transport (*see also* TRANSPORTATION AND SLAUGHTER). Transportation is stressful to pigs. The strain may be severe or even fatal if pigs also experience exposure to extreme temperatures; long durations without water, food, and rest; mixing with alien pigs; overcrowding; and slippery floors. Regulations concerning animal transport are being gradually imposed, but unacceptable practices are still common.

Slaughter (see also TRANSPORTATION AND SLAUGHTER). Most industrialized countries require instantaneous stunning of pigs before slaughtering. It is the preslaughter handling and housing of pigs rather than the slaughter itself that causes considerable suffering because of its large scale, total anonymity, and the tendency among the personnel to depreciate the suffering.

Selected Bibliography. Fraser, A. F., and D. M. Broom, *Farm Animal Behaviour and Welfare*, 4th ed. (Wallingford: CAB International, 1993); Grandin, T. (Ed.), *Livestock Handling and Transport* (Wallingford: CAB International, 1993); Phillips, C., and D. Piggins (Eds.), *Farm Animals and the Environment* (Wallingford: CAB International, 1992); Recommended Code of Practice for Care and Handling of Pigs, Publication 1771/E, Agriculture Canada, Ottawa, 1984; Sainsbury, D., *Farm Animal Welfare: Cattle, Pigs, and Poultry* (London: Collins, 1986).

MAREK ŠPINKA

PITTMAN-ROBERTSON ACT. *See* HUNTING.

PLUTARCH

Plutarch (c. 46–c. 120) was a Greek philosopher famous for his *Lives*. His defense of the Pythagorean diet led him to expound the philosophical basis of vegetarianism.* Instead of asking why vegetarians abstain from meat, we should ask why flesh eaters consume animals: "For the sake of a little flesh we deprive them of sun, of light, of the duration of life to which they are entitled by birth and being" (*Moralia*, 535–579; Magel, *Keyguide*, 72). His other essays, "Whether Land or Sea Animals Are Cleverer" (*Moralia*, 309–479) and "Beasts Are Rational" (*Moralia*, 487–533), defend animal intelligence, their ties of kinship with humans, and especially their right to be treated justly.

Selected Bibliography. Magel, Charles, *Keyguide to Information Sources in Animal Rights* (London: Mansell Publishing; Jefferson, NC: McFarland and Company, 1989); Martinengo-Cesaresco, Evelyn, Plutarch the Humane, in *The Place of Animals in Human Thought* (London: T. Fisher Unwin, 1909); Plutarch, *Moralia*, trans. by H. Cherniss and W. C. Helmbold (London: William Heinemann; Cambridge, MA: Harvard University Press, 1968); Plutarch, On Eating Flesh [extracts], in Tom Regan and Peter Singer (Eds.), *Animal Rights and Human Obligations*, 1st ed. (Englewood Cliffs, NJ: Prentice-Hall International, 1976), 111–117; Sorabji, Richard, *Animal Minds and Human Morals: The Origins of the Western Debate* (London: Duckworth, 1993); Tsekourakis, D., Pythagoreanism or Platonism and Ancient Medicine? The Reasons for Vegetarianism in Plutarch's Moralia, in *Aufstieg und Niedergang der Römischen Welt*, 2.36.1., 366–393.

ANDREW LINZEY

POLYISM

Polyism is a term used to describe the failure to care for or empathize with animal suffering (*see* EMPATHY FOR ANIMALS) because of the large numbers involved, in contrast with the situation when only a few members of that species may be in a similar environment. Very often it is associated with the difficulty or impracticality of doing anything else—allowing fish* to die on the decks of trawlers due to being unable to absorb oxygen from air, or tolerating lameness, fractured limbs, and death* in the intensive farming (see FACTORY FARMING) of broiler chickens* kept in sheds in tens of thousands.

DAVID B. MORTON

PORPHYRY

Porphyry (232–309) was perhaps the strongest animal advocate in the Greek world. A devoted pupil of Plotinus, he wrote influential commentaries on Plato, Plotinus, and Aristotle. His work *On Abstinence from Animal Food* attacks not only animal sacrifice and meat eating but also culling animals, maintaining that such action is unnecessary because nature is a self-regulating system. Like Theophrastus* and Plutarch,* he rejected the denial of animal rationality and their kinship with us that were features of Aristotle's philosophy. In *Against the Christians*, of which only fragments survive, he argues that Jesus was "not much of a saviour" since he allowed the swine to plunge over the cliff to their death (commentary on Matthew 8:28–34; Sorabji, *Animal Minds*, 181), though by endorsing Christianity's preference for spiritual, rather than blood, sacrifice, he confirms that the early church rejected animal sacrifice. For Porphyry, God was a spiritual being who could only be properly worshipped through spiritual sacrifices.

Selected Bibliography. Porphyry, *Abstinence from Animal Food*, ed. Esme Wynne-Tyson, trans. T. Taylor (London: Centaur Press, 1965); Sorabji, Richard, *Animal Minds and Human Morals: The Origins of the Western Debate* (London: Duckworth, 1993).

ANDREW LINZEY

POULTRY. *See* CHICKENS.

PREDATION

Predation refers to animals killing other animals for food. Most animal rights* philosophers argue that humans should stop being predators; they claim that we have a moral obligation to become vegetarians (see VEGE-TARIANISM). Questions that arise relate to other animal predators: Is there a moral obligation for them to stop preying on other animals, and is there a moral obligation for humans to interfere with other animal predators?

There are two lines of reasoning that lead to the conclusion that, like humans, other animals should not be predators. First, the animals who are killed for food suffer both the fear of being hunted and the pain* of being killed, often in gruesome ways. These animals also suffer the loss of the rest of their lives, which could have been happy lives. Since one of the basic goals of one view of morality (*see* UTILITARIANISM) is to reduce suffering* and increase happiness, the world would be a morally better place without predation.

The second line of reasoning to this conclusion starts from the idea that, like humans, animals have a right to life. It would be wrong to kill humans for food, because that would violate their right to life. Similarly, a rabbit's or a gazelle's right to life is violated when it is killed for food by a fox or a lion. Violations of rights are morally very serious matters; they should not happen. Thus predation should stop.

Both of these lines of reasoning have been put forward in attempts to discredit animal rights* philosophy. A standard way of discrediting a proposal is to show that following it would lead to an absurd conclusion. Since the organization of nature depends on one thing living by killing another, including some animals preying on others, the idea that this process involves a violation of rights or is otherwise fundamentally immoral and should be stopped is an absurd idea for many people, including most, if not all, animal rights advocates. Consequently, these advocates have responded to the two arguments that animals should not be predators by attempting to show that the absurd conclusion does not follow from animal rights principles. They have made the following two counterarguments.

First, it is argued that animal rights principles concern only how humans should treat animals; they do not concern how nonhuman animals should treat each other. Ideas of moral rights and obligations arise only in situations where there are beings who can recognize rights and obligations and regulate their behavior accordingly—they are moral agents (*see* MORAL AGENCY AND ANIMALS). However, it can be argued that although foxes cannot recognize the rights of rabbits, humans can, and since we know that foxes kill rabbits, the situation is not limited to just those nonhuman animals. The situation is like the obligation adults have to prevent young children from being cruel to animals. Young children cannot recognize moral rights and obligations; nonetheless, it is still wrong for them to torment and kill rabbits. Adults who see what the children are doing should step in to protect rabbits from being killed by the children. Similarly, humans can have an obligation to protect rabbits from being killed by foxes, even though the foxes cannot understand moral concepts.

The second counterargument is stronger. Stopping predation would not reduce suffering and increase happiness. The only way to stop predation would be to kill all the predators. Also, the populations of many animals previously killed by predators would then increase dramatically. These extra animals would then die of disease and starvation; thus they would suffer.

Consequently, although animals suffer pain and loss when they are killed by other animals, that is a lesser evil than would occur if we were to try to prevent predation.

Since most, but not all, humans can live on a vegetarian diet, we can eliminate the suffering and death caused by most human predation. However, the natural order is not one that can exist without suffering and death. Most predation by nonhuman animals is necessary for the survival of life on earth, and so it cannot be eliminated.

Selected Bibliography. Callicott, J. Baird, Animal Liberation: A Triangular Affair, in *In Defense of the Land Ethic* (Albany: State University of New York Press, 1989); Rollin, Bernard, Must We Police Creation? and Don't Animals Kill Each Other? in *Animal Rights and Human Morality* (Buffalo, NY: Prometheus Books, 1981); Rolston, Holmes, III, Values Gone Wild, in *Philosophy Gone Wild* (Buffalo, NY: Prometheus Books, 1986); Salt, Henry S., *Animals' Rights* (London: George Bell and Sons, 1892); Sapontzis, S. F., Saving the Rabbit from the Fox, in *Morals, Reason, and Animals* (Philadelphia: Temple University Press, 1987).

STEVE F. SAPONTZIS

Nonrightist's View

A major issue with which animal rightists (*see* ANIMAL RIGHTS) have difficulty is the ubiquitous torture inflicted by predators, especially when large mammals kill animals that can be regarded as attractive. Witnessing the pain* of these animals can be disturbing. With insects, such as ants killing a beetle, the predation can be regarded as instinctive and as involving little in the way of pain. When an owl or a fox chomps on a mouse, they can be seen as providing control over what could otherwise be a disastrous overpopulation of vermin.

Animal rightists seem to find it difficult to support enthusiastically the efforts of government and other agencies that result in increasing the number of those predators that are listed as endangered species.* These wolves, condors, chimpanzees,* lions, and the like exact grisly tolls. It seems a reasonable question to ask how one can argue for a universal reducing of the pain of animals when so much of it is engendered by the activities of animal predators. Of course, wild animals can suffer from diseases and die of starvation before becoming prey or carrion, but animal rightists have been known to ignore the fact that predation is prevalent in the activities of many animals.

It would also not make sense to claim that human beings have none of the propensities of predatory primates; the story of civilization is, to no small extent, about how society has developed means of coping with homicidal tendencies, and we still have much to do with regard to our problems of war and crime. How society is to regard animal life involves the use of a principle that we are capable of appreciating fear,* pain, or suffering* of large captive animals and have been ready to carefully develop protective regulations so as to avoid the cruelty. But it is important to deal with the matter selectively: for instance, we do not discourage bloody surgery because so many people would find it revolting to watch, nor do we engage in criticism of the use of animal skins by Eskimos living in cold climates. The killing of microbes is an important human activity. There have been unbridled attacks on furriers and gastronomes who have been spitefully harassed because some do not approve of their uses of animals.

Many rightists believe that they have developed a workable theme that includes care for almost every living nonhuman creature. They have to blind themselves to the wide range of predatory activity going on all the time around us on earth. They do not approve of hunting,* no matter how circumspect. There are those who believe every sentient (*see* SENTIENTISM) organism should have rights as humans do. Some leave the impression that they would rather have starving people die than be granted approval to eat an animal. A prescription that might help reorient some animal rightists and might even help them think kindly about the selective use of animals by human beings is to ponder the reality of predation and how it kills the generalized animal rights theme.

Selected Bibliography. Davis, H., and D. Balfour (Eds.), *The Inevitable Bond* (New York: Cambridge University Press, 1992); Kerasote, T., *Bloodties: Nature, Culture, and the Hunt* (New York: Random House, 1993); Lansdell, H., Laboratory Animals Need Only Humane Treatment: Animal "Rights" May Debase Human Rights, *International Journal of Neuroscience* 42 (1988): 169–178; Lansdell, H., The Three Rs: A Restrictive and Refutable Rigmarole, *Ethics and Behavior* 3 (1993): 177–185; Lutherer, L. O., and M. S. Simon, *Targeted: The Anatomy of an Animal Rights Attack* (Norman: University of Oklahoma Press, 1992); Marquardt, K., *AnimalScam: The Beastly Abuse of Human Rights* (Washington, DC: Regnery Gateway, 1993); Pluhar, E. B., *Beyond Prejudice: The Moral Significance of Human and Nonhuman Animals* (Durham, NC: Duke University Press, 1995); Rollin, B. E., *Animal Rights and Human Morality* (Buffalo, NY: Prometheus Books, 1981); Rowan, A. N., and B. E. Rollin, Animal Research—For and Against: A Philosophical, Social, and Historical Perspective, *Perspectives in Biology and Medicine* 27 (1983): 1–17.

HERBERT LANSDELL

PREFERENCE AND MOTIVATION TESTING

In a preference test, experimenters give animals a choice of two or more different options or environments and then monitor the animals' behavior to determine which alternative they select. Preference testing has been used in many ways in animal welfare* research. Animals' preferences have been established for air temperature, for type and level of light, and for common materials used in cage or pen design. The methods have also been used to assess how strongly animals seek to avoid aspects of animal handling such

Preference and Motivation Testing: A preference experiment in which pigs were given free access to two pens, one with a bare concrete floor and one bedded in straw. Photo courtesy of the Department of Agriculture and Agri-Food Canada, Government of Canada.

as noise, vibration, and various forms of restraint. Such knowledge has allowed more effective design of animal housing and handling equipment.

On the surface, determining the preferences of animals seems like a simple task, but the simplicity is more apparent than real. For example, the preference of pigs* for straw-bedded pens turns out to be remarkably complex. Pigs strongly prefer straw when they are actively foraging; they are indifferent to straw when they are using a food or water dispenser; and they either select or avoid a bedded floor as a resting area depending on whether the environment is cool or warm. Furthermore, mature sows take a sudden interest in straw when they are building a nest just before giving birth. To characterize the animals' preferences, we need a comprehensive study that asks how this preference varies with the animal's age, reproductive state, and ongoing behavior and with fluctuations in the environment.

Experiments also need to identify how animals' preferences are affected by their previous experience. In the short term, animals may show a temporary avoidance of, or attraction to, novel options; these temporary reactions should not be used to infer longer-term preferences.

Various methods have been used to assess the strength of an animal's preferences or its degree of motivation to obtain the preferred option. In some experiments, animals are trained to press a lever or peck a key to obtain a reward such as a larger cage or access to social companions; then the "price" can be increased by requiring more and more lever presses for the same reward. This method can be used to compare, for example, the animal's motivation for a larger cage versus its motivation to eat or drink. In other experiments, animals have been required to push against a weighted door to gain access to a preferred cage. This method literally measures how much effort they will expend to obtain a given reward.

In using preference research to asses and improve animal welfare, we normally assume that animals will prefer those environments or options that promote their health and psychological well-being.* This is often true, but there are exceptions. For example, many fish* species avoid being harmed by aquatic pollutants such as copper simply by swimming away from contaminated water, but the same species may fail to avoid other contaminants such as phenol even at levels that cause serious damage or death. Problems may also arise if a choice requires a level or type of cognitive ability that the animal does not possess (see ANIMAL COGNITION). We cannot, for example, expect animals to weigh up the short-term and long-term benefits of making a particular choice.

Because they seem intuitively simple, preference tests are potentially very influential. As noted by Ian Duncan in a criticism of some of the early preference testing, the argument that the animal itself prefers a given option is very convincing in public discussion of animal welfare. It is important, therefore, that animal welfare scientists ensure that preference-testing methods are used and interpreted appropriately so that misleading conclusions are avoided.

Selected Bibliography. Dawkins, M. S., *Animal Suffering* (London: Chapman and Hall, 1980); Dawkins, M. S., From an Animal's Point of View: Motivation, Fitness, and Animal Welfare, *Behavioral and Brain Sciences* 13 (1990): 1–61; Duncan, I. J. H., The Interpretation of Preference Tests in Animal Behavior, *Applied Animal Ethology* 4 (1978): 197–200; Duncan, I. J. H., Measuring Preferences and the Strength of Preferences, *Poultry Science* 71 (1992): 658–663; Fraser, D., Preference and Motivational Testing to Improve Animal Well-being, *Lab Animal* 25 (1996): 27–31.

DAVID FRASER

PRIMATT, HUMPHRY

Humphry Primatt (c. 1725– c. 1780) was an 18th-century divine and historically influential zoophile. His work *The Duty of Mercy and the Sin of Cruelty to Brute Animals* (1776) is the first systematic theology of the status of animals using arguments derived from reason and revelation. Of particular significance is his anticipation of the modern argument for equal consideration of interests based on sentiency (see SENTIENTISM). "Pain is pain, whether it be inflicted on man or on beast; and the creature that suffers it, whether man or beast, being sensible of it whilst it lasts, suffers evil" (1992 edition, 21). Also significant is his sophisticated theological interpretation of the generosity of God as the basis for human moral generosity toward animals (*see* GENEROSITY PARADIGM). Primatt was an inspiration to Arthur Broome,* who founded the first society for the prevention of cruelty to animals in 1824 and who published an abridged edition of Primatt's work in 1831. Primatt served various churches in Suffolk and Norfolk and became doctor of divinity at Aberdeen University in 1773. *The Duty of Mercy* is presumably based on his doctoral dissertation and is his only known work.

Selected Bibliography. Primatt, Humphry, *The Duty of Mercy and the Sin of Cruelty to Brute Animals*, 2nd rev. ed. by Arthur Broome (Edinburgh: T. Constable, 1832).

ANDREW LINZEY

Q

QUALITY OF LIFE. *See* WELL-BEING OF ANIMALS.

R

REGULATION 3254/91. *See* TRAPPING.

RELIGION AND ANIMALS

Animal Theology

Animal theology relates Christian thinking to contemporary debates about the status and rights of the nonhuman animals (*see* ANIMAL RIGHTS). It seeks to address and redress the failure of historical theology to take seriously alternative insights that lie largely silent within the Christian tradition (*see* RELIGION AND ANIMALS, Christianity). Systematic theology has largely proceeded on the basis of the virtual nonexistence of animals. Historically, animals have been the outcasts of theology, defined as beings with no mind, reason, immortal soul, or moral status (*see* MORAL STANDING OF ANIMALS). Basic questions about their status and significance have simply not been addressed. The question raised by animal theology is whether Christian doctrine is necessarily speciesist (*see* SPECIESISM) and whether it can incorporate animal-centered concerns into mainstream thinking. Modern theologians argue variously that even conservative theological understandings can be enhanced and deepened by the adoption and development of these insights.

In terms of traditional doctrine, there are three main areas. The first is *creation.* Much theological emphasis has been laid on the special creation of humans to the detriment of the nonhumans. But the "specialness" of humanity in creation can be read another way: as support for the special role of humanity in looking after the world, not as the master but as the servant species. The second is *incarnation*. Traditional doctrine affirms that God became human in the person of Jesus Christ. While this is frequently taken as a vindication of human uniqueness, some church fathers have argued that the incarnation is the raising up of all fleshly substance (*ousia*) to be with God: the Word becoming flesh affirms all flesh, animal and human. The third is *redemption*. While much traditional interpretation excludes animals directly or indirectly from the sphere of God's redemptive purposes, it can be argued that notions of ultimate justice specifically require animal immortality (*see* RELIGION AND ANIMALS, Theodicy). Viewed from this threefold perspective, God creates, unites, and redeems all living beings, and the focus of this divine work is not just the human species but specifically sentient (*see* SENTIENTISM), fleshly creatures.

Apart from the plausibility of these reinterpretations, there is one reason why theology needs to take animals more seriously. It lies in the traditional claim that the *Logos* is the source of all life, because if so, it must follow that a theology based on the *Logos* must be able to render an account not just of the human species but the entire created universe. In other words, the implicit promise of traditional theology is that it will deliver us from humanocentricity (*see* ANTHROPOCENTRISM).

Selected Bibliography. Linzey, Andrew, *Animal Theology* (London: SCM Press; Urbana: University of Illinois Press, 1995); Linzey, Andrew, *Christianity and the Rights of Animals* (London: SPCK; New York: Crossroad, 1987); McDaniel, Jay B., *Of God and Pelicans: A Theology of Reverence for Life* (Louisville: Westminster/John Knox Press, 1989); Pinches, Charles, and Jay B. McDaniel (Eds.), *Good News for Animals? Christian Approaches to Animal Well-being* (Maryknoll, NY: Orbis Books, 1993); Webb, Steve, *Pet Theories: A Theology of Animal Passion* (New York: Oxford University Press, 1997).

ANDREW LINZEY

Buddhism

The Buddhist tradition is a varied series of religious phenomena, and few valid generalizations are possible. Attitudes toward other animals, however, are one of the few areas where generalizations can be made. Generally the Buddhist tradition was unconcerned with any systematic exploration of the physical world, including the realities of other animals. It accepted most of the views of other animals that were important in the cultures and subcultures where Buddhism developed.

At its core Buddhism is a salvation-like concern (usually referred to as "liberation") for the individual. Theoretically, each individual Buddhist attempts to discover about himself or herself the basic features of existence experienced by Gotama, the historical Buddha of the 5th or 6th century before the current era (B.C.E.). The core of this experience was that each living being has, in the end, no lasting self. Similarly, there is no lasting deity or creator of the earth. Instead, all is in process and subject to change.

The unifying elements in the tradition are (1) reverence of some kind for Gotama and (2) a strong, consistent, hermitlike tradition that adheres to time-honored rules of conduct. It is this tradition that has provided a relative unity and stability in the moral code.

The Buddhist monastic code (known as the *Vinaya*) reveals that early Buddhists accepted the view that all animals other than humans belong to one realm that is lower than that of human beings. Yet it is clear that Buddhists did not know other animals well despite their claim that all nonhuman animals, from the simplest of karmic forms on up to the most complex such as the large-brained social mammals, form a single kingdom that does not include humans. As noted later, the tradition displayed poor awareness of the elaborate realities of the lives of other animals, lumping them together in a group below humans in the hierarchy of the universe.

In one very important way, however, Buddhism was clearly revolutionary with regard to the moral significance of other animals, for Buddhism, along with Jainism (*see* RELIGION AND ANIMALS, Jainism), was important in opposing the sacrifice* of other animals that was part of the brahminical tradition in India. Similarly, the tradition spread important precepts, or moral undertakings, that affirmed that killing other sentient beings was a violation of the most basic moral norms of the universe. The first precept in the tradition is "I undertake to abstain from the destruction of life." This is an ethical commitment that the tradition has from its very beginnings identified as part of the core of religious living. Society for a Buddhist, then, is not to be taken in the narrow sense of human society, but in a broader sense of a community comprising all living or sentient beings.

There is another, less favorable side to the Buddhist view of other animals, however. The way in which early Buddhists talked about other animals reveals that they thought about them in rather negative ways. For Buddhists, any animal other than a human was in an inferior position and could, if it lived a perfect life, be reborn as a human. This reliance on reincarnation as an explanation of the justice of the present state of any being also functioned as a justification of many of the social divisions of the day, although Gotama resisted the notion that humans in the lower social divisions were less important than high-status individuals. But rich humans were deemed to have been rewarded for past good deeds, and lame, stupid, and unfortunate humans were deemed to be paying for past bad acts. Below even the most unfortunate and morally corrupt humans were all other animals.

The Buddhist tradition's attitude toward uses of other animals reveals these same hierarchical notions of life. Elephants, whose natural history was poorly known by Buddhists, were used with Buddhist approval. Indeed, Gotama himself understood that use of elephants was a morally based reward. Rich humans were entitled to ride around on elephants, having lived past lives in such a way as to justly deserve this reward. Sadly, the Buddhist scriptures contain many indications that elephants suffered during captivity, being deprived of their naturally complex social lives with other elephants (*see* CIRCUSES AND CIRCUS ELEPHANTS).

Selected Bibliography. Chalmers, R. (Trans.), *Further Dialogues of the Buddha (Translated from the Pali of the Majjhima Nikaaya)*, 2 vols., Sacred Books of the Buddhists series, 5 and 6 (London: Humphrey Milford/Oxford University Press, 1926–1927); Gombrich, Richard, The Buddhist Way, in Heinz Bechert and Richard Gombrich (Eds.), *The World of Buddhism: Buddhist Monks and Nuns in Society and Culture* (London: Thames and Hudson, 1991), 9–14; Gombrich, Richard, *Theravada Buddhism: A Social History from Ancient Benares to Modern Colombo* (London and New York: Routledge, 1988); Keown, Damien, *Buddhism and Bioethics* (London: Macmillan; New York: St. Martin's Press, 1995); Keown, Damien, *The Nature of Buddhist Ethics* (London: Macmillan, 1992); Schmithausen, Lambert, *Buddhism and Nature: The Lecture Delivered on the Occasion of the EXPO 1990: An Enlarged Version with Notes* (Tokyo: International Institute for Buddhist Studies, 1991); Story, Francis, *The Place of Animals in Buddhism* (Kandy, Ceylon: Buddhist Publication Society, 1964); Waldau, Paul, Buddhism and Animal Rights, in Damien Keown (Ed.), *Buddhism and Contemporary Issues* (Oxford: Oxford University Press, 1997); Williams, Paul, *Mahayana Buddhism: The Doctrinal Foundations* (London and New York: Routledge, 1994).

PAUL WALDAU

Christianity

Many of the important ideas that have governed our understanding and treatment of animals arise from Christian and Jewish (*see* RELIGION AND ANIMALS, Judaism) sources or from reaction to, development of, or opposition to them. Many zoophiles (animal lovers or, more broadly, those who care for animals) maintain that Christian indifference has been one of the main causes of the low status of animals. Within the Christian tradition in almost every period of history there were both strong negative and positive ideas and attitudes toward animals (*see* ATTITUDES TOWARD ANIMALS, Pre-Christian Attitudes, Changing Attitudes throughout History). Though it is true that largely negative ideas have predominated, it would be false to suppose that subtraditions have not sustained alternative viewpoints and sometimes radical ones.

There are three major negative tendencies. The first may be called instrumentalism, the view that animals are here for human use. St. Thomas Aquinas, interpreting Aristotle, held that in the created hierarchy that God had made animals were the intellectual inferiors of humans and were made essentially for human use. According to this view, the purpose of animals was primarily, if not exclusively, for the service of human subjects.

Second, and allied to instrumentalism, there has been a consistent humanocentricity (*see* ANTHROPOCENTRISM) that has effectively defined animals out of the moral picture. This has been achieved largely through the emphasis upon certain perceived differences between humans and animals. Animals are judged as beings with no reason or immortal soul and incapable of friendship with human subjects. From this it has been deduced that humans have no direct duties to animals because they are not moral subjects of worth in themselves. Many contemporary secular theories, for example, contractualism, owe their origin to this developing Scholastic view that animals do not form part of a moral community with human beings.

The third tendency may be described as dualism—the way Western culture has made distinctions and separations between, for example, the rational and nonrational, flesh and spirit, and mind and matter. Animals are still viewed as being on the wrong side of these desirable attributes, the most important of which has been rationality. As Scholastic philosophy and theology began to stress the centrality of rational intellect (and since it was almost universally accepted that animals had none), it followed that animals had no moral status. Rationality became, and in many ways still is, the key to moral significance (see MORAL STANDING OF ANIMALS).

But in order to see the broader picture, we need to set alongside these negative tendencies a range of positive insights, many of which are clearly biblical in origin. Three are presented here. The first centers on the notion of "dominion" found in Genesis 1:28 (see DOMINIONISM). Although dominion has often been interpreted as little less than tyranny, in original context it meant that humans had a God-given responsibility to care for the earth (confirmed by the fact that the subsequent verses command a vegetarian diet and envisage a world in Sabbath harmony). A rival interpretation of dominion as "stewardship" or responsibility can be traced back to the earliest Christian writers and came to the fore in the emergence of 18th- and 19th-century zoophily. The second concerns the notion of "covenant" found in Genesis 9. Against the prevailing notion that humans and animals are utterly separate, the idea of God's covenant with all living creatures kept alive the sense of a wider kinship. The third positive insight is preserved in the notion of moral generosity (see GENEROSITY PARADIGM), which came to prominence in the emergence of humanitarian movements of the 19th century. According to this perspective, we owe animals charity, benevolence, and merciful treatment. Cruelty* was judged incompatible with Christian discipleship: to act cruelly, or even to kill wantonly, was ungenerous, a practical sign of ingratitude to the Creator (see BROOME, ARTHUR; PRIMATT, HUMPHRY; SHAFTESBURY). The Christian tradition, which had in many ways supported, defended, and provided the ideological justification for the abuse of animals in previous centuries, came to spearhead a new movement for animal protection.

Selected Bibliography. Clarke, P. A. B., and Andrew Linzey (Eds.), *Political Theory and Animal Rights* (London and Winchester, MA: Pluto Press, 1990); Gunton, Colin E., *Christ and Creation: The Didsbury Lectures* (London: Paternoster Press, 1992); Joranson, Philip N., and Ken Butigan (Eds.), *Cry of the Environment: Rebuilding the Christian Creation Tradition* (Santa Fe: Bear and Company, 1984); Linzey, Andrew,

Christianity and the Rights of Animals (London: SPCK; New York: Crossroad, 1987); Linzey, Andrew, and Dan Cohn-Sherbok, *Celebrating Animals in Judaism and Christianity* (London: Cassell, 1997); Linzey, Andrew, and Tom Regan (Eds.), *Animals and Christianity: A Book of Readings* (London: SPCK, 1989; New York; Crossroad, 1989); Murray, Robert, *The Cosmic Covenant: Biblical Themes of Justice, Peace, and the Integrity of Creation* (London: Sheed and Ward, 1992); Santmire, H. Paul, *The Travail of Nature: The Ambiguous Ecological Promise of Christian Theology* (Philadelphia: Fortress Press, 1985); Thomas, Keith, *Man and the Natural World: A History of the Modern Sensibility* (New York: Pantheon Books, 1983).

ANDREW LINZEY

Disensoulment

Disensoulment is the stripping away of the spirit powers or souls of animals and of the sanctity of the living world. This process occurred over the centuries as early herders and farmers intensively exploited animals and nature and needed new myths and other psychic levers to resolve their very old beliefs in animals as First Beings, teachers, tribal ancestors, and the souls of the living world (*see* ANIMAL PRESENCE).

In the ancient Middle East, the cradle of Western culture, where animal husbandry was the key to nation and wealth building, agrarian societies invented misothery* and other ideas that aided the debasement of animals. There, the builders of the bustling city-states preached misothery in their arts and in their rising, new agrarian religions. In these, the essential message was to debase animals and nature and to elevate human beings over them. The effect, spiritually speaking, was to turn the world upside down: before domestication,* the powerful souls or supernaturals (or "gods") were animals, and primal people looked up to them; after domestication, the gods were "humanoid," and people looked down on animals. In primal culture, all beings had souls, of which the greatest was the tribe's totem animal; in agriculture, humans alone have souls, and god is in human form. Animal-using agrarians stripped animals of their souls and powers and put them in what they perceived to be their proper place: far beneath—and in the service of—humankind.

Selected Bibliography. Campbell, Joseph, *The Way of the Seeded Earth*, vol. 2 of *Historical Atlas of World Mythology* (San Francisco: Harper and Row, 1988); Eisler, Riane, *The Chalice and the Blade* (San Francisco: Harper and Row, 1987); Fisher, Elizabeth, *Woman's Creation* (Garden City, NY: Anchor Press/Doubleday, 1979); Lerner, Gerda, *The Creation of Patriarchy* (New York: Oxford University Press, 1986).

JIM MASON

Hinduism

Hinduism, the oldest of the major religious traditions, is not a single religion, but an umbrella under which one finds very different kinds of beliefs. These include, among others, Vaishnavism, Shaivism, Shaaktism, and

Tantrism, each of which in turn is a complex religious tradition that has many forms of its own. The term "Hinduism" was coined by European scholars in the 19th century as a description of native beliefs, other than Buddhism and Islam (*see* RELIGION AND ANIMALS, Buddhism, Islam), that occurred in the Indian subcontinent. These beliefs are internally diverse, such that nontheistic beliefs coexist with theistic and devotional beliefs.

In Hinduism there is no single view of other animals. The different views are dominated by two general beliefs that govern the ways in which other animals are conceived. First, human beings, though recognized to be in a continuum (*see* CONTINUITY) with other animals, are considered the model of what biological life should be. A corollary of this first belief is the claim that the status "human" is far above the status of any other animal. The second general belief is that any living being's current position in the cycle of life (created by repeated incarnations) is determined by the strict law of karma. Belief in reincarnation is the hallmark of most, though not all, Hindus' beliefs. These two beliefs have resulted in other animals being viewed with uncertainty. Positively, other animals have been understood to have souls just as do humans. Negatively, they have been understood to be inferior to any human, a corollary of which is the belief that the existence of other animals must be particularly unhappy, at least compared to human existence.

Importantly, humans are by no means considered equal to one another in classical Hinduism, for according to the *sanatana dharma* (the eternal law or moral structure of the universe) men are not born equal. Like other animals, they are born into that station in life for which their past karma has fitted them. Inequalities required by the social system are not viewed as unjust; rather, they are simply the result of good or bad deeds performed in former lives. A common claim is that those who act morally are assured of a good rebirth in higher social classes, while wrongdoers are assured of being reborn into the wombs of outcasts or, worse yet, a nonhuman animal.

Despite all this, the tradition has often exhibited great sensitivity to other animals. In the Srima Bhagavantam, the believer is told, "One should treat animals such as deer, camels, asses, monkeys, snakes, birds and flies exactly like one's own children" (7.14.9; Prime, 51). A contemporary Hindu who is an environmental ethicist argues, "All lives, human or nonhuman, are of equal value and all have the same right to existence" (Dwivedi, 203). More generally, the economics of village life in India provide many examples of coexistence with other animals and environmentally sensitive ways of living.

The tradition has vast sources, and some do support the view that humans have no special privilege or authority over other creatures, but instead have more obligations. This argument relies on the belief that many Hindu deities, such as Rama and Krishna (closely associated with monkeys and cows, respectively), have been incarnated as other animals. In addition, the deities worshipped in India include Ganesh, an elephant-headed god, and Hanuman, the monkey god.

This sensitive side in Hindus' awareness of other animals is often symbolized by the image of sacred cows wandering the streets of India unmolested and free; yet the realities for animals in Hindu societies have been and continue to be far more complicated. The traditional respect for other animals has been affected greatly by economic factors that inhibit transmission of ancient values that encourage respect for other animals. Nowadays, the pace of modern development is leaving behind the strong emphasis that almost all Hindu scriptures place on the notion that benefits can be received by not killing or harming other animals. Thus, while there is throughout the Hindu tradition a culturally significant sense of the continuity of all life, the already-pronounced sense of discontinuity between humans and all other animals threatens to change for the worse.

One important ancient form of the tradition (sometimes known as brahminical religion) was challenged by the Buddhist and Jain (*see* RELIGION AND ANIMALS, Jainism) traditions because it was characterized by a heavy emphasis on animal sacrifice.* This practice stemmed from the ancient scriptures known as the Vedas. The Jains and Buddhists challenged these sacrifices as cruel and unethical and had a great effect on the later Hindu views of the decency of intentionally sacrificing other animals. *Ahimsa*, the historically important emphasis on nonviolence, has now become a central feature of the tradition.

Hindu social codes, embodied in the Laws of Manu, reflect the one-dimensional view of other animals as completely inferior to humans. This belief that all other animals are qualitatively inferior to any human is also reflected in the myths of the origin of other animals, which one important myth (the Purusa Sukta in the important Rig Veda) attributes to remnants of a primal male (*purusa*) sacrificed by the gods. Thus in the Hindu tradition, as with the Buddhists and with Plato (*Timaeus*) in the West, other animals derive their origins from, and are a degenerate form of, elevated humanity.

Selected Bibliography. Basham, A. L., *The Sacred Cow: The Evolution of Classical Hinduism*, ed. Kenneth G. Zisk (London: Rider, 1990); Chapple, Christopher Key, *Nonviolence to Animals, Earth, and Self in Asian Traditions* (Albany: State University of New York Press, 1993); Dwivedi, O. P., Satyagraha for Conservation: Awakening the Spirit of Hinduism, in J. Ronald Engel and Joan Gibb Engel (Eds.), *Ethics of Environment and Development: Global Challenge, International Response* (London: Bellhaven Press, 1990), 202–212; Hardy, Friedhelm, *The Religious Culture of India: Power, Love, and Wisdom* (Cambridge: Cambridge University Press, 1994); Prime, Ranchor, *Hinduism and Ecology: Seeds of Truth* (London: Cassell, 1992); Zaehner, R. C., *Hinduism* (New York: Oxford University Press, 1966).

PAUL WALDAU

Islam

Islam, along with Judaism and Christianity (*see* RELIGION AND ANI-MALS, Christianity, Judaism), is a member of the larger category of Abrahamic religious traditions. This group of traditions is dominated by an ethical anthropocentrism,* although the human-centeredness of each tradition is modified by important insights into the moral dimension of (1) other animals' lives and (2) humans' instrumental use of other animals. Each of the Abrahamic traditions treads the delicate balance between true theocentrism and a reasoned but hidden speciesism.*

In the Islamic tradition, much emphasis is given to the importance of humans as the center of the universe, with other animals having been put here for the benefit of humans. But recognition of a moral dimension of other animals, as well as in humans' treatment of creatures who are deemed creatures of Allah (the Arabic equivalent to the English word "God"), does play an important role in the tradition. For example, there are passages in the principal Islamic scripture, the Qur'an, as well as in other important writings of the tradition such as the Hadith (the traditional collection relating the actions and sayings of Mohammed and his companions) and the Shari'ah (or "way," the body of legal provisions), that recognize that other animals are not solely for human use and have their own importance as Allah's creatures.

Mohammed himself commented, "Whoever is kind to the creatures of Allah, is kind to himself," and he compared the doing of good or bad deeds to other animals with similar acts done to humans. Other animals might be said to have a high profile in the tradition, for there are many Surahs (chapters in the Qur'an) named after animals ("The Cow," "The Cattle," "The Bees," and "The Elephants").

Negative views of other animals appear in some Qur'anic passages; negative views also appear in the beliefs of various sects that infidels after death become other animals or that hell is full of noxious nonhuman animals. The practice of public, ceremonial slaughter of other animals for food (*dhabh*), which occurs at the end of Ramadan, the traditional month of fasting, and at other times when the meat is used for a celebrative feast and often distributed to the poor, reflects the basic belief that humans are the vice-regent (*Khalifah*) of Allah and other animals are for their use. Rules designed to make the killing more humane moderate the metamessage that humans are the only animals that really matter.

Environmental concern, which can benefit other animals even if other animals are not the direct concern (*see* ENVIRONMENTAL ETHICS), is another possible route by which interpretations of core passages in the tradition can favor other animals. The tradition offers both legal and ethical reasons for protecting the environment, although there is no agreement that Islam is, at its core, inclined to such protections. Some have argued that concern for nature is anything but conspicuous in the Qur'an, while others have argued that Muslims have a strong tradition of earth sensitivity.

It is difficult to find bases for animal rights* in the Islamic tradition because its perception of the moral norm of the universe has been interpreted in such an anthropocentric manner (*see* ANTHROPOCENTRISM). Islamic ethics are based on two principles, the first of which is a general understanding of human nature, and the second of which is a combination of religious and legal grounds. Because the Islamic tradition's fundamental ethical values are held to be revealed, accurate, and unalterable, the mechanisms that are available to foster acceptance of the increasing knowledge about perceptions of other animals' lives and abilities are not yet well developed. The existing patterns of reasoning continue to start from the view that it is humans who really matter. Accordingly, the possibilities of developing views that other animals do not need to serve human interests in any way are few.

Selected Bibliography. Deen (Samarrai), Mawil Y. Izzi, Islamic Environmental Ethics, Law, and Society, in J. Ronald Engel and Joan Gibb Engel (Eds.), *Ethics of Environment and Development: Global Challenge, International Response* (London: Bellhaven Press, 1990), 189–198; Deen (Samarrai), Mawil Y. Izzi, Islamic Ethics and the Environment, in Fazlun Khalid and Joanne O'Brien (Eds.), *Islam and Ecology* (New York and London: Cassell, 1992), 25–35; Khalid, Fazlun, and Joanne O'Brien (Eds.), *Islam and Ecology* (New York and London: Cassell, 1992); Masri, B. A., *Animals in Islam* (Petersfield, England: Athene Trust, 1989); Masri, B. A., *Islamic Concern for Animals* (Petersfield, England: Athene Trust, 1987); Nasr, Seyyed Hossein, *Man and Nature* (London: Unwin Paperbacks, 1990); Williams, John Alden, *Islam* (New York: Washington Square Press, 1963).

PAUL WALDAU

Jainism

One of the world's oldest religions, Jainism is also distinguished as one of the faiths that cares the most about nonhuman animals. It is a religion without God that yet holds that our souls can become gods through liberation. It is said that our souls accumulate karman particles through both good and bad actions, which make good or bad things, respectively, happen to us in turn. The goal is to cease all passions and actions that generate good and bad karma, as these particles literally make us too heavy to leave the realm of rebirth. The soul that has escaped the cycle of rebirth ascends to a permanent resting place at the very apex of the Jaina universe. The key to achieving divine liberation is by practicing *ahimsa*, or avoiding injury to all life. The positive side of this is a reverence for all life (*See* RELIGION AND ANIMALS, Reverence for Life) or a universal love for all creatures.

If one acts badly in a lifetime, one could be reborn as a primitive being. There are simple one-sense beings with only a sense of touch (e.g., plants and microscopic *nagodas*, which come in the form of earth bodies, water bodies, fire bodies, and wind bodies), two-sense beings who also have taste (e.g., worms, leeches), three-sense beings who can also see (e.g., ants, moths), four-sense beings who smell things as well (e.g., bees, flies, mosquitoes), and five-sense beings who hear in addition to the other senses (e.g., fish, dolphins, elephants, or any being born in a womb). There are rational and nonrational five-sense beings.

Inflicting injury on these creatures is wrong because of the suffering* caused, and also because it produces passions in the killer leading to karma and rebirth. The Jains condemn all animal sacrifices,* build animal shelters,* and never hunt or fish. Farming, which injures insects, is permitted because the harm is unintentional, but Jain monks beg with a bowl so crumbs will not attract insects that would be crushed underfoot. Monks brush their path to sweep away small life forms they might otherwise step on. It is prohibited to breed destructive animals and considered noble to allow oneself to be bitten by a snake rather than kill it. Jains are vegetarians (*see* VEGETARIANISM), but consume milk.

Selected Bibliography. Dundas, Paul, *The Jains* (New York: Routledge, 1992); Gopalan, S., *Outlines of Jainism* (New York: Halsted Press, 1973); Jain, Jyotiprasad, *Religion and Culture of the Jains* (New Delhi: Bharatiya Jnanpith, 1975); Jaini, Padmanabh S., *The Jaina Path of Purification* (Berkeley: University of California Press, 1979); Mardia, K. V., *The Scientific Foundations of Jainism* (Delhi: Motilal Banarsidass, 1990).

DAVID SZTYBEL

Judaism

From ancient times Judaism has expressed concern for the welfare of animals. This principle is referred to as *tsa'ar ba'alei chayim* (do not cause sorrow to living creatures). Since the animal kingdom is part of God's creation, human beings are to exercise responsibility for their care. Thus the Book of Genesis declares that humankind is to dominate all living things (1:26–28). Here the concept of dominance is interpreted as stewardship: humans are to ensure that all living creatures are treated humanely. Such an attitude is exemplified in the Torah, which lists various laws governing the treatment of animals. The Book of Deuteronomy, for example, states that when an ox is threshing grain, it should be allowed to eat what has been beaten out (25:4). Again, Deuteronomy 22:1–3 states that all Israelites are to look after domestic animals that have been lost. Such kindness toward the beasts of the field should be extended to other living things. Specific legislation is also put forth to ensure that animals will be protected in other circumstances.

Following such biblical commands, the rabbis of later centuries emphasized the need for animal welfare*; in their view, all living things are part of the created order and therefore require special consideration. Maimonides, a 12th-century Jewish philosopher, stated, "It is . . . prohibited to kill an animal with its young on the same day . . . for the pain of the animals under such circumstances is very great." Such a concern not to cause animals pain* is reflected in the various prescriptions regarding killing of animals for food. In the Jewish tradition, meat eating is regarded as giving in to human weakness; in this light, animals must be spared pain when they are slaughtered. Only a properly qualified slaughterer is permitted to engage in such an activity; he is to be a pious and sensitive person. The knife used must be sharp and clean without imperfections so that animals are slaughtered as painlessly as possible; the act of slaughter should render the animal senseless. Although arguably more humane methods of slaughter have been introduced in the modern world involving prestunning (*see* TRANSPORTATION AND SLAUGHTER), this ancient practice was intended to cause as little suffering as possible. Such concern about animal welfare is reflected in a variety of incidents in which the rabbis expressed the importance of preventing cruelty to animals. These acts of compassion were perceived as equivalent to prayers. According to tradition, vegetarianism* is the ideal state that existed in the Garden of Eden and will prevail in the Messianic Age. Increasingly, Jews from across the religious spectrum are embracing this form of consumption.

The primary source dealing with animal experimentation is the commentary of Rabbi Moses Isserles in the Code of Jewish Law. Here he states that animal experiments are permissible only if they advance human welfare. The principle of *tsa'ar ba'alei chayim* would rule out such scientific procedures for inessential human needs and would encourage the pursuit of alternative methods of research.

The principle of compassion for all living creatures similarly applies to hunting.* Judaism categorically condemns all forms of hunting for pleasure, including fox hunting, bullfights, dogfights, and cockfights. In the same spirit, the Jewish tradition is opposed to killing animals for their pelts: hence the Jewish faith would condemn such practices as using bone-crushing leg-hold traps (*see* TRAPPING) to capture wild animals or clubbing baby seals and skinning them while alive.

Selected Bibliography. Berman, Louis, *Vegetarianism and the Jewish Tradition* (Hoboken, NJ: Ktav, 1981); Cohen, Noah, *Tsa'ar Ba'ale Hayim: The Prevention of Cruelty to Animals: Its Bases, Development, and Legislation in Hebrew Literature* (Spring Valley, NY: Feldheim, 1979); Kalechofsky, Roberta (Ed.), *Judaism and Animal Rights* (Marblehead, MA: Micah Publications, 1992); Kalechofsky, Roberta, and Richard H. Schwartz, *Vegetarian Judaism: A Guide for Everyone* (Marblehead, MA: Micah Publications, 1988); Phillips, A., Animals and the Torah, *Expository Times*, June 1995.

DAN COHN-SHERBOK

Judaism and Animal Sacrifice

During biblical times animal sacrifice (*zebach*) was practiced as part of Jewish religious observance. Animals were offered to God as an institutionalized means of relief from the impurity generated by human violations of moral rules or purity taboos. The animals selected for sacrifice were those that were deemed useful to humans, and both anthropomorphism* and anthropocentrism* can be seen in the description of these animals, and not others, as "pleasing to God." The well-known "Thou shall not kill" was not thereby violated because, in the Hebrew tradition, this moral rule is interpreted as "Thou shall not kill unlawfully." Methods for lawful killing are defined by the Torah ("law"), which contains the written code with 613 laws of ethical human behavior, and by the later oral tradition and rabbinical commentary. The practice of animal sacrifice was discontinued after the destruction of the second temple by the Romans in 70 C.E.

Another view of sacrifice appears in the tradition's self-criticism, although in this criticism there has been little emphasis on the obvious point that it was cruel to the individual animals. Maimonides, a 12th-century Jewish philosopher, argued that sacrifices were a concession to barbarism. Some modern theologians continue to argue that sacrifice "in its way" represented respect for animal life. A more balanced observation is that sacrifice does not necessarily involve a low view of the sacrificed animals' lives (Linzey, *Christianity and the Rights of Animals*, 41). This is plausible given that the tradition contains powerful passages recognizing that the blood of humans and other animals is sacred (for example, Leviticus 17:10). Ultimately, Judaism moved away from this practice, though Orthodox Jewish prayer books to this day ask for a reestablishment of the temple sacrifices, and there remain rules governing ritual slaughter (*shechita*) by a specially trained religious functionary (*shochet*).

The occurrence of these instrumental uses of other animals and ultimate rejection of the old sacrificial practices are of limited value in assessing Judaism's views of other animals, as they deal with only a few domestic animals (*see* DOMESTICATION). There were many other complex animals with which the Jewish tradition was unfamiliar.

Selected Bibliography. Clark, Bill, "The Range of the Mountains Is His Pasture": Environmental Ethics in Israel, in J. Ronald Engel and Joan Gibb Engel (Eds.), *Ethics of Environment and Development: Global Challenge, International Response* (London: Bellhaven Press, 1990), 183–188; Kalechofsky, Roberta, *Judaism and Animal Rights: Classical and Contemporary Responses* (Marblehead, MA: Micah Publications, 1992); Linzey, Andrew, *Christianity and the Rights of Animals* (New York: Crossroad, 1987); Maimonides, *The Guide for the Perplexed*, trans. M. Friedlander (New York: Dover Publications, 1956); Murray, Robert, *The Cosmic Covenant: Biblical Themes of Justice, Peace, and the Integrity of Creation* (London: Sheed and Ward, 1992); Schwartz, Richard H., *Judaism and Vegetarianism* (Marblehead, MA: Micah Publications, 1988).

PAUL WALDAU

Reverence for Life

Reverence for life is a concept pioneered by the Alsatian theologian and philosopher Albert Schweitzer* in 1922. According to Schweitzer, ethics consists in experiencing a "compulsion to show to all will-to-live the same basic reverence as I do to my own." The relevance of Schweitzer's thought to modern debates about animals is immense. According to Schweitzer, other life forms have a value independent of ourselves, and our moral obligation follows from the experience and apprehension of this value. This insight is essentially religious in character and therefore basic and nonnegotiable. Schweitzer was undoubtedly prophetic. "The time is coming," he wrote, "when people will be astonished that mankind needed so long a time to learn to regard thoughtless injury to life as incompatible with ethics."

Selected Bibliography. Linzey, Andrew, *Animal Theology* (London: SCM Press; Urbana: University of Illinois Press, 1995); Linzey, Andrew, Moral Education and Reverence for Life, in David A. Paterson (Ed.), *Humane Education: A Symposium* (London: Humane Education Council, 1981), 117–125; Schweitzer, Albert, The Ethics of Reverence for Life [extract], in Andrew Linzey and Tom Regan (Eds.), *Animals and Christianity: A Book of Readings* (London: SPCK, 1989; New York: Crossroad, 1989), 118–120, 121–133; Schweitzer, Albert, *Reverence for Life*, trans. R. H. Fuller, foreword by D. E. Trueblood (London: SPCK, 1970).

ANDREW LINZEY

Saints

There is a remarkable range of material linking Christian saints with animals. The stories of St. Francis of Assisi preaching to the birds and St. Anthony of Padua preaching to the fishes are well known. Much less well known are the stories, to take just a few examples, of St. Columba and the crane or St. Brendan and the sea monster. Most scholars and theologians have dismissed this wealth of material as legend or folklore, but its significance, historically and theologically, can be noted. First, it is testimony to a widespread positive tradition within Christianity that has linked spirituality with a benevolent and sensitive regard for animals. The underlying rationale for this study of saints appears to be that as individuals grow in love and communion with their Creator, so too ought they to grow in union and respect for animals as God's creatures. Something like two-thirds of canonized saints East and West apparently befriended animals, healed them from suffering,* assisted them in difficulty, and celebrated their life through prayer and preaching. Second, despite the negative tradition within Christianity that has frequently downgraded animals, regarding them, at its very worst, as irrational instruments of the Devil, literature on these saints makes clear God's benevolent concern for other than human creatures and the common origin of all life in God. Third, because of this common origin in

God, it necessarily follows that there is a relatedness, a kinship between humans and nonhumans. According to St. Bonaventure, St. Francis was able to call creatures "by the name of brother or sister because he knew they had the same source as himself." Fourth, many of these stories prefigure a world of peaceful relations between humans and animals where human activity is no longer injurious or detrimental to other creatures. St. Brendan's voyage, for example, culminates in the discovery of a new Eden-like land characterized by the absence of predatory nature (see PREDATION) and widespread vegetarianism.* Such stories are testimonies to a substratum within Christianity that is inclusive of concern for animal life. The ideas they embody of respect, generosity (see GENEROSITY PARADIGM), and kinship between species reflect the themes that mainstream Scholastic tradition has almost entirely failed to incorporate into its thinking.

Selected Bibliography. Butler, Alban, *Lives of the Saints*, revised by Herbert Thurston and Donald Attwater, 4 vols. (New York: P. J. Kenedy and Sons, 1946); Linzey, Andrew, and Dan Cohn-Sherbok, *Celebrating Animals in Judaism and Christianity* (London: Cassell, 1997); Low, Mary, *Celtic Christianity and Nature: Early Irish and Hebridean Traditions* (Edinburgh: Edinburgh University Press, 1996); Sorrell, Roger D., *St. Francis of Assisi and Nature: Tradition and Innovation in Western Christian Attitudes toward the Environment* (New York: Oxford University Press, 1988); Waddell, Helen, *Beasts and Saints* (1947), rev. ed. by Esther De Vaal (London: Darton, Longman and Todd, 1995).

ANDREW LINZEY

Theodicy

Theodicy comes from the Greek words *theos* (god) and *dike* (justice) and is a branch of theology concerned with exploring and defending the justice of God in relation to physical and moral evil. Theodical issues are frequently at the heart of debates about animal rights* and animal welfare* and are used both positively and negatively in encouraging or discouraging concern for animal suffering.* A great deal of historical theology has utilized theodical arguments negatively in ways that seem to satisfy the claim that God is just and good but at the expense of animals. The first negative type solves the problem of animal pain* by effectively denying its existence. Historically, Cartesianism (see DESCARTES, RENÉ) has played a vital part in the development of this argument, but it has not lacked modern adherents. For example, Charles Raven argued that "it may be doubted whether there is any real pain without a frontal cortex, a fore-plan in mind, and a love which can put itself in the place of another; and these are the attributes of humanity." Clearly there can be no problem of animal pain to solve if such pain is illusory. The second negative type admits of some animal pain but minimizes its significance morally. For example, John Hick holds that animal pain is necessarily different from human pain because animals cannot anticipate death.* "Death is not a problem to the animals. . . . We may indeed say of them 'Death is not an injury rather life a privilege.' " Clearly, if death is not a "problem" to animals, then the moral significance of killing is necessarily reduced. The third negative type also admits of the existence of animal pain but denies its significance theologically. For example, Peter Geach holds that God is essentially "indifferent" to animal pain. "The Creator's mind, as manifest in the living world, seems to be characterized by mere indifference to the pain that the elaborate interlocking teleologies of life involve." This appeal to the world as it now exists has historically been one of the major theodical arguments against animal welfare. In the crisp summary of Samuel Pufendorf: "For it is a safe conclusion from the fact that the Creator established no common right between man and brutes that no injury is done brutes if they are hurt by man, since God himself made such a state to exist between man and brutes." Such an argument finds its contemporary and largely secular expression in an ecological form of theodicy that maintains that since nature is essentially predatory (*see* PREDATION), we should abide by nature's "rules." Nature's perceived "law" is baptized into "natural" or "moral" law.

Alongside these negative types, there are positive ones too. Here are three examples. The first is that animal pain and predation, far from being the Creator's will, are actually contrary to it. C. S. Lewis,* for example, held that both animal pain and carnivorousness were the result of "Satanic corruption" of the earth before the emergence of human beings (see *The Problem of Pain*). It follows that humans therefore have a duty not to imitate such malevolent distortion and to fight against it. The second is that while the Creator allows pain in creation (both animal and human) as an inevitable corollary of the freedom allowed to creation itself, such pain will eventually be transformed by a greater joy beyond death. Keith Ward, for example, holds that "immortality, for animals as well as humans, is a necessary condition of any acceptable theodicy" and that "necessity, together with all the other arguments for God, is one of the main reasons for believing in immortality." Such a prospect both maintains the ultimate justice of God and justifies the alleviation of pain (as an anticipation of God's final will) in the present. The third form of positive theodicy maintains that the God revealed in the suffering of Jesus suffers with all innocents, whether human or animal, in this world and will redeem all such suffering. From this perspective, Andrew Linzey concludes that the "uniqueness of humanity consists in its ability to become the servant species," that is, "co-participants and co-workers with God in the redemption of the world." Far from being indifferent to suffering, God is seen as manifest within it, beckoning human creatures to active compassion to remove the causes of it.

However we may judge the satisfactoriness of these negative or positive theodicies, it is inevitable that ethical concern for animals will continue to be influenced by one or more of them in one form or another. Concern for

animal suffering rarely stands by itself as a philosophical position and requires the support of some form of meta-ethical framework in which the
problem of a specific injustice can be properly recognized and addressed only
within the context of a sufficiently comprehensive vision of ultimate justice
for all.

Selected Bibliography. Geach, Peter, *Providence and Evil* (Cambridge: Cambridge
University Press, 1977); Hick, John, *Evil and the God of Love* (1966), Fontana ed.
(London: Collins, 1967); Kingston, A. Richard, Theodicy and Animal Welfare, *The-
ology* 70(569) (November 1967): 482–488; Lewis, C. S., *The Problem of Pain* (London:
Geoffrey Bles, 1940); Linzey, Andrew, *Animal Theology* (London: SCM Press;
Urbana: University of Illinois Press, 1995); Linzey, Andrew, and Tom Regan (Eds.),
Animals and Christianity: A Book of Readings (London: SPCK, 1989; New York: Cross-
road, 1989), which includes extracts from Hick, Geach, Lewis, and Ward in The
Problem of Animal Pain, 39–78, and The Question of Animal Redemption, 81–109;
Raven, Charles E., *The Creator Spirit* (London: M. Hopkinson, 1927); Ward, Keith,
Rational Theology and the Creativity of God (Oxford: Blackwell, 1982).

ANDREW LINZEY

Theos-Rights

Theos-rights denotes God's (theos) own rights as Creator to have what is
created treated with respect. According to this perspective, rights are not
awarded, negotiated, or granted, but *recognized* as something God-given.
Comparatively little attention has been devoted to the theological basis of
animal rights (*see* RELIGION AND ANIMALS, Animal Theology), though
it offers a coherent theoretical basis for the intrinsic value of (especially)
sentient (*see* SENTIENTISM) beings. Whereas in secular ethics, rights are
usually correlative of duties, for example, if A has a duty toward B, it usually
follows that B has a right against A, in theological ethics the reverse may be
claimed. For example, Dietrich Bonhoeffer maintains that "we must speak
first of the rights of natural life, in other words of what is given to life and
only later of what is demanded of life." Rights thus may be characterized as
what are given to creatures by their Creator to which humans owe a primary
obligation. The value of theos-rights lies conceptually in the way in which
it frees ethical thinking from humanocentricity. As Andrew Linzey writes:
"According to theos-rights what we do to animals is not simply a matter of
taste or convenience or philanthropy. When we speak of animal rights we
conceptualize what is objectively owed to animals as a matter of justice by
virtue of their Creator's right. Animals can be wronged because their Creator
can be wronged in his creation." Although some Christians oppose the language of rights altogether as unbiblical or contrary to creation construed as
"grace," the notion of rights has a long history in theological ethics. Thomas
Tryon was probably the first to use it in a specifically theological context
relating to animals (1688), but it continues to be used in modern contexts

as well. For example, Cardinal John Heenan stressed that "animals have very positive rights because they are God's creatures. . . . God has the right to have all creatures treated with proper respect."

Selected Bibliography. Bonhoeffer, Dietrich, *Ethics*, 2nd ed. (London: SCM Press; New York: Macmillan, 1971); Heenan, John, Foreword to Ambrose Agius, *God's Animals* (London: Catholic Study Circle for Animal Welfare, 1970), 2–3; Linzey, Andrew, *Animal Theology* (London: SCM Press; Urbana: University of Illinois Press, 1995); Linzey, Andrew, *Christianity and the Rights of Animals* (London: SPCK; New York: Crossroad, 1987), 68–98; Tryon, Thomas, Complaints of the Birds and Fowls of Heaven to Their Creator, in *The Country-Man's Companion* (London: Andrew Sowle, 1688).

ANDREW LINZEY

REPRODUCTIVE CONTROL

Controlling reproduction is often considered desirable when populations of animals become large or are in competition with increasing human populations, and lethal methods to reduce populations are thought unacceptable by many people. For example, greater and more diverse segments of the public want to be involved in controlling populations of wildlife. No longer accepting that killing is the only option, the public is demanding humane methods of population control. In one study of deer, residential property owners favored contraception over trapping* and transferring, hunting,* or allowing nature to take its course. Methods to control reproduction in wildlife should not only be safe for the species targeted, but for all other species who may eat contraceptives placed in their habitat. Researchers who develop wildlife contraceptives have challenging problems to overcome. Because many of their "experimental subjects" are animals on public lands, they must receive wide support from the public for any investigational study. Such support will likely be obtained only if treated animals are evaluated on a long-term basis, assuring the public that each proposed method of reproductive control is safe and humane. Because the experimental subjects (the wildlife) and the research laboratory (the parks and forests) belong to the public, such public acceptance is crucial in order for those who want to control reproduction to succeed.

There is also a surplus of pets (dogs* and cats*; *see* COMPANION ANIMALS AND PETS) in the United States. The most common method for controlling reproduction in pet animals is the surgical neuter operation, called a spay procedure in female animals and a castration in male animals. The spay procedure consists of surgically removing both ovaries and the uterus (ovariohysterectomy). The castration procedure is performed on male dogs and cats and consists of surgically removing both testes. Spaying and castrating do not influence behaviors such as play behavior, fear-related aggression, or friendliness. Behaviors such as urine marking or spraying, roaming, mounting, and male-male aggression are reduced through surgical neutering.

In order to control reproduction, it is desirable to neuter animals prior to puberty to prevent unwanted litters of puppies or kittens. Recent research suggests that it is safe to neuter as early as 6 to 8 weeks of age. In a survey of 500 pet-owning households, nearly 20% of all neutered pets had been allowed to produce offspring before sterilization. If neutering occurs after a pet has had one or two litters, the problem of pet overpopulation continues to thrive. For example, if cats are allowed to reproduce for only one year, 2 cats can be the progenitors of over 170,000 cats in seven years. If cats are allowed to continue to reproduce, 2 cats can be the progenitors of over 400,000 cats in seven years. Therefore, for any surgical or nonsurgical method of preventing pregnancy to maximally control pet reproduction, the method should be used prior to puberty, which can occur prior to 6 months of age in some dogs and cats.

Drugs or newer technologies may provide for nonsurgical approaches for controlling animal populations. Unfortunately, the cost of such alternatives may be based on the price of a surgical neuter operation as drug or biotechnology companies strive to make profits. Therefore, although nonsurgical methods to control reproduction may be less invasive and perhaps more humane, they may not be used by a wider group of pet owners if cost remains a significant barrier.

Attempting to control reproduction in animals is frequently accompanied by other economic and political factors. As one source of puppies decreases (litters of puppies in humane shelters*), consumers will likely find alternative sources of pets. RU 486, the controversial drug that terminates human pregnancies, is seemingly effective and safe in dogs. However, because of the controversy surrounding the use of RU 486 for people, it remains doubtful whether veterinarians* will have access to this drug.

Selected Bibliography. Hetts, S., *Behavioral Effects of Spaying and Neutering: The Case for Early Neutering* (Englewood, CO: American Humane Association, 1996); MSPCA Spay/Neuter Survey Summary (Boston: MSPCA, 1991); Olson, P. N., *The Case for Early Neutering* (Englewood, CO: American Humane Association, 1996); Olson, P. N., and S. D. Johnston, New Developments in Small Animal Population Control, *Journal of the American Veterinary Medical Association* 202(6) (1993): 904–909; Stout, R. J., and B. A. Knuth, *Effects of a Suburban Deer Management Communication Program, with Emphasis on Attitudes and Opinions of Suburban Residents* (Ithaca, NY: Human Dimensions Research Unit, Department of Natural Resources, Cornell University, 1995); Theran, P., Early-Age Neutering of Dogs and Cats, *Journal of the American Veterinary Medical Association* 202(6) (1993): 914–917.

PATRICIA OLSON

REPTILES

The class Reptilia includes turtles, squamates (lizards, snakes, and relatives), crocodilians, and two recognized species of the highly protected tuatara. Reptiles (other than birds, which many experts now also classify as reptiles) share several traits, including being ectothermic (dependent on external sources of heat) and covered with hard plates, scales, or bony shells. Reptiles live in almost all habitats except year-round subfreezing or deep-sea environments. Reptiles have adapted to many conditions, exploit a wide range of food items with diverse foraging methods, and have evolved diverse social systems. All tuataras, turtles, and crocodilians lay eggs, the latter also showing highly developed nest guarding and posthatching parental care. Many squamate reptiles, which constitute about 95% of all reptile species, give live birth.

The abilities of reptiles to learn, suffer (see SUFFERING OF ANIMALS), communicate, play, and socialize are generally underestimated, even by many herpetologists—scientists who study them. This mistake is made by not realizing that although reptiles do not have complex facial or vocal repertoires, tactile, chemical, and whole-body visual displays are common and important in communication. The metabolic rate of reptiles is about 10% that of mammals and birds, and thus their behavior is often slow (for example, land turtles) or sporadic, although there are many exceptions. Furthermore, reptiles are often ecologically specialized and critically dependent upon having proper temperature, humidity, diets, lighting, substrates, perches, retreats, and other captive arrangements to stimulate normal activity (see ENRICHMENT FOR ANIMALS). Knowing their natural behavior aids greatly in providing appropriate captive conditions for reptiles.

Reptiles are growing in popularity as pets, especially green iguanas, box turtles, boas, and pythons. A major problem is that the behavioral, nutritional, environmental, medical, and psychological needs* of reptiles are very different from ours and those of our common companion animals* (dogs,* cats,* rodents). This leads to many problems and the premature deaths of literally thousands of animals each year. For example, reptiles can go much longer without food than other vertebrates, and many slowly starve to death or succumb to poor nutrition, insufficient temperatures for digesting food, or lighting with inadequate ultraviolet radiation.

Reptiles possess many traits that are useful in answering important questions in animal biology and behavior. Snakes possess chemosensory abilities more acute than those of most other terrestrial vertebrates. Reptiles can be both short- and long-lived, have behavior patterns that can be measured and recorded easily, and are important ecological components of many habitats where they occur. Many species are affected by habitat loss or changes due to human activity. Many reptiles are also killed directly by people; others are exploited for food, skins, and the pet trade in numbers that threaten the survival of many species, including once-common species of turtles in North America. Social, foraging, and antipredator (defensive) behavior can differ greatly within and between closely related forms, especially in squamates. Thus it is very difficult to generalize across species, raising problems in maintaining many species in captivity, developing effective conservation plans, and studying their behavior and understanding the way they experience their lives.

Selected Bibliography. Breen, J. F., *Encyclopedia of Reptiles and Amphibians* (Neptune City, NJ: T.F.H. Publications, 1974); Burghardt, G. M., Of Iguanas and Dinosaurs: Social Behavior and Communication in Neonate Reptiles, *American Zoologist* 17 (1977): 177–190; Greenberg, N., G. M. Burghardt, D. Crews, E. Font, R. Jones, and G. Vaughan, Reptile Models for Biomedical Research, in A. Woodhead (Ed.), *Nonmammalian Animal Models for Biomedical Research* (Boca Raton, FL: CRC Press, 1989), 290–308; Schaeffer, D. O., K. M. Kleinow, and L. Krulisch (Eds.), *The Care and Use of Amphibians, Reptiles, and Fish in Research* (Bethesda, MD: Scientists Center for Animal Welfare, 1992); Warwick, C., F. L. Frye, and J. B. Murphy (Eds.), *Health and Welfare of Captive Reptiles* (London: Chapman and Hall, 1995).

GORDON M. BURGHARDT

RESEARCH ANIMALS. *See* LABORATORY ANIMAL USE.

REVERENCE FOR LIFE. *See* RELIGION AND ANIMALS; SCHWEITZER, ALBERT.

RIGHTS. *See* ANIMAL RIGHTS.

RODEOS

Rodeo, a tradition from the days of the American trail and range cowboy, is extremely popular throughout certain areas of the western United States. Rodeo is an integral part of traditional life for many people in the Great Plains, where there is historical continuity between the cattle frontier, ranching, and the modern "cowboy sport" that developed from it. The origins of rodeo can be traced to the Wild West show as well as to the sports and contests that were first held by early-day working cowboys for their own amusement. Rivalry between cowhands as to who could ride the wildest bronco for the longest time or rope the liveliest calf or the biggest steer led to riding and roping matches. Ultimately these events attracted enthusiastic spectators and developed into full-scale rodeo.

In standard rodeo, the program is divided into two categories of contests: bucking or rough-stock events and timed or cattle-ranch–oriented events.

Various other special events and exhibitions are often added, and barrel racing for women is typically included. In the rough-stock events, cowboys compete for the best score in riding bucking broncos or bulls for eight seconds. Timed events consist of contests in which cowboys compete for the shortest time in accomplishing tasks that are based on ranch work, such as calf roping, steer roping, and steer wrestling. Broncos and bulls are the cowboys' opponents in the bucking contests, whereas cowboys' mounts in timed events have the role of partners in the subduing of various types of cattle.

While individuals for whom rodeo is a way of life do not generally view their sport as particularly inhumane, many people who are outsiders to the ranch and rodeo complex point out that some events of rodeo involve cruelty* to horses and cattle. This controversy highlights an important but often overlooked factor in evaluating the treatment of animals: the presence of publicly displayed, as opposed to privately inflicted, cruelty. In rodeo, whatever brutality toward animals is involved is generally displayed for the audience to see. Painful procedures such as shocking with electric prods, tightening of flank straps, and spurring of broncos are visible to anyone who stands near the chutes or watches the rodeo. Audiences who attend some other equine sporting events, on the other hand, typically see only the performance itself. The bronco undergoing eight seconds of obvious pain* while in the arena could suffer far less, overall, than the show horse who, isolated in a stall, may suffer for months and years from being confined in head and tail sets and having weighted shoes and painful devices applied to feet and legs to alter or enhance its gaits.

Judgments about cruelty involve cultural and psychological factors that vary among individuals. Rodeo contestants feel that the confinement of horses characteristic of eastern horse management is a much greater evil than their own seasonal use of broncos who are likely to spend the remainder of the year free on the range. These examples demonstrate that measurements of inhumane treatment must take into account all phases of the animals' lives, hidden and revealed. Because such great exertion and so many complex tasks are demanded of horses, and because of their sensitive nature and remarkable willingness to submit to trainers and riders and obey the human will, equine animals are particularly vulnerable to overexploitation and abuse involving injury, pain, and even death.* Cattle used in rodeo also may suffer trauma and pain and are sometimes killed in the context of the sport. Their treatment in rodeo is directly related to their role in society as meat animals, since producing beef is the purpose of cattle ranching.

Selected Bibliography. Lawrence, Elizabeth Atwood, *Rodeo: An Anthropologist Looks at the Wild and the Tame* (Chicago: University of Chicago Press, 1984); Rollins, Philip Ashton, *The Cowboy: An Unconventional History of Civilization on the Old-Time Cattle Range* (New York: Ballantine, 1973); St. John, Bob, *On Down the Road: The World of the Rodeo Cowboy* (Englewood Cliffs, NJ: Prentice-Hall, 1977); Slotkin, Richard, *Gunfighter Nation: The Myth of the Frontier in Twentieth-Century America* (New

York: HarperCollins, 1993); Westermeier, Clifford P., *Man, Beast, Dust: The Story of Rodeo* (Denver: Dieter, 1947).

ELIZABETH ATWOOD LAWRENCE

ROYAL SOCIETY FOR THE PREVENTION OF CRUELTY TO ANIMALS (RSPCA) AND EARLY BRITISH LEGISLATION

At the beginning of the 19th century the English would have been surprised to hear themselves praised for special kindness to animals. City streets were crowded with horses and dogs* that served as draft animals and beasts of burden, as well as with herds of cattle and sheep being driven to slaughter. Many of these animals were obviously exhausted or in pain,* as were many of the horses and donkeys used for riding. Popular amusements included cockfighting, dogfighting, rat killing, bull running, and the baiting of wild animals. By the end of the century, however, officials of such organizations as the Royal Society for the Prevention of Cruelty to Animals (RSPCA), founded by Arthur Broome,* routinely claimed that kindness to animals was a native English trait.

This shift in opinion reflected real changes. The 19th century saw a series of administrative and legal breakthroughs with regard to the humane treatment of animals, as well as steadily widening public support for animal welfare* and the laws and societies dedicated to protecting animals from cruelty and abuse. Although the first animal-protection bill to be introduced in Parliament failed miserably in 1800, in 1822 a pioneering piece of legislation was enacted. Known as Martin's Act, after its originator and chief advocate Richard Martin,* it aimed to "prevent cruel and improper treatment of Cattle," which included most farm and draft animals, but not bulls or pets. Later legislation (subsequent acts were passed in 1835, 1849, and 1854) periodically extended protection until all domesticated (*see* DOMESTICATION) mammals were covered, as well as some wild mammals in captivity.

When the Society for the Prevention of Cruelty to Animals (SPCA) was founded in 1824, one of its primary goals was to ensure that the provisions of the new legislation actually took effect. The SPCA funded its own special corps of constables and instructed civilian sympathizers how to arrest wrongdoers encountered in the streets. Despite the initial obstacles it faced, the SPCA (RSPCA beginning in 1840, when Queen Victoria granted the society permission to prefix "Royal" to its name) was successful on every front. As legal protections for animals expanded, so did the society's membership, in both numbers and social prestige. It boasted a series of royal patrons, and the aristocracy was heavily represented on its governing board.

By the 1900s the RSPCA epitomized respectable philanthropy, the kind of charity routinely remembered in the wills of the prosperous. With such

SMITHFIELD MARKET.—THE DROVER'S GOAD.—(SEE PAGE 42.)

Royal Society for the Prevention of Cruelty to Animals: This 19th-century drawing depicts the type of treatment of animals that led to the formation of the Humane Society. *Source: Illustrated London News,* mid-19th century.

powerful backing, the size of the RSPCA increased from its initial complement of only a few men to 8 officers by 1855, 48 by 1878, and 120 by 1897. In its first year of operation the society conducted 147 successful prosecutions under Martin's Act; by the end of the century successful prosecutions peaked at over 8,000 per year before horses, the most frequent victims of prosecuted offenses, were replaced by motor vehicles.

One reason that cab horses and draft horses figured so prominently in RSPCA prosecutions was that there were many of them, and they were abused in plain sight on the public streets. But another was that their abusers were apt to belong to the part of human society where the middle- and upper-class members of the RSPCA expected to encounter depraved behavior. Indeed, it is likely that some humanitarians viewed the animal protection laws as a useful supplement to existing legal and social mechanisms for controlling unruly humans. When animals suffered at the hands of the genteel, the RSPCA and kindred organizations found it more difficult to prosecute or, often, even to acknowledge that a problem existed. For this reason, such sports as steeplechasing and fox hunting (indeed, hunting* of all kinds) were subjects of contention within the mainstream Victorian humane movement. The hardest case of all in these terms was posed by vivisection (*see* ANTI-VIVISECTIONISM), an exclusively middle-class and upper-middle-class pursuit. Although John Colam, then the secretary of the RSPCA, offered strong testimony against the use of vivisection in teaching when he testified before a royal commission on vivisection in 1876, few of his constituents shared his strong views. As a consequence, committed antivivisectionists withdrew from the mainstream humane movement, and, at least for several years, they languished while it prospered.

Selected Bibliography. Fairholme, Edward G., and Wellesley Pain, *A Century of Work for Animals: The History of the R.S.P.C.A., 1824–1924* (New York: E. P. Dutton, 1924); Ritvo, Harriet, *The Animal Estate: The English and Other Creatures in the Victorian Age* (Cambridge, MA: Harvard University Press, 1987); Salt, Henry, *Animals' Rights Considered in Relation to Social Progress* (1892; reprint, Clark's Summit, PA: Society for Animal Rights, 1980); Thomas, Keith, *Man and the Natural World: A History of the Modern Sensibility* (New York: Pantheon Books, 1983); Turner, James, *Reckoning with the Beast: Animals, Pain, and Humanity in the Victorian Mind* (Baltimore: John's Hopkins University Press, 1980).

HARRIET RITVO

ROYAL SOCIETY FOR THE PREVENTION OF CRUELTY TO ANIMALS (RSPCA) REFORM GROUP

Frustrated by the ineffectiveness of the Royal Society for the Prevention of Cruelty to Animals (RSPCA)* in dealing with the modern cruelties of factory farming,* animal exploitation, and the increasingly internationalized abuse of wildlife, some members of the RSPCA, led by Brian Seager, John Bryant, and Stanley Cover, formed the Reform Group in 1970. They supported the attempt by Vera Sheppard to persuade the RSPCA to oppose fox hunting and other cruel sports and succeeded in 1972 in securing the election to the RSPCA Council of five Reform Group supporters, including Bryant, Seager, Andrew Linzey, and Richard Ryder. Over the next eight years, until the end of the decade, the Reform Group faction succeeded in changing the world's oldest and largest animal welfare* organization beyond recognition. In 1976 Ryder was made vice chairman and was then chairman of the RSPCA Council from 1977 until 1979. During these years of reform the society not only came out against cruel sports but, for the first time, developed comprehensive animal welfare policies across the board and elevated the welfare of farm, laboratory, and wild animals to a priority status equal with the welfare of pets. The reformers set up staff departments to deal with these areas of abuse and revived the society's campaigning function, which had been allowed to lapse since the Edwardian era. Publicity, parliamentary, and scientific facilities were established, and the society even gave its support to Lord Douglas Houghton's* successful initiative, the General Election Coordinating Committee for Animal Welfare, to persuade all major British political parties to officially include, for the first time, animal welfare policies in their election platforms in 1979. Before the end of Ryder's term of office (which was followed by a temporary reversal of the society's performance initiated by conservatives), an undercover plainclothes section of the RSPCA's inspectorate was established, and, perhaps most important, the society initiated the establishment and funding of a powerful political lobby for animals in the European Community, subsequently to be named the Eurogroup for Animal Welfare.

Selected Bibliography. Ryder, Richard, *Animal Revolution* (Oxford: Blackwell, 1989).

RICHARD D. RYDER

S

SACRIFICE. *See* LABORATORY ANIMAL USE; RELIGION AND ANIMALS.

SAINTS. *See* RELIGION AND ANIMALS.

SALT, HENRY STEPHENS

Henry Stephens Salt (1851–1939) was a pioneering 19th-century animal rights* advocate whose prescient work *Animals' Rights* (1892) anticipates virtually all of the important modern arguments in favor of animals' interests. While this and Salt's other works concerning vegetarianism* and animals' rights were little read in his time, Salt nevertheless exerted extraordinary influence on such contemporaries as Edward Carpenter, Mohandas Gandhi,* John Howard Moore,* William Morris, Sydney Olivier, George Bernard Shaw,* Count Leo Tolstoy,* and other prominent reformers. The Humanitarian League, which he founded with Fabian Socialists and other acquaintances in 1891, attacked a range of 19th-century cruelties and is regarded as the first modern animal rights organization. Salt and his colleagues campaigned not only against the violation of animals' rights but also against the oppression and torment of human beings in such contexts as warfare, criminal justice, labor relations, hospitals, military and school discipline, and colonialism.

Selected Bibliography. Hendrick, George, *Henry Salt: Humanitarian Reformer and Man of Letters* (Urbana: University of Illinois Press, 1977); Hendrick, George, and

Willene Hendrick (Eds.), *The Savour of Salt: A Henry Salt Anthology* (Fontwell, Sussex: Centaur Press, 1989); Salt, Henry S., *Animals' Rights Considered in Relation to Social Progress* (Clarks Summit, PA: International Society for Animal Rights, 1980); Winsten, Stephen, *Salt and His Circle* (London: Hutchinson, 1951).

BERNARD UNTI

SCHOPENHAUER, ARTHUR

Arthur Schopenhauer (1788–1860) was a German philosopher who provided a sharp critique of Immanuel Kant's* view that duties to animals are only indirect duties to humankind. He rejected as "revolting and abominable" Kant's notion that animals are beings without reason—indeed, only "things"—and can therefore be used as means to humans' ends. He castigated religious systems for failing to appreciate the profound similarities that humans share with animals and therefore for failing to take moral account of them. Schopenhauer's critique influenced other philosophers, notably Friedrich Nietzsche,* and laid the basis for the modern rejection of theological notions of human uniqueness and for the claim that animals deserve protection for their own sakes.

Selected Bibliography. Schopenhauer, Arthur, *On the Basis of Morality*, trans. E. F. J. Payne (Indianapolis: Bobbs-Merrill Company, 1965), extract in Tom Regan and Peter Singer, *Animal Rights and Human Obligations*, 1st ed. (Englewood Cliffs, NJ: Prentice-Hall, 1976), 124–128.

ANDREW LINZEY

SCHWEITZER, ALBERT

Albert Schweitzer (1875–1965), an Alsatian theologian, missionary, and humanitarian, first publicly formulated his concept of "reverence for life" in the Dale Lectures at Mansfield College, Oxford (1922), which were subsequently published in English as *The Decay and Restoration of Civilisation* and *Civilisation and Ethics* (1923). Schweitzer's concept broke new ground in European ethics by expressly including the nonhuman within the sphere of human responsibility. He conceived of "reverence" (*Ehrfurcht*) in largely "mystical" terms (see Linzey, *Animal Theology*) in which individuals perceive as a revelation the divinely given worth of other creatures. In this way he anticipated the work of modern animal rights* theorists who appeal to the "intrinsic" or "inherent" value of sentient creatures. Schweitzer's thought is prophetic: "The time is coming, however, when people will be astonished that mankind needed so long a time to learn to regard thoughtless injury to life as incompatible with ethics" (*Civilisation and Ethics*, 215). He was awarded the Nobel Peace Prize in 1952. The Animal Welfare Institute's highest honor, a medal awarded annually, is named after Schweitzer.

Selected Bibliography. Linzey, Andrew, *Animal Theology* (London: SCM Press, Urbana: University of Illinois Press, 1995); Schweitzer, Albert, *Civilisation and Ethics* (1923), trans. C. T. Campion (London: Unwin Books, 1967); Schweitzer, Albert, *The Decay and Restoration of Civilisation* (London: A. C. Black, 1923); Schweitzer, Albert, *Out of My Life and Thought: An Autobiography*, trans. C. T. Campion (London: Allen and Unwin, 1933); Schweitzer, Albert, *Reverence for Life*, trans. R. H. Fuller, foreword by D. E. Trueblood (London: SPCK, 1970).

ANDREW LINZEY

SENTIENTISM

Sentientism, a term coined by Andrew Linzey in 1980, denotes an attitude that arbitrarily favors sentients over nonsentients. The term is historically parallel to that of "speciesism"* coined by Richard Ryder in 1970. Although Linzey was one of the early advocates of sentiency as the basis of rights, he subsequently warned against claiming too much for any one form of classification as the basis of moral standing* or rights. Raymond Frey specifically argues that sentiency as the basis of rights "condemns the whole of nonsentient creation, including the lower animals, at best to a much inferior status or . . . at worst possibly to a status completely beyond the moral pale."

The issue is how to recognize the value and moral relevance of sentiency as a criterion while avoiding falling into the error of previous generations who have isolated one characteristic or ability—for example, reason, language, culture, or friendship—and used it as a barrier to wider moral sensibility. There is a need to be aware that all moral categories and distinctions are themselves liable to change as our own moral sensibilities develop and our scientific understanding increases.

Selected Bibliography. Frey, R. G., What Has Sentiency to Do with the Possession of Rights? in David A. Paterson and Richard D. Ryder (Eds.), *Animals' Rights: A Symposium* (London: Centaur Press, 1979), 106–111; Linzey, Andrew, *Animal Rights: A Christian Assessment* (London: SCM Press, 1976); Linzey, Andrew, Moral Education and Reverence for Life, in David A. Paterson (Ed.), *Humane Education: A Symposium* (London: Humane Education Council, 1981), 117–125; Schweitzer, Albert, *Civilization and Ethics* (1923), trans. C. T. Campion (London: Unwin Books, 1967).

ANDREW LINZEY

Individual Interests

Simply put, individual interests are individual stakes in life. More precisely, individual interests are defined as relationships between an individual and his or her opportunities to maximize positive experiences and to minimize negative experiences over his or her lifetime. Since the capacity for having positive or negative experiences is equivalent to being sentient, only sentient beings have individual interests. It is in an individual's interest to use all present and future opportunities, whether he or she is aware of their existence or not. Individuals, whether human or nonhuman, are usually not aware and probably cannot be aware of all opportunities that would be in their interest to pursue. An individual may have interest in something without taking an interest in it. For example, a bored (see ANIMAL BOREDOM) individual does not take interest in anything but still has an interest in whatever would alleviate its boredom. The moral concept of having an (individual) interest ("A is in the interest of X") ought to be clearly distinguished from the psychological concept of taking an interest ("X is interested in A"). The ability to take interest is dependent upon individuals having wants and desires. While having an interest in something does not necessarily imply taking an interest in it, the converse is not true: wanting or desiring something does imply having an interest in obtaining or avoiding it. The scope of individual interests is, therefore, dependent on the diversity of psychological interests, which is, in turn, dependent on cognitive capacities.

A major step in the evolution of animal cognition* that led to an expansion of psychological and individual interests was the emergence of the capacity to form a value-laden mental representation of an external situation. This capacity is clearly present in many mammals and birds and probably in some other animals. Another major step in the evolution of individual interests was the emergence of reflective self-consciousness (see ANIMAL COGNITION, Conscious Experience) and self as a major source of positive and negative experiences.

There is a controversy over whether animals have an interest in life as opposed to interests in specific experiences. This controversy stems from the ambiguity of the terms "interest" (as discussed here), "self-consciousness" (or "self-awareness"), and "life." Perceptual self-consciousness implies an experiential awareness of one's own body and the distinction between the body and the environment. Reflective self-consciousness is an ability to reflect upon oneself, which implies having a *concept* of oneself. The basic, restrictive meaning of life is the life in itself, the very existence of an individual, which enables it to experience anything at all. The broad meaning of life includes individual existence and all that matters to the individual in its lifetime. Only a reflectively self-conscious individual can have some concept of, and thus take an interest in, one's own individual existence in itself, that is, may not want to die no matter what experience is to be expected. Since most animals (with a few exceptions, especially "higher" primates) do not appear to show evidence of reflective self-consciousness, they cannot take interest in life itself. However, a good life, which means an existence with predominantly positive experiences, is obviously in an animal's interest. Whether life in either sense is in an animal's interest depends, therefore, on its expected quality (see WELL-BEING OF ANIMALS).

Selected Bibliography. Elzanowski, A., The Moral Career of Vertebrate Values, in M. H. Nitecki and D. V. Nitecki (Eds.), *Evolutionary Ethics* (Albany: State University of New York Press, 1993), 259–276; Frey, R. G., *Interests and Rights* (Oxford: Clarendon Press, 1980); Rollin, B. E., *Animal Rights and Human Morality* (Buffalo, NY: Prometheus Books, 1992); Sapontzis, S. F., *Morals, Reason, and Animals* (Philadelphia: Temple University Press, 1987); Teutsch, G. M., *Mensch und Tier: Lexikon der Tierschutzethik* (Man and animal: Lexicon of animal protection ethics) (Göttingen: Vandenhoeck und Ruprecht, 1987).

ANDRZEJ ELZANOWSKI

SEWELL, ANNA

Anna Sewell (1820–1878) was the author of *Black Beauty* (1877), the most influential anticruelty novel of all time. A lifelong invalid, Sewell wrote the book in her fifties, dictating it to her mother from her sickbed. She sold the book outright for a negligible amount and did not live to see its enormous success and impact. The popularity of *Black Beauty* has been linked to the abolition of the bearing rein and to the wider success of the humane movement worldwide. Within two weeks of receiving a copy in February 1890, Massachusetts Society for the Prevention of Cruelty to Animals (MSPCA) founder George Angell* had arranged for a pirated edition. Soon the book was selling at the rate of 250,000 copies per year. New editions of the book continue to appear, and the story of Black Beauty has attracted the talents of a number of illustrators and cinematographers.

Selected Bibliography. Chitty, Susan, *The Woman Who Wrote Black Beauty* (London: Hodder and Stoughton, 1971).

BERNARD UNTI

SHAFTESBURY (7TH EARL OF), LORD ANTHONY ASHLEY COOPER

Anthony Ashley Cooper, Lord Shaftesbury (1801–1885), was a British evangelical philanthropist active in many social causes, including factory reform, the abolition of child labor, and mental health. His animal advocacy is less well known. Together with Henry Manning,* Frances Power Cobbe,* and George Hoggan he founded the world's first antivivisection* society, the Victoria Street Society for the Protection of Animals from Vivisection, in 1875 and became its first president. Although he was an advocate of total abolition, he supported the 1876 Cruelty to Animals Act because "while he believed restriction might be effective, he feared that abolition would be a dead letter" (*Hansard*, 1876, 1016). Although Shaftesbury subsequently spoke in favor of abolitionist legislation in 1878, it was unsuccessful. Shaftesbury's moral credo is encapsulated in these lines: "I was convinced that God had called me to devote whatever advantages He might have bestowed upon me to the cause of the weak, the helpless, both man and beast, and those who have none to help them" (letter, April 30, 1881). Shaftesbury exercised great influence in the movement until his death. Cobbe wrote of him: "Lord Shaftesbury never joined the Victoria Street Society, it was the Society which joined Lord Shaftesbury" (*In Memoriam*, 3).

Selected Bibliography. Cobbe, Frances Power, *In Memoriam, The Late Earl of Shaftesbury, K. G., First President of the Victoria Street Society* (London: 1885); Kramer, Molly Beer, and Andrew Linzey, Vivisection, in Paul Barry Clarke and Andrew Linzey (Eds.), *Dictionary of Ethics, Theology, and Society* (London and New York: Routledge, 1996), 870–874; Lord Shaftesbury, letter, April 30, 1881, cited and discussed in Roberta Kalechofsky, *Between the Species: A Journal of Ethics* 6 (3) (Summer 1990): 160; Lord Shaftesbury, Speech in the House of Lords on the Cruelty to Animals Bill, May 22, 1876, *Hansard* (London: HMSO, 1876), 1016–1030; Vyvyan, John, *In Pity and in Anger: A Study of the Use of Animals in Science* (London: Michael Joseph, 1969).

ANDREW LINZEY

SHAW, GEORGE BERNARD

George Bernard Shaw (1856–1950) was an Irish-born author, playwright, pamphleteer, and essayist. An outstanding humanitarian of his age, Fabian Socialist, vegetarian, and antivivisectionist (*see* ANTIVIVISECTIONISM), he was a scathing critic of all forms of animal abuse. His "Shavian" wit was used to devastating effect on opponents. On vivisection, he argued that the plain logic of such experimentation would be to include human subjects in research too since an unlimited right to know would justify boiling human infants to find out what boiled babies taste like ("These Scoundrels"). When H. G. Wells eulogized Pavlov's experiments with dogs,* Shaw replied, "And from twenty-five years of this sort of thing all that the world learned was how a dog behaved with half its brains out, which nobody wanted to know, and, what was perhaps important, what sort of book a physiologist could write without having any brains at all" (cited in Pearson, *Bernard Shaw*, 274). Angered by rabbit coursing near his home, he wrote of sport hunters: "To kill in gratification of a lust for death is at least to behave villainously. . . . But to kill, being all the time quite a good sort of fellow, merely to pass away the time . . . is to behave like an idiot or a silly imitative sheep"(*Prefaces*, 148). But it was Shaw's unrepentant vegetarianism* that most disturbed his contemporaries, since he spoke of meat eating as "cannibalism with its heroic dish omitted" (Pearson, 64). The heart of Shaw's philosophical position on animals was straightforward: humanitarianism is about the extension of "fellow-feeling"; it is illogical not to extend such sympathy to animals. Shaw's circle included his close friend Henry Salt,* for whose anthology *Killing for Sport* (1915) he wrote a preface, and his wife, Kate Salt, who provided both secretarial support and inspiration, not least of all as the model for the female lead in *Candida*. In 1925 Shaw was awarded the Nobel Prize for Literature.

Selected Bibliography. Pearson, Hesketh, *Bernard Shaw* (London: Collins, 1942; Four Square ed., 1964); Shaw, George Bernard, *The Complete Plays* (London: Odhams Press, 1936); Shaw, George Bernard, *The Dynamitards of Science* [pamphlet] (London: London Anti-Vivisection Society, 1900); Shaw, George Bernard, Preface to Henry S. Salt (Ed.), *Killing for Sport: Essays by Various Writers* (London: George Bell, 1915), xi–xxxiv; Shaw, George Bernard, *Prefaces* [to his plays] (London: Constable and Company, 1934); Shaw, George Bernard, *Shaw on Vivisection*, ed., G. H. Bowker (London: George Allen and Unwin, 1949); Shaw, George Bernard, These Scoundrels: Vivisection—The "Science" of Imbeciles, *Sunday Express*, August 7, 1927.

ANDREW LINZEY AND BERNARD UNTI

SHELTERS

Animal shelters in the United States range from small buildings in rural areas to progressive, state-of-the-art facilities, many providing crucial services that go beyond the basic "sheltering" of animals. Responsible shelters today provide humane care and treatment of all animals needing protection; seek to return lost or stray animals to their owners; seek responsible, lifelong homes for animals without owners; and provide a humane death for unwanted animals when necessary.

Generally speaking, shelters tend to fall within one of three categories: (1) municipal animal control agencies, run by governmental entities in cities and towns; (2) private, nonprofit agencies governed by a board of directors; and (3) private, nonprofit agencies with a governmental contract to provide animal care and control services. Most communities have at least one (and often several) animal shelters.

Terms such as "humane society," "society for the prevention of cruelty to animals (SPCA)," and other similar names are generic, meaning that any organization can use them. Thus, organizations with these names vary dramatically in focus and services provided. In addition, there is no national governing (or oversight) organization that dictates standards or policies for these agencies. Two national organizations, the Humane Society of the United States (HSUS)* and the American Humane Association (AHA), offer guidelines and recommendations for animal shelters. Although local agencies are under no obligation to follow these recommendations, many of them do.

Working with limited human and financial resources, shelters have both legal and ethical responsibilities to provide responsible animal control and sheltering services for both the animals and people in their community.

However, the needs of the animals in any given community far exceed those of simply sheltering them from the elements. Beyond that, responsible shelters invest energy into three general areas to fulfill their mission: (1) preventing cruelty and/or suffering of animals; (2) enforcing animal-protection laws; and (3) instilling humane principles into society. To achieve the goal of protecting both the animals and people within their community, responsible shelters accept every animal brought in; never charge a fee for surrendered animals; maintain a clean, comfortable, safe, and healthy environment for animals; hold stray animals a minimum of five operating days, including a Saturday; screen prospective adopters using adoption standards; use sodium pentobarbital (the most humane method), administered by well-trained, compassionate individuals, when euthanasia* is necessary; and ensure that all adopted animals are sterilized to prevent future births.

A little over a hundred years ago, there was no protection for abandoned or abused animals. Sometimes, animals found roaming the streets were gathered up by city workers and taken to "impoundment" lots (or "pounds") where they were held for a brief time to give a rare owner the opportunity to claim his or her "property." All that changed in 1870 when Caroline White* refused to accept the inhumane practices at her local pound and took over the responsibility for the care of unwanted animals in Philadelphia, Pennsylvania, thus forming the first real "sheltering" organization in the United States.

While private, nonprofit sheltering programs began to take wing, municipal animal control services continued to consist primarily of "catching and killing" animals. As a result, pounds and shelters began to flourish simultaneously in the United States. Each lacked standards, policies, and a unified response to animal care and control issues. There remained a great divide between the services, operations, and missions of these two differing types of agencies, which led to increased strife within the animal community.

In the early 1960s, Phyllis Wright worked hard to unite municipal animal control agencies and private animal shelters. She showed them that they should, in fact, have the same mission in mind. At that time, professional standards for shelter operations and animal control programs had not yet been established. Joining the Humane Society of the United States (HSUS) in 1969, Wright helped to create these standards by becoming a national liaison to shelter workers, animal control professionals, and governmental agencies involved in animal issues. She helped to create national training opportunities for shelter workers that succeeded in "professionalizing" the movement into what it is today.

Both municipal animal control agencies and private shelters have expanded their focus to include diverse issues such as wildlife rehabilitation, humane education (*see* HUMANE EDUCATION MOVEMENT), fostering programs, obedience training, and pets in housing. These shelters also understand that the root of the problem is not the animals themselves, but people. Therefore, public support, understanding, and resources are crucial to resolving the many problems associated with animals.

Selected Bibliography. Allen, R. D., and W. Westbrook (Eds.), *The Handbook of Animal Welfare, Biomedical, Psychological, and Ecological Aspects of Pet Problems and Control* (New York: Garland STPM Press, 1979); *Animal Sheltering* (published by the Humane Society of the United States) 19(1) (January–February, 1996); Curtis, P., *The Animal Shelter* (New York: E. P. Dutton, 1984); Humane Society of the United States, How to Organize a Humane Society, 1985; Local Animal Control Management, *MIS Report* 25 (9) (September 1993): 1–20.

SALLY FEKETY

SILVER SPRING MONKEYS

In 1981 the Institute for Behavioral Research (IBR) in Silver Spring, Maryland, was raided by police as a result of accusations of cruelty to animals. This was the first time in American history that a scientific research laboratory had been raided by police as a result of alleged cruelty to animals, and it quickly became a landmark case that set legal and political precedents across the United States.

The research at the IBR, led by Edward Taub, was funded by the National Institutes of Health (NIH) and focused on somatosensory deafferentation (removing sensation) research in primates, in which all sensation was surgically abolished from one or both forelimbs. The extent to which the animals then used their limbs (or could use them) was evaluated. It was believed that voluntary movement was impossible in the absence of sensory feedback, a conclusion disproved by the research at IBR.

In the early summer of 1981, an animal activist named Alex Pacheco asked Taub for a job at IBR. Taub told Pacheco that there was no paying job available at the institute, but that he was welcome to work at the laboratory on a volunteer basis. Taub was not aware that Pacheco was one of the founding members of People for the Ethical Treatment of Animals (PETA). During his five months at IBR, Pacheco took photographs of the conditions in the facility. In addition, while Taub was away on vacation, he brought five scientists (two zoo veterinarians* and three animal activists, two of whom were primatologists) into the facility to witness the conditions in the laboratory.

On September 22, 1981, in response to the affidavits of the five scientists alleging grossly unsanitary conditions and inadequate care and the photographs provided by Pacheco, the Montgomery County police raided IBR, confiscating the primates and seizing laboratory records. Taub was subsequently charged with cruelty to animals. In November 1981, Taub was found guilty of providing inadequate veterinary care to six of the seventeen primates. The other 113 charges were dismissed. Taub appealed the conviction, demanding a second trial before a jury, and was found guilty on a single count of inadequate veterinary care. He appealed to the Maryland Supreme Court, which dismissed the case because, it argued, the Maryland anticruelty statute did not apply to federally funded research. The NIH subsequently determined that the IBR facilities and program violated several aspects of NIH animal research policies, and it first suspended and then terminated Taub's funding.

The case has had a tremendous impact on the animal research debate and on resulting public policy. At the time of the police raid, Congress had scheduled hearings on several animal research bills. The news coverage of the raid and the publicity generated by Pacheco's photographs refocused the congressional hearings. NIH also found its own policies too vague to deal adequately with the events and initiated a major revision of its animal research policies. The research community, particularly the American Psychological Association and the Society for Neuroscience, was very concerned about the case and rallied behind Taub to defend him from his critics. In contrast, two laboratory animal veterinarians testified for the prosecution that the conditions pictured at IBR were grossly substandard for the care of primates. Subsequently, PETA and NIH fought over the fate of the Silver Spring monkeys, especially the deafferented animals, which ended up at the Delta Primate Research Center. The monkeys continued to be the focus of court battles well into the 1990s until the last animal was euthanized (*see* EUTHANASIA) because of failing health.

Selected Bibliography. Guillermo, K. S., *Monkey Business: The Disturbing Case That Launched the Animal Rights Movement* (Washington, DC: National Press Books, 1993).

ANDREW N. ROWAN

SIZEISM

Sizeism, a form of speciesism,* specifically relates to the failure to empathize with (*see* EMPATHY FOR ANIMALS) or give small animals the same consideration that would be given to larger animals. Although there are no good physiological reasons to doubt that small animals feel pain,* scientific procedures are carried out on them that would not be carried out on larger animals without an anesthetic, for example, amputation of digits, docking* of tails, castration, cardiac puncture, and intracerebral injections.

DAVID B. MORTON

SLAUGHTER. *See* TRANSPORTATION AND SLAUGHTER.

SPECIES-ESSENTIALISM

Essentialism is the claim that every member of a real kind shares some one quality with all and only others of that kind. What is now in doubt is that such kinds can ever be identified with biological species. One can question whether it is necessary, to be a dog,* to share some quality with all (and only) dogs and whether it is necessary to suppose that there are "pure" dogs, having no other qualities than dogs require. Biologists typically blame Aristotle or his followers for "species-essentialism," for supposing that there are real, discrete biological kinds, such that there are "perfect" specimens of each such kind. The truth is that Aristotle insisted that there were no absolute divisions in nature: we could conveniently classify living things, but would always find that there were hybrids and intermediates in any system.

Aristotle was correct: the existence of cross-species hybrids and the supposed existence of ancestral species from which several modern species have evolved show that nature is a continuum (see CONTINUITY; DARWIN, CHARLES). A species is a set of interbreeding populations, not a natural kind. There need be no one quality that every member of a species shares with all and only the others. Not all members even resemble all their conspecifics (members of the same species) more than they resemble creatures of other species. Nor is there any "perfect specimen" of a given species: any member of a species, however unusual, is equally and perfectly a member. Nothing says that any individual can have fertile intercourse with any conspecific of the other sex, nor that every individual of that species shares any one particular character with every other, nor that its failure to have some feature shared by most is any real defect. Some groups, closed off from others, will be highly uniform; others will not, yet the differences do not grow into true species differences unless the group happens to split up. Sometimes one species will turn into two only because some crucial, intermediate population has perished (without any change in any other population). It is not even entirely true that genetic information cannot pass between real species: occasional hybrids aside, viral infection transfers genetic material.

With respect to the human species, it turns out not to be a "natural kind": it is just the set of interbreeding populations. There may have been (and there may yet be) more than one such "human" species: what the individuals concerned were (or will be) like (and what our duties might be toward them) cannot be settled by deciding on their species.

Selected Bibliography. Clark, S. R. L., Is Humanity a Natural Kind? in T. Ingold, (Ed.), *What Is an Animal?* (London: Routledge, 1994); Douglas, Mary, *Natural Symbols* (Harmondsworth: Penguin, 1973); Gotthelf, A., and J. G. Lennox (Eds.), *Philosophical Issues in Aristotle's Biology* (Cambridge: Cambridge University Press, 1987);

Mayr, Ernst, *Animal Species and Evolution* (Cambridge, MA: Belknap Press of Harvard University Press, 1963); Sober, Elliott, *From a Biological Point of View* (Cambridge: Cambridge University Press, 1994).

STEPHEN R. L. CLARK

SPECIESISM

The term *speciesism* was first coined by Richard Ryder in 1970. In 1985 the *Oxford English Dictionary* defined speciesism as "discrimination against or exploitation of certain animal species by human beings, based on an assumption of mankind's superiority." This definition marked the official acceptance of "speciesism" into the language. Peter Singer did much to establish its use. Two chapters in his classic work *Animal Liberation* include the term in their titles.

Two slightly different, but not often clearly distinguished usages of "speciesism" should be noted. A human may seek to justify discrimination against, say, an armadillo on the grounds that the armadillo cannot talk, is not a moral agent (*see* MORAL AGENCY AND ANIMALS), has no religion, or is not very intelligent (*see* ANIMAL COGNITION, Intelligence); such an attitude is often described as speciesist. But, more strictly, it is when the discrimination or exploitation against the armadillo is justified solely on the grounds that the armadillo is of another species that it is speciesist. This latter usage should perhaps be called *strict speciesism*. A strict speciesist might argue, for example, that painful experiments are allowable on intelligent and communicative chimpanzees* but not upon human beings of any sort, even brain-dead ones; here, the speciesist regards the species difference itself as the all-important criterion.

By drawing the parallel between speciesism, sexism, and racism, campaigners have been able to attract the attention, and often the support, of liberals, democrats, and others who might otherwise have remained indifferent to the interests of nonhumans. Thus, although the concept has proved useful on the philosophical level, for example, as a means to address the subject without any commitment to the idea of "rights," it has had value on the psychological and political levels also.

Selected Bibliography. Ryder, Richard D., *Animal Revolution: Changing Attitudes towards Speciesism* (Oxford: Basil Blackwell, 1989); Ryder, Richard D., Experiments on Animals, in Stanley Godlovitch, Roslind Godlovitch, and John Harris (Ed.), *Animals, Men, and Morals: An Enquiry into the Maltreatment of Non-Humans* (London: Victor Gollancz, 1971); Ryder Richard D., *Victims of Science: The Use of Animals in Research* (London: Davis-Poynter, 1975); Ryder, Richard D. (Ed.), *Animal Welfare and the Environment* (London: Gerald Duckworth, 1992); Singer, Peter, *Animal Liberation* (New York: New York Review of Books, 1990).

RICHARD D. RYDER

Historical Views

The term "speciesism" has become a valuable tool in describing how humans have thought of and treated other animals. As a concept, speciesism is an attempt to describe an attitude that has been the primary justification for the many ways in which humans have deprived other animals of basic moral protections such as life, liberty, and freedom from purposeful infliction of avoidable harm.

The exclusion of other animals' interests has taken different forms, such as justifications that even the minor interests of humans (such as cosmetic appearance, recreation, or convenience) outweigh the major interests of other animals. Overriding the interests of other animals has traditionally been supported by claims that other animals exist for humans. Aristotle made such a claim (*Politics* 1.8) in the fourth century B.C.E., and three centuries later Cicero made even more anthropocentric (*see* ANTHROPOCENTRISM) claims (*De Natura Deorum* 2.14). The claim is still made in great earnestness, as in the 1994 Catholic Catechism passage that says, "Animals, like plants and inanimate things, are by nature destined for the common good of past, present and future humanity" (paragraph 2415).

It is not merely the inclusion of all humans that is the target of antispeciesism advocates. Rather, the problem is the exclusion of all other animals solely because they are not members of the human species. It is the concentration on the species line as the border of moral considerability that has led to the charge that membership in the human species has been the real criterion for determining which animals are valued.

What have stimulated and continue to drive the charge of speciesism are justifications of many avoidable, nonessential human activities. Instrumental use of other animals, sport hunting,* factory farming,* testing of cosmetics, biomedical experiments, roadside animal shows, and recreational animal parks involve intentional, but avoidable damage to other animals' interests.

Speciesism is a valuable tool for describing the terrain we are in with regard to our understanding of the moral status of other animals. As Gary Francione points out, other animals are property in contemporary legal systems (*see* LAW AND ANIMALS), and speciesist exclusions are the foundation of such thinking. The continuation of such views is also a central feature of the most influential secular and religious institutions in Western culture, thereby anchoring the anthropocentrism of traditional ethics. One recent attempt to breach the species barrier is the Great Ape Project,* which, in the interest of many other animals, focuses on humans' closest genetic cousins as the first step in dismantling the traditional prejudices that draw their life from the practice of defining moral considerability in terms of membership in the human species.

Selected Bibliography. *Catechism of the Catholic Church* (London: Gregory Chapman, 1994); Cavalieri, Paola, and Peter Singer (Eds.), *The Great Ape Project: Equality beyond Humanity* (London: Fourth Estate, 1993); Cone, James H., *Black Theology and Black Power* (New York: Seabury Press, 1969); Francione, Gary L., *Animals, Property, and the Law* (Philadelphia: Temple University Press, 1995); Midgley, Mary, *Animals and Why They Matter* (Athens: University of Georgia Press, 1983); Ryder, Richard, *Animal Revolution: Changing Attitudes towards Speciesism* (Oxford: Basil Blackwell, 1989); Ryder, Richard, *Victims of Science: The Use of Animals in Research* (London: Davis-Poynter, 1975); Singer, Peter, *Animal Liberation* (New York: New York Review of Books, 1990).

PAUL WALDAU

Biological Classification

Speciesism is the attribution of weight to species membership in evaluating the ethical treatment of individuals. When we say that all and only human life is sacred, we are embodying speciesism in a basic moral principle. When we treat nonhuman animals as mere means to our ends, while condemning the same attitude in the case of human beings, we are incorporating speciesism into our practices.

Recently, speciesism has been equated with racism and sexism as a form of arbitrary discrimination. Some philosophers have pointed out that if we reflect on the human rights theory, we can realize that we have already settled similar questions of relevance. People generally believe that race and sex membership should play no role in our morality. To be consistent, the same judgment should be made in the case of species membership. On this view, the very idea of human equality tells us that speciesism is ethically objectionable.

However, one should explain what is wrong with racism and sexism. An answer seems evident. Races and sex are biological classifications. As such, they are concerned with purely physical characteristics such as skin color and reproductive role, rather than with psychological properties such as the capacity for being harmed or benefited. Since ethics is an autonomous theoretical subject, endowed with its own standards of justification, criteria coming from different disciplines have no bearing on it.

Against this, it can be said that there is a correspondence between race or sex and the possession, or lack, of some characteristics that are morally relevant, so that group membership may be appealed to as a mark of this difference. This can be called the "correspondence approach." Thus, for example, racists often claim that members of other races are less intelligent than members of their own race. However, even if the claim were true, this approach would not work. First, if the underlying reference is to other characteristics, drawing a line through race membership is uselessly confusing. Second, what we shall find will be overlap, not mutual exclusion, between races, and to treat individuals not on the basis of what is allegedly "normal" for their group would be irrational.

Thus it seems that racism and sexism are in fact arbitrary discriminations. Many have disputed that we can say the same for speciesism. Since it is undeniable that species is a biological characteristic just as race and sex are, the objections to the parallel have focused on the correspondence approach. While seen as unacceptable in the case of humans, this approach has claimed to be sensible in the case of other animals, because the gulf between us and them allegedly is so large as to prevent overlap.

However, since the work of Charles Darwin,* we have given up the idea of a gulf between us and the other animals: we see the animal world as composed of a multitude of organisms that resemble one another in some ways, but differ in others, and we hold that differences among species should be viewed as differences in degree rather than in kind. Moreover, if some people want to stick an arrangement of beings in a linear, ascending scale, they still have to be concerned with the presence within our species of disabled, disturbed, or brain-damaged individuals (*see* MARGINAL CASES).

All in all, it seems that racism, sexism, and speciesism are arbitrary discriminations. If this conclusion is sound, we can only preserve our belief that there are no morally relevant barriers within our species at the price of abandoning the belief that there is a morally relevant barrier around our species.

Selected Bibliography. Johnson, Edward, Species and Morality (Ph.D. dissertation, Princeton University, July 1976; Ann Arbor, MI: University Microfilms International, 1977); Pluhar, Evelyn, Speciesism: A Form of Bigotry or a Justified View?" *Between the Species* 4(2) (Spring 1988): 83–96; Rachels, James, *Created from Animals: The Moral Implications of Darwinism* (Oxford: Oxford University Press, 1990); Singer, Peter, *Practical Ethics*, 2nd ed. (Cambridge: Cambridge University Press, 1993); Tooley, Michael, *Abortion and Infanticide* (Oxford: Oxford University Press, 1983).

PAOLA CAVALIERI

SPECIMENS

Natural history museums house scientific specimens of animals in their collections. Specimens serve as the essential permanent records of biodiversity. They are used to study systematic biology, taxonomy, distribution, ecology, physiology, behavior, wildlife management, and conservation. Much of our information about wild animals is based on these museum collections. Many species of mammals (such as mice,* shrews, and bats) are not seen unless they are captured and cannot be identified without examining the skull. Many species of birds can only be correctly identified with a specimen in hand.

These research collections are looked after with great care, so that they will continue to provide information well into future generations. A specimen of a mammal usually consists of a skin and a skull; it may also consist of other parts, including the skeleton or parts thereof, the body or body parts preserved in alcohol, or tissue samples for genetic analyses. A single specimen provides a wealth of information on the individual (for example, breeding condition, diet, molt) in addition to documenting when this particular animal existed in a particular locale. For example, a black-footed ferret specimen from Denver, collected in the 1940s, lets us know that Denver used to provide a good habitat for these animals. The natural phenomenon of bird navigation, involving five billion birds per year in North America that travel an average distance of 2,000 kilometers, is documented with specimens. Ornithological collections often contain eggs of birds. The negative impact of DDT on birds was first made known by studying eggshells in museum collections and comparing them with present-day eggs. Modern techniques even allow reconstruction of the genetic information of an individual. Studies of chemical composition of hair, feathers, or shells provide indications of levels of chemical pollutants at the time of capture.

In recent years, concern has been raised about the morality of killing animals for this collecting of specimens. In response to this concern, and as a consequence of space limitations in museums, the labor intensity of caring for collections, and declining populations in the wild, collecting has become far more conservative. In addition, alternatives to collecting in certain situations are being discussed. These include photographs of animals to document distributions of easily identifiable animals, and blood and/or tissue samples to access genetic information.

In an attempt to balance a land ethic that values individuals, populations, and ecosystems (see ENVIRONMENTAL ETHICS), an animal rights* ethic that values individuals, and the interests of science that value knowledge, Robert Loftin has outlined criteria for justifiable collecting of specimens. These criteria include necessity, importance, novelty, least damage, mercy, maximum information, no long-term impact, and no jeopardy to endangered species.*

Although it is true that some individuals are killed during collecting, specimens can be used for purposes that lead to the conservation of the species the specimens represent. We can only conserve and protect populations, species, and ecosystems based on our knowledge of what was there prior to the present.

Selected Bibliography. Banks, R. C. (Ed.), *Museum Studies and Wildlife Management: Selected Papers* (Washington, DC: Smithsonian Institution Press, 1979); Bogan, M. A., R. B. Finley, and S. J. Petersburg, The Importance of Biological Surveys in Managing Public Lands in the Western United States, in *Management of Amphibians, Reptiles, and Small Mammals in North America*, General Technical Report, U.S.D.A. Forest Service, RM-166 (1988), 254–261; Finley, R. B., Jr., The Value of Research Collections, *BioScience* 37 (1987): 92; Finley, R. B., Jr., and M. A. Bogan, Studies of Biological Diversity: The U.S. Fish and Wildlife Service Experience, *Association of Systematics Collections Newsletter* 20 (1992): 110–111; Loftin, R. W., Scientific Collecting, *Environmental Ethics* 14 (1992): 253–264; Yates, T. L., Value and Potential of the Collection Resource, in H. H. Genoways, C. Jones, and O. L. Rossolimo (Eds.), *Mammal Collection Management* (Lubbock: Texas Tech University Press, 1987), 9–17.

CARRON A. MEANEY

STEREOTYPIES IN ANIMALS

A stereotypy is a repeated, relatively invariant sequence of movements that has no obvious function. It is the repetition of the same behavior pattern that makes the stereotypy so obvious to an observer, and the abnormality is also indicated by the distinction from useful repetitive behaviors such as breathing, walking, or flying. Among the most striking abnormal behaviors shown by some animals in zoos* and in confined conditions on farms are stereotypies such as route tracing, bar biting, tongue rolling, or sham chewing. Georgia Mason described a female mink, in a 75 × 37.5 × 30-cm cage on a mink farm, who would repeatedly rear up, cling to the cage ceiling with her forepaws, and then crash down on her back.

Stereotypies can be shown by humans with neurological disorders, by those with some degree of mental illness, and by those in situations where they have little or no control over aspects of their interaction with their environment. People with no illness may show stereotypies when confined in a small cell in prison or when exposed to situations like waiting for an important interview or waiting for their wife to give birth.

The causes of stereotypies in nonhuman animals seem to be very similar to those in humans. Frustrated individuals, especially those unable to control their environment for a long period, are the most likely to show the behavior. Individuals treated with particular drugs, especially psychostimulants such as amphetamine and apomorphine, may show stereotypies, but it is not clear what this tells us about the causation of stereotypies. Many stereotypies seem to be related to oral movement or to locomotion, so the control systems for such movements are clearly susceptible to being taken over by whatever causes repetition. The age of the individual and the amount of time in the housing condition can affect the stereotypies shown, for example, horses changing from crib biting to wind sucking or from side-to-side pacing to head weaving and confined sows changing from bar biting to sham chewing. Movements can also become more complex with age.

In most cases we do not know whether a stereotypy is helping the individual to cope (*see* ANIMAL WELFARE, Coping) with the conditions, has helped in the past but is no longer doing so, or has never helped and has always been just a behavioral abnormality. None of the studies that demonstrate a relationship between the extent of occurrence of stereotypies and opioid receptor blocking or opioid receptor density measurement tells us with certainty whether or not stereotypies have any analgesic or calming function. But in all cases the stereotypy indicates that the individual has some difficulty in coping with the conditions, so it is an indicator of poor welfare. Some stereotypies must indicate worse welfare than others, but any individual showing them has a problem.

Stereotypies are sometimes ignored by those who keep animals and may be taken to be normal behavior by those people if they see only disturbed animals. For example, zoo keepers may see route tracing by cats or bears, laboratory staff may see twirling around drinkers by rodents, and farmers may see bar biting or sham chewing by stall-housed sows without realizing that these indicate that the welfare of the animals is poor. A greater awareness of the importance of stereotypies as indicators of poor welfare is resulting in changes in animal housing. More complex environments that give the individual more control and hence result in the occurrence of fewer stereotypies are now being provided in good animal accommodation (see ENRICHMENT FOR ANIMALS). These environments also give opportunities for a larger proportion of the full behavioral repertoire to be expressed, and for the patterns of movements in the repertoire to be varied. The consequent reduction in frustration and increase in the proportion of an individual's interactions with its environment that are under its control improve its welfare.

Selected Bibliography. Broom, D. M., Stereotypies as Animal Welfare Indicators, in D. Smidt (Ed.), Indicators Relevant to Farm Animal Welfare, *Current Topics in Veterinary Medicine and Animal Science* 23 (1983): 81–87: Broom, D. M., and K. G. Johnson, *Stress and Animal Welfare* (London: Chapman and Hall, 1993); Lawrence, A. B., and J. Rushen (Eds.), *Stereotypic Animal Behaviour: Fundamentals and Applications to Welfare* (Wallingford: CAB International, 1993); Mason, G. J., Stereotypies: A Critical Review, *Animal Behaviour* 41 (1991): 1015–1037; Ödberg, F., Abnormal Behaviours (Stereotypies), *Proceedings of the First World Congress on Ethology Applied to Zootechnics, Madrid* (Madrid: Industrias Graficas España, 1978), 475–480.

DONALD M. BROOM

STRESS

On most occasions when people say that they are stressed, or that some other individual, whether human or not, is stressed, they mean that their environment is having an adverse or harmful effect on them. Hans Selye, a physiologist, emphasized that the secretion of glucocorticoids from the adrenal cortex is a widespread, nonspecific stress response. However, since then, others have shown that exposure to high temperature, hemorrhage, prolonged close confinement, a nonnutritive diet, or dehydration elicits no adrenal cortex response or reduced glucocorticoid production, but some adaptive, useful activities such as courtship, copulation, and hunting for food do elicit glucocorticoid production. Hence it is not useful to define the term *stress* with reference to increased adrenal activity.

Scientists have used the word "stress" to refer to even minor perturbations of homeostasis (or balance), and one defined it as "any displacement from the optimum state," so that stress seemed to be the effect of almost any stimulus. Brief exposure to the warm sun, which elicits simple physiological and behavioral responses, would be called stress using such a definition. To use "stress" for a circumstance in which such regulatory responses occur is unnecessary and misleading.

The necessity for consideration of psychological as well as physical effects of the environment on individuals has been emphasized in many discussions of stress. It is of interest in this context that many of those discussing stress in domestic or wild animals have tended to emphasize physical problems, while people discussing themselves concentrate on coping difficulties of a mental nature. At least in the more complex animals, both must be important, and stress does not refer to a single coping system (*see* ANIMAL WELFARE, Coping).

The ultimate measure of distress* for animals is impairment of biological fitness—how many offspring they produce who then go on to reproduce. If an individual is adversely affected by his environment to such an extent that he is less able to pass on his genes to the next generation because he dies or is unable to produce as many offspring, then his fitness is reduced. In many cases it is not easy to be sure that fitness is reduced, but it can be confidently predicted on the basis of previous knowledge.

In order to take account of the functioning of coping systems and each of the points made earlier, stress is defined as an environmental effect on an individual that strains his control systems and reduces his fitness or appears likely to do so. A distinction is therefore made between a minor disturbance to an individual's equilibrium that may necessitate the use of energy to correct it and would not be referred to as stress and greater effects that are sufficient to reduce fitness.

Stress may result from a variety of kinds of effects, but Selye was right to emphasize that particular changes in physiology and immune-system function are common to many individuals and circumstances. A variety of harsh conditions can result in immunosuppression, increased pathology, and sometimes general failure of body function and then death. There is an overlap between the concept of stress and that of the welfare of an individual. If the individual is stressed, his welfare will be poor. However, stress refers to failure to cope with the environment, and poor welfare also includes the situation in which the individual has difficulty in coping with his environment without fitness reduction.

Selected Bibliography. Broom, D. M., and K. G. Johnson, *Stress and Animal Welfare* (London: Chapman and Hall, 1993); Mason, J. W., Psychoendocrine Mechanisms in a General Perspective of Endocrine Integration, in L. Levi (Ed.), *Emotions: Their Parameters and Measurement* (New York: Raven Press, 1975), 143–82; Moberg, G. P., Biological Response to Stress: Key to Assessment of Animal Well-Being?

in G. P. Moberg (Ed.), *Animal Stress* (Bethesda, MD: American Physiological Society, 1985), 27–49; Selye, H., The Evolution of the Stress Concept, *American Scientist* 61 (1973): 692–699; Trumbull, R., and M. H. Appley, A Conceptual Model for the Examination of Stress Dynamics, in M. H. Appley and R. Trumbull (Eds.), *Dynamics of Stress: Physiological, Psychological, and Social Perspectives* (New York: Plenum Press, 1986).

DONALD M. BROOM

SUBJECTIVITY OF ANIMALS

To be interested in animal welfare* is to assume that animals are capable of having subjective (or personal) feelings and thoughts. Only if we assume that animals can feel fearful, frustrated, unhappy, or bored does it make sense to want to improve their situation. However, for scientists working in the field of animal welfare, the problem is whether, and how, we can be certain that animals have such kinds of experiences.

To consider this problem, we should take a closer look at what is meant by the term "subjective." First, this term refers to inner experience; it indicates that animals (and humans) are beings with their own individual view of the world and their own needs* and desires. Second, the term refers to human knowledge and has the assumption of "antiobjective"; to call a statement "subjective" is to claim that it is based on private opinion and has no broader general validity among different people.

Unfortunately, these two meanings of the term "subjective" are frequently tied into one. Many assume that because feelings are of an inner, subjective nature, they therefore are not open to reliable, objective assessment, only to biased personal judgment. However, the two meanings of subjectivity should not be tied, but be carefully pried apart. With appropriate criteria, objective, unbiased investigation of subjective experience in animals may well be possible.

Various approaches to the study of subjective experience in animals have been put forward in recent years. One of the first and most influential ideas was to let animals "vote with their feet": when given a choice of environments, animals will spend most of their time in the environment they presumably like best (*see* PREFERENCE AND MOTIVATION TESTING). Another proposal was to test how hard animals are prepared to work for various kinds of rewards: to gain access to litter, for example, chickens* are willing to peck a key many times. Such studies indicate what animals prefer and like; however, they do not tell us what animals experience when they do not get what they like. We do not know whether they then suffer, and if so, how much. One approach is to test whether "out of sight is out of mind"; if animals can form mental images of their experiences and remember them (for example, companionship, litter, or the provision of food), we can ask whether they miss these experiences when they are absent, and suffer as a result. Another approach is to assess an animal's suffering* through detailed study of its expression. The animal's body posture, eyes, ears, tail, and the overall manner in which it relates to the environment are all expressive traits indicating how the animal feels (*see* ANIMAL BOREDOM). Although this approach does not address the cause of an animal's distress,* it does allow rapid diagnosis of serious subjective affliction.

We do not need absolute proof to take a phenomenon seriously and study it. After all, the existence of human suffering has, strictly speaking, not been scientifically proven either. That science as yet cannot explain why and how subjective experience exists does not mean that its existence is uncertain or unavailable for description and analysis. Careful description of phenomena is the start of scientific explanation, not the result.

Several philosophers have provided helpful starting points, arguing that an animal's perspective is closely linked to the species-typical way it interacts with the environment. To understand why subjective experience exists, perhaps the brain is the best focus of study; but to investigate what it is, what are the range and diversity of experience of which animals are capable, behavior in all its richly expressive aspects provides the best starting point. People who closely interact with animals in mutual partnership, such as dog* and horse trainers, zoo* keepers, and pet owners (*see* COMPANION ANIMALS AND PETS), develop an intimate acquaintance with the expressive repertoire of their animals and learn to understand it well. The science of animal welfare can be stimulated through development of various approaches, as indicated earlier.

Selected Bibliography. Dawkins, M. S., From an Animal's Point of View: Motivation, Fitness, and Animal Welfare, *Behavioral and Brain Sciences* 13 (1990): 1–61; Duncan, I. J. H., and J. C. Petherick, The Implications of Cognitive Processes for Animal Welfare, *Journal of Animal Science* 69 (1991): 5017–5022; Hearne, V., *Adam's Task: Calling Animals by Name* (London: Heinemann, 1986); Nagel, T., What Is It Like to Be a Bat? *Psychological Review* 83 (1974): 435–451; Wemelsfelder, F., The Scientific Validity of Subjective Concepts in Models of Animal Welfare, *Applied Animal Behaviour Science* 53 (1–2) (1997): 75–88.

FRANÇOISE WEMELSFELDER

SUFFERING OF ANIMALS

Suffering is a general term used in referring to animals who may be experiencing adverse physiological and mental states such as pain, * discomfort, fear,* distress,* frustration, boredom (*see* ANIMAL BOREDOM), torment, or grief. It is possible for an individual to suffer without pain—for example, an individual who constantly fears something—and to experience pain without suffering—for example, when one pinches oneself. In humans, suffering is recognized as having the dimension of mental processing involving awareness of self in relation to that physical state and reflects the integration of earlier experiences and future desires with the adverse state(s) being experienced. There is increasing evidence that animals other than humans have this ability, particularly the great apes, some other nonhuman primates, and perhaps other mammals (and even other vertebrates), but to date there is little empirical evidence for this.

Assessment of suffering is difficult in animals because they cannot directly communicate through a common language, and so it is based on careful observations of animal behavior and clinical signs. Such signs can be observed accurately and are either nonparametric or parametric. Nonparametric signs are observable as being present or absent but are not measurable on a continuum, as with parametric signs. Examples of nonparametric signs include harsh coat, runny eyes, hangdog look, eyes half open, diarrhea, lameness, hopping lame, and changes in behavior such as changes from docility to aggression or from quiet to vocalizing on approach. Parametric signs are measurable on a continuum and include body weight, body temperature, heart rate, or rate of breathing. Such an assessment of animal suffering is only possible when the normal physiological parameters and behavior of that individual animal or strain (breed) or species are well known. When these parameters have been established, one can estimate fairly objectively how far an animal has deviated from normality and what an animal may be feeling, and so begin to assess the level of suffering. Generalizing from human experiences in a similar condition to nonhuman animals also guides one to look for signs an animal may show, but has to take into account relevant biological differences between humans and animals. This approach has been termed critical anthropomorphism.*

Selected Bibliography. DeGrazia, D., and A. Rowan, Pain, Suffering, and Anxiety in Animals and Humans, *Theoretical Medicine* 12 (1991): 193–211; Fitzgerald, M., Neurobiology of Foetal and Neonatal Pain, in Patrick Wall and Ronald Melzack (Eds.), *Textbook of Pain*, 3rd ed. (London: Churchill Livingstone, 1994), 153–163; Institute of Laboratory Animal Resources and National Research Council, Committee on Pain and Distress in Laboratory Animals, *Recognition and Alleviation of Pain and Distress in Laboratory Animals* (Washington, DC: National Academy Press, 1992); International Association for the Study of Pain, Guidelines on Painful Experiments: Report of the International Association for the Study of Pain Subcommittee on Taxonomy, *Pain* 6 (1979): 249–252; Melzack, R., and P. Wall, *The Challenge of Pain* (Harmondsworth, UK: Penguin Books, 1982); Morton, D. B., Recognition and Assessment of Adverse Effects in Animals, in N. E. Johnson (Ed.), *Proceedings of Animals in Science Conference: Perspectives on Their Use, Care, and Welfare* (Melbourne, Australia: Monash University, 1995), 131–148; Morton, D. B., and P. H. M. Griffiths, Guidelines on the Recognition of Pain, Distress, and Discomfort in Experimental Animals and an Hypothesis for Assessment, *Veterinary Record* 116 (1985): 431–436.

DAVID B. MORTON

SYMPATHY FOR ANIMALS

Sympathy for animals has been the obvious motivating force behind the animal-protection movement. Stories of how animals are treated on factory farms (*see* FACTORY FARMING) or in scientific laboratories (*see* LABO-RATORY ANIMAL USE; SILVER SPRING MONKEYS) have been important for getting people to change their attitudes toward eating meat or toward scientific research using animals. Discussions of sympathy often fail to note that there are two different, if related, senses of "sympathy." These can be illustrated by contrasting two phrases: "sympathy for" and "sympathy with." Sympathy for (or toward) X always involves one experiencing something of the feelings that one imagines X has in the situation X is in. In addition, the feelings targeted are always negative, although they lead, in sympathy for, to feelings of generosity as well, as when we give aid because we feel sorry for a beggar or for someone who has to do something unpleasant.

The ability to feel sympathy for appears to be based on the capacity to have sympathy with: a broad ability to respond in a mirroring way to the emotions and, more generally, to the inner mental life of other conscious beings. For A to have sympathy with some being B is for A to think and feel in the same way as B does, and to do so on the basis of A's perception of B's situation. This mirroring response occurs unconsciously, for the most part, and at best is a set of "as-if" feelings and thoughts. Sympathy with, however, is not necessarily considerate or generous and does not necessarily lead to sympathy for. Hunters, farmers, anglers, animal trainers, and guards in prison may all have a sensitive understanding of the objects of their attentions—and have it by an inner mirroring of the mental life of the other—and yet feel neither sympathy for nor benevolence toward these objects.

The philosopher David Hume based his notion of sympathy on a natural or innate common sentiment among humanity that leads to a sort of sympathetic contagion in which one person's emotion tends to cause the same emotion in other people. Hume was aware that such sympathy extends beyond the human sphere to include our responses to nonhuman animals. The notion of "empathy"* is an alternative explanation for some of the same psychological phenomena. Although usually used as a synonym for "sympathy," "empathy" was originally invented early in the 20th century to describe and explain the experience of projecting one's feelings onto works of art, and it has been extended to describe psychological abilities and experiences that some people have to identify with others. Empathy is thought to involve different mechanisms than sympathy, most notably a projection of the self onto or into the "other."

Although some scientists are doubtful about the validity of human feelings of sympathy for or with other animals, cognitive psychologists have begun to investigate the idea that there is some mechanism of sympathy that explains how humans understand each other. Their idea is that we do it by means of an inner simulation, that we produce an experiential modeling of the operations of other minds, rather than understanding them by making inferences about what is going on in other minds through applying a general cognitive theory.

Even if sympathy is for the most part acceptable, there are still questions about its variability and whether we can appropriately extend sympathy to more distant life forms. Sympathy for animals seems to vary enormously cross-culturally, cross-historically, and even within a given individual's life. Some people pamper their pets while being cruel and heartless to other similar animals. The answer may be to educate sympathies so that they are based on the best theory of the animal in question. We learn, for example, that chimpanzees* "grin" when they are aggressive, not when they are amused. This assumes, however, that our sympathy-with feelings can be educated and are sufficiently flexible to encompass a wide range of beings.

Selected Bibliography. Chismar, D., Empathy and Sympathy: The Important Difference, *Journal of Value Inquiry* 22 (1988): 257–266; Fisher, J. A., Taking Sympathy Seriously; A Defense of Our Moral Psychology toward Animals, *Environmental Ethics* 9 (1987): 197–215; Gordon, R. M., Sympathy, Simulation, and the Impartial Spectator, *Ethics* 105 (1995): 727–742; Midgley, M., The Mixed Community, in *Animals and Why They Matter* (Athens: University of Georgia Press, 1983); Morton, D. B., G. M. Burghardt, and J. A. Smith, Critical Anthropomorphism, Animal Suffering, and the Ecological Context, in S. Donnelly, and K. Nolan (Eds.), Animals, Science, and Ethics [Special issue], *Hastings Center Report* 20 (1990): 13–19.

JOHN ANDREW FISHER

T

TAIL DOCKING. *See* DOCKING.

THEODICY. *See* RELIGION AND ANIMALS.

THEOPHRASTUS

Theophrastus (371–286 B.C.) was a Greek philosopher who asserted a close mental kinship between humans and nonhuman animals. Born in Eresus on Lesbos, he studied at Athens under Aristotle, eventually succeeding him as head of the school (Lyceum) from 322. Although much of his output is now lost, sizable portions of his *On Piety* were preserved by Porphyry* (Sorabji, *Animal Minds*, 175) and make clear his view that we owe animals justice, and also that it is wrong to sacrifice animals and, explicitly, to eat meat. Theophrastus is modern in his insistence not only that it is wrong to cause suffering to animals, but also that killing is unjust because it robs animals of their life. Unlike his teacher Aristotle, who held that animals could not form part of the moral community because they were incapable of rational friendship, Theophrastus maintained that animals enjoy kinship with humans and therefore deserve moral solicitude.

Selected Bibliography. Cole, Eve, Theophrastus and Aristotle on Animal Intelligence, in William Fortenbaugh and Dimitri Gutas (Eds.), *Theophrastus: His Psychological, Doxographical, and Scientific Writings*, Rutgers University Studies in Classical

Humanities, 5 (1992); Fortenbaugh, William, Pamela Huby, Robert Sharples, and Dimitri Gutas (Eds.), *Theophrastus of Eresus: Sources for His life, Writings, Thought, and Influence*, pt. 2 (Leiden, 1992), 404–437; Sorabji, Richard, *Animal Minds and Human Morals: The Origins of the Western Debate* (London: Duckworth, 1993).

ANDREW LINZEY

THEOS-RIGHTS. *See* RELIGION AND ANIMALS.

THERAPEUTIC USE OF ANIMALS. *See* ANIMAL-ASSISTED THERAPY.

TOLSTOY, LEO NIKOLAYEVICH

Leo Nikolayevich Tolstoy (1828–1910) was a Russian aristocrat, novelist, and writer. Like Mohandas Gandhi,* he was deeply committed to the principle of nonviolence, which he also extended to the animal world. He translated Howard Williams's *The Ethics of Diet* into Russian with an accompanying essay "The First Step" (1892), in which he commends vegetarianism* as a step toward achieving the moral perfection required by Christ's teaching as illustrated by the Sermon on the Mount. Tolstoy corresponded with the Humanitarian League and eventually became a member. Although he was influenced by Orthodox spirituality, he was deeply critical of the established Orthodox Church, complaining that it legitimized violence and cruelty. His many novels illustrate the need for a spiritual life inclusive of respect for animals: nowhere is this more powerfully stated than in the opening section of *Resurrection* (1904), where humans are pictured in their own physical and moral prison, unable to grasp that "every man and every living creature has a sacred right to the gladness of the springtime" (9).

Selected Bibliography. Sarolea, Charles, *Count L. N. Tolstoy: His Life and Work* (London: Thomas Nelson and Sons, 1932); Tolstoy, Leo, *A Confession and Other Religious Writings*, trans. with an introduction by Jane Kentish (Harmondsworth: Penguin Books, 1987), especially The Law of Love and The Law of Violence, 152–221; Tolstoy, Leo, The First Step (1892), in *Recollections and Essays*, trans. with an introduction by Aylmer Maude, 4th ed. (London: Oxford University Press, 1961), 123–135; Tolstoy, Leo, The First Step [extract], in Andrew Linzey and Tom Regan (Eds.), *Animals and Christianity: A Book of Readings* (London: SPCK; New York: Crossroad, 1989), 194–197; Tolstoy, Leo, *The Gospel of Humaneness: Selections from the Writings of Count Leo Tolstoy*, Vegetarian Jubilee Library (London: Ideal Publishing Company, 1897); Tolstoy, Leo, *Resurrection* (1904), trans. Vera Traill, foreword by Alan Hodge (New York: New American Library, 1961).

ANDREW LINZEY AND BERNARD UNTI

TRANSPORTATION AND SLAUGHTER

Many abuses that occur during transport and slaughter of animals are due to poor management. The single most important factor that determines how animals are treated during transport and slaughter is the attitude of management. Employees who handle thousands of animals can become numb and desensitized. To be most effective, a good manager must be involved enough in day-to-day activities to prevent detachment but must not become so involved in daily operations that desensitization occurs. A combination of well-designed equipment, trained employees, and dedicated, caring management results in transport and slaughter that is done with a minimum of discomfort.

Abusive treatment of "downers," sick or crippled animals that can't stand up, is the number one transport problem. Crippled animals are sometimes dragged or thrown. Good husbandry practices, such as hoof trimming, gentle handling, and selling cattle and sows when they are still fit to travel, can prevent most downers. Poorly managed dairies and farms are likely to have the highest percentage of downer animals.

A major welfare problem in the dairy industry is abuse of newborn calves. Some poorly managed dairies transport calves off the farm before they are old enough to walk, resulting in high death losses. Some dairies neglect to feed new calves the mother's colostrum, which helps them fight sickness. Other problem areas are transport of horses in double-deck cattle trucks and rest-stop requirements. Double-deck cattle trucks have adequate head room for cattle, but tall horses are likely to be injured when they hit the ceiling. Some horses and cattle are transported for many hours without rest stops. However, too many rest stops can increase stress.* The stress of loading and unloading has to be balanced against the benefits of rest. Welfare during transport can be improved with air-ride truck suspensions and improved ventilation systems. An air-ride suspension provides a much smoother ride for the animals.

Another serious welfare problem during transport is death losses due to genetic weakness in animals. Pigs* and poultry selected for superlarge muscles are weaker and die more often during transport than conventional animals. Overselection for leanness also results in nervous, excitable pigs and cattle who are more likely to become stressed during handling and transport.

Slaughter of cattle, pigs, horses, and other farm mammals is covered by the Humane Slaughter Act.* Poultry (*see* CHICKENS) are not covered. This act only applies to animals on the premises of the slaughter plant. Transport outside the plant premises is not covered. The Humane Slaughter Act requires that livestock be rendered insensible to pain* prior to slaughter by either captive bolt stunning, electric stunning, or CO_2 gas. The law is enforced in each slaughter plant by a U.S. Department of Agriculture (USDA) veterinarian* who is in charge of inspection. With captive bolt stunning, the animal is shot with a gun that drives a steel bolt into its forehead, and it is killed instantly. Captive bolt seems to be painless when done correctly. Captive bolt guns require very careful maintenance to maintain maximum hitting power. Poor captive bolt maintenance is one of the major welfare problems in a poorly managed plant.

Most pigs in the United States are rendered unconscious with cardiac arrest electric stunning. An electric current at 1.25 amps and about 250 to 300 volts is passed through both the heart and the brain of the pigs. The pig is electrocuted instantly and does not feel the shock when it is done correctly. Proper placement of the electrodes is essential. The pig's brain must be in the current path, and sufficient amperage (current) must pass through the brain to induce grand mal epileptic seizure. The two main welfare problems are placement of the electrodes in the wrong location and use of less than 1.25 amps. In CO_2 stunning, the pigs are anesthetized with gas, which seems to be very humane for certain genetic types of pigs. However, some genetic lines of pigs react very badly and become very agitated when they first come into contact with the gas.

Ritual slaughter is controversial and is exempt from the Humane Slaughter Act. In kosher (Jewish) (see RELIGION AND ANIMALS, Judaism and Sacrifice) and halal (Muslim) (see RELIGION AND ANIMALS, Islam) slaughter, fully conscious cattle, sheep, goats, and poultry are slaughtered without preslaughter stunning. The animal's throat is cut while it is fully conscious. In evaluating this procedure, one must separate the variable of the method used to restrain the animal from the actual throat cut. Cruel, stressful methods of restraint, such as hanging live cattle upside down by one back leg, are probably much more distressful to the animal than the throat cut. A properly done cut with a very sharp knife appears to cause little reaction from the animal. Many Muslim religious authorities will accept preslaughter stunning, but stunning is not permitted prior to kosher slaughter.

Effective, well-designed equipment is available for handling and holding cattle during slaughter. Systems with curved chutes with high, solid sides help prevent the animals from becoming frightened by using principles of animal behavior. Most large slaughter plants hold cattle during stunning in a conveyor restrainer system. When these systems are operated properly, the animals will quietly follow each other. For cattle, slaughtering is often less stressful than handling on the farm for vaccinations. The systems for cattle work much better than the systems for pigs. Cattle by instinct line up and walk up a single-file chute. Pigs resist moving in single file. Danish researchers are working on a new low-stress pig-handling system where pigs are stunned in groups.

Broken wings and legs on poultry can be greatly reduced when handlers are given payment bonuses for keeping injuries low. Bruises on cattle or pigs will be much lower if producers have to pay for them. One of the best ways

Temple Grandin demonstrates to an employee how to quietly turn a steer by shaking plastic streamers. Photo courtesy of Temple Grandin.

A curved shute with high solid sides utilizes behavior principles to keep cattle calm at the slaughter plant. Photo by Temple Grandin.

to reduce animal injuries is to make each person who handles or transports an animal financially accountable for damage.

Selected Bibliography. Grandin, T., Animal Handling, in E. O. Price (Ed.), *Farm Animal Behavior. Veterinary Clinics of North America, Food Animal Practice* 3(2) (1987): 323–338; Grandin, T., Euthanasia and Slaughter of Livestock, *Journal of the American Veterinary Medical Association* 204 (1994): 1354–1360; Grandin, T., Farm Animal Welfare during Handling, Transport, and Slaughter, *Journal of the American Veterinary Medical Association* 204 (1994): 372–377; Grandin, T. (Ed.), *Livestock Handling and Transport* (Wallingford, Oxon, UK: CAB International, 1993); Livestock Conservation Institute, 1910 Lyda Drive, Bowling Green, Kentucky.

TEMPLE GRANDIN

TRAPPING

The majority of trapped animals are captured for their fur skins, which are sold, or for management purposes, oftentimes referred to as "animal damage control." Others are trapped for biological studies.

Professional trappers are very few in number. In the United States there are only about 2,000 individuals who earn a living by hunting* and/or trapping (U.S. Department of Commerce, 1990). The vast majority of trappers are under the age of 20 and are involved in the activity for "recreation." The number of trappers fluctuates dramatically based on the price trappers are able to collect for the animals' pelts. Oftentimes the cost of trapping supplies exceeds the economic return.

There are three principal types of traps: limb-restraining, killing, and confinement. Traps are set on land, in the shallows with a slide to drag the animal into the water and drown it, or underwater to kill by drowning, strangulation, or a sharp blow to the neck.

Steel-jaw leghold traps are the most commonly used traps for catching animals for the fur trade. This limb-restraining device is used in all three set locations previously described. Steel-jaw traps have been condemned internationally as inhumane and have been banned in 88 countries, but are still used in the United States and Canada. When the trap is triggered, the jaws slam together with tremendous force upon the limb of whatever animal has set off the device. The jaws of the trap are standard steel, or they may contain sharp teeth or a small strip of hard rubber (called "padding" by the fur industry).

Scientists and veterinarians* have documented the injuries caused to leghold-trapped animals from being caught and from their violent struggle to escape the painful capture. Traumas include broken bones, severed tendons and ligaments, fractured teeth, and severe soft-tissue damage. Gangrene of the affected appendage can begin within as little as half an hour after being trapped. Animals may chew off their own limb to escape, an act termed

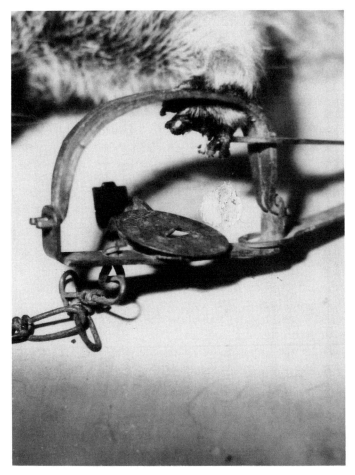

Trapping: A raccoon chewed at its captured foot in an attempt to escape. *Source*: Animal Welfare Institute.

"wring-off" by trappers. Nontarget animals, including companion dogs* and cats,* raptors, and deer, often get trapped.

Footsnares are another type of limb-restraining trap. Footsnares can greatly reduce the amount of pain* and injury caused to trapped animals as compared with leghold traps. Use of these traps virtually eliminates broken bones and broken teeth. The footsnares must incorporate a coating or tubing around the snare cable to prevent the snare from biting into the animal's limb, and the snare must have a means to prevent tourniquet-type tightening of the noose so that circulation is not cut off to the limb.

Killing traps are intended to kill the animal before the trapper returns, but many fail to kill rapidly. Necksnares, which are supposed to kill by stran-gulation, are examples of traps that cause a slow death for the animals; these traps are also known to catch many nontarget animals such as deer.

Mechanical killing traps are intended to strike the animal on the back of the neck, causing irreversible unconsciousness until death. The conibear is the most commonly used killing trap, but frequently it fails to cause a quick death. Animals not killed outright are tortured by this trap's grip on their heads or body. A number of other types of killing traps have shown more ability to kill immediately and to catch the target species.

Confinement traps, such as cage and box traps, are devices that hold the animal without gripping any appendages. These traps are popular because generally they cause no pain and little or no injury to animals who are caught. Most often the cages are covered in wire mesh, but there are also traps with solid sides of metal, plastic, or even logs. In fact, perhaps the least cruel trap available is the log box trap handmade of native materials. Cap-tured animals are sheltered from the weather, and there is no steel for them to damage their teeth on.

Regulation of trapping of furbearing animals differs in each of the 50 states. A small number of states have made tremendous strides in reducing trapping cruelty. Most states still allow trapping practices that result in an enormous amount of animal suffering.*

The law with the greatest impact on reducing the suffering of trapped animals throughout the world is Regulation 3254/91, adopted unanimously by the European Union in 1991. This law prohibited use of steel-jaw leghold traps in the 15 nations of the European Union beginning January 1, 1995. A second part to the law prohibits import into the European Union of fur from 13 species of wildlife (badger, beaver, bobcat, coyote, ermine, fisher, lynx, marten, muskrat, otter, raccoon, sable, and wolf) if the country has not ended use of steel-jaw leghold traps or adopted "internationally agreed hu-mane trapping standards." Currently, no such international standards exist.

Millions of rats and mice* are trapped throughout the world. The majority of them are caught in "snap traps" that are supposed to kill the animal rapidly with a lethal blow. Often they do not kill as intended, but strike in a nonlethal location, leaving the animal to die a slow death. Glue traps are becoming more common. Powerful glue adheres to the mouse, rat, or other small animal who enters the cardboard box so that it cannot extricate itself, nor can it be rescued by pulling it loose. Often the box and its still-living occupant are thrown away. Animals caught in these traps die of dehydration, starvation, or asphyxiation. Box traps are available for catching rats and mice, and these are much less cruel as long as they are checked frequently.

CATHY LISS

u

URBAN WILDLIFE

For most, if not all, of urban history there have been wild animals living in close proximity to humans. These *synanthropes* are far less studied than their counterparts in other habitats. By the late 1960s the significance of urban wildlife began to be recognized, and the first of a number of national conferences in the Americas focused on the many emerging issues associated with this field.

Today, urban wildlife is recognized as a subdiscipline of the larger field of urban ecology. From a perspective involving animal welfare* issues, three areas of concern regarding urban wildlife can be visualized: human-wildlife conflicts, the benefits and positive values associated with urban wildlife, and the suitability of urban (and suburban) habitats for wildlife.

Conflicts between people and wild animals in cities are not new. In the 1st century A.D., the historian Josephus described specially constructed metal spires that were installed on rooftops in Jerusalem to repel birds. The devastation of the plagues in medieval Europe was caused by fleas whose host was the Norway rat (*Rattus norvegicus*) and may have been the most destructive conflict (for humans) ever to occur. Municipal shelters* and animal control agencies historically have had little to do with wildlife, and state and federal wildlife agencies have traditionally focused their attention on agricultural and farming issues. Private individuals (animal rescuers and rehabilitators) and nature centers were often the only resource available to guide urbanites on resolving conflicts with wildlife until quite recently, when private businesses—nuisance wildlife control officers

(NWCOs) or pest control officers (PCOs)—began practicing in many metropolitan areas.

Many of the practitioners of urban wildlife damage control adhere to traditional wildlife-management perspectives. These regard wildlife as a renewable resource and emphasize management from a utilitarian (see UTILITARIANISM) and materialistic perspective. The control of problem wildlife populations occurs through hunting,* trapping,* or other activities that result in the destruction of animals. Yet urban wildlife managers face quite different issues than their traditionalist counterparts and a public that typically eschews hunting as a management procedure.

Studies indicate that urban populaces have strong humanistic and moralistic feelings about animals. Such concerns, coupled with rising environmental awareness, have led to favoring new approaches to conflict resolution. The field of integrated pest management encompasses environmentally responsible strategies for solving problems with "pest" species; the objective of the strategies is to harmonize the relationship of humans to other species.

The positive values that humans derive from an association with wild animals are the subject of much general speculation and discussion, but little focused study or research. Improved psychological and even physical health is associated with contact with natural environments and with wild animals themselves. Better environmental health has long been associated with greater juxtaposition of natural areas with human-built environments. Such areas support wildlife species that, because of their position at higher trophic levels, are sensitive and fairly precise indicators of environmental quality. Recently, much attention has been paid to the role of specific species like beavers in modifying environments to control natural processes that are regarded as injurious to humans, such as flooding.

Most human influences on global ecosystems arise from urban populations. The demands, requirements, and decisions of urban populations control the global ecosystem. Wildlife is a preferred component of natural systems, one in which humans typically vest more interest and attention than in physical environments or even other living communities. How the quality of the human environment is improved and enhanced by wildlife is an issue that will engage much attention as human populations become increasingly urban.

Selected Bibliography. Adams, L. W., *Urban Wildlife Habitats* (Minneapolis: University of Minnesota Press, 1994); Fitter, R. S. R., *London's Natural History* (London: Collins, 1990; Gilbert, O. L., *The Ecology of Urban Habitats* (London: Chapman and Hall, 1989); Hone, J., *Analysis of Vertebrate Pest Control* (Cambridge: Cambridge University Press, 1994); Platt, R. H., R. A. Rowntree, and P. C. Muick (Eds.), *The Ecological City* (Amherst: University of Massachusetts Press, 1994).

JOHN HADIDIAN

UTILITARIANISM

The term "utilitarianism" is often used to describe any ethical stance that judges whether an action is right or wrong by considering whether the consequences of the action are good or bad. In this broad sense of the term, "utilitarianism" is equivalent to what is sometimes called "consequentialism." It is opposed to rule-based ethical systems, according to which an action is right if it is in conformity with moral rules and wrong if it is in violation of these rules, irrespective of its consequences.

An example may help to make this more concrete. Is it wrong to break a promise? Those who base ethics on a set of moral rules and include "keep your promises" among these rules would say that it is. On the other hand, a utilitarian would ask: what are the consequences of keeping the promise, and what are the consequences of breaking it? In some situations the good consequences achieved by breaking the promise would clearly outweigh the consequences of keeping it.

This gives rise to a further question: what kind of consequences are relevant? According to the classic version of utilitarianism, first put forward in a systematic form by the English philosopher and reformer Jeremy Bentham, what ultimately matters is pleasure or pain.* Thus classic utilitarians judge acts right if they lead to a greater surplus of pleasure over pain than any other act that the agent could have done. Bentham included in his calculations the pleasures and pains of all sentient (*see* SENTIENTISM) beings. In rejecting attempts to exclude animals from moral consideration (as virtually everyone did in his day) Bentham (*Introduction*, 17.1.4) wrote: "The question is not, Can they *reason?* nor Can they *talk?* but, Can they *suffer?*"

Nowadays there are many who continue to call themselves utilitarians who, while still holding that the rightness of an act depends on its consequences, think that the idea that pleasure and pain are the only consequences that should count is too narrow. They argue that some people may prefer other goals—for example, a writer might be able to achieve a life of luxury by working for an advertising agency, but may prefer the long and lonely work of writing a serious novel. Bentham could claim that she thinks that she will get more lasting pleasure from writing the novel, but it is also possible that she simply considers writing something of lasting literary value to be more worthwhile, irrespective of how much pleasure it is likely to add to her life and the lives of others, than writing advertising copy. Considering such cases has led to the development of a form of utilitarianism known as "preference utilitarianism." Preference utilitarians judge acts to be right or wrong by attempting to weigh up whether the act is likely to satisfy more preferences than it frustrates, taking into account the intensity of the various preferences affected. On this view, too, animals will count as long as they are capable of having preferences, and an animal who can feel pain or distress* can be presumed to have a preference to escape that feeling.

Utilitarianism has great appeal because of its simplicity, and because it avoids many of the problems of other approaches to ethics, which can always require you to obey a rule or follow a principle, even though to do so will have worse consequences than breaking the rule or not following the principle. On the other hand, this very flexibility may also mean that the utilitarian reaches conclusions that are at odds with conventional moral beliefs. Hence one of the most popular ways of attempting to refute utilitarianism is to show that it can, in appropriate circumstances, real or imaginary, lead to the conclusion that it is right to break promises, tell lies, betray one's friends, and even kill dear old Aunt Bertha in order to give her money to a worthy cause. To this some utilitarians respond by retreating to some form of a "two-level" view of morality, based on the idea that at the level of everyday morality we should obey some relatively simple rules that will lead us to do what has the best consequences in most cases, while in some special circumstances, and when assessing the rules themselves, we should think more critically about what will lead to the best consequences. Other, more tough-minded utilitarians say that if our common moral intuitions clash with our carefully checked calculations of what will bring about the best consequences, then so much the worse for our common moral intuitions.

Selected Bibliography. Bentham, Jeremy, *An Introduction to the Principles of Morals and Legislation* (1789; New York: Hafner, 1948); Hare, R. M., *Moral Thinking* (Oxford: Clarendon Press, 1981); Mill, John Stuart, *Utilitarianism* (1863; London: Dent, 1960); Sidgwick, Henry, *The Methods of Ethics*, 7th ed. (London: Macmillan, 1907); Smart, J. J. C., and Bernard Williams, *Utilitarianism, For and Against* (Cambridge: Cambridge University Press, 1973).

PETER SINGER

Utilitarian Assessment of Animal Experimentation

Many defenders of animal experimentation claim that the practice is justified because of its enormous benefits to human beings. Utilitarians can judge conflicts between members of different species by saying that the moral worth of an action would be the product of the moral worth of the creature that suffers, the seriousness of the wrong it suffers, and the number of such creatures that suffer.

Many defenders of research often speak as if utilitarian (cost-benefit) calculation is easy. Frequently they cast the public debate as if the choice to pursue or forbid animal experimentation were the choice between "your baby or your dog." However, this way of framing the question can be grossly misleading. The choice has not been, nor will it ever be, between your baby and your dog. Single experiments (and certainly single experiments on single animals) do not confirm biomedical hypotheses. Only a series of related experiments can confirm such hypotheses. Animal experiments are part of a scientific framework that considers whether the practice of animal experimentation is sufficiently beneficial to justify its costs.

Whatever the precise details of this utilitarian analysis, animal experimentation clashes with the moral codes (a) against doing evil to promote some good and (b) against inflicting suffering* on one creature of moral value to benefit some other creature of moral worth. That is, we do an evil to animals to provide goods for humans. Moreover, the evil we do (inflicting suffering on animals) is definite, while the good we promote (preventing the suffering of humans) is only possible. Additionally, the creatures that suffer will not be the ones who benefit from that suffering. Dogs* pay the cost of experimentation; humans reap the benefits.

The force of these codes of conduct is deep in our ordinary morality. Although undergoing a painful bone-marrow transplant to save the life of a stranger is noble, we think that requiring a person to undergo that procedure would be wrong. Abandoning these codes of conduct, though, would mean that nonconsensual moral experiments on humans could be justified if the benefits to humans were substantial enough. It would also require abandoning the idea of the moral separateness of creatures, a view central to all Western conceptions of morality. For instance, virtually everyone would be opposed to requiring people to give up one of their good kidneys to save someone else's life. Thus, even if we assume that animals have less value than humans, this latter imbalance means that researchers must show staggering benefits of experimentation to justify the practice morally.

Moreover, when determining the gains relative to the cost of animal experimentation, we must include not only the costs to animals (which are direct and substantial), but also the costs to humans (and animals) of misleading experiments. For instance, we know that animal experiments mislead us about the dangers of smoking. By the early 1960s, researchers found a strong correlation between lung cancer and smoking. However, since efforts to induce lung cancer in nonhuman animal models* had failed, the government delayed acting.

Furthermore, since we should include possible benefits (since no benefits are certain) on the scales, we must also include possible costs. For example, some researchers have speculated that AIDS was transferred to the human population through an inadequately screened polio vaccine given to 250,000 Africans in the late 1950s. Although the hypothesis is likely false, something like it might be true. We know, for instance, that one simian virus (SV_{40}) entered the human population through inadequately screened vaccine. Therefore, it is difficult to know how researchers could possibly claim that there would be no substantial ill effects of future animal experimentation. These possible ill effects must be counted.

Finally, and perhaps most important, the moral calculation cannot look simply at the benefits of animal experimentation. It must look instead at the benefits that only animal research could produce. To determine this utility, the role that medical intervention played in lengthening life and improving health, the contribution of animal experimentation to medical intervention, and the benefits of animal experimentation relative to those of nonanimal research programs have to be ascertained.

Selected Bibliography. Bailar, J., III, and E. Smith, Progress against Cancer? *New England Journal of Medicine* 314 (1986): 1226–1231; Brinkley, J., Animal Tests as Risk Clues: The Best Data May Fall Short, *New York Times*, national ed., March 23, 1993, C1, C20–C21; Cohen, Carl, Animal Experimentation Defended, in S. Garattini and D. W. van Bekkum (Eds.), *The Importance of Animal Experimentation for Safety and Biomedical Research* (Dordrecht: Kluwer Academic Publishers, 1990); Elswood, B. F., and R. B. Stricker, Polio Vaccines and the Origin of AIDS [letter to the editor], *Research in Virology* 144 (1993): 175–177; LaFollette, H., and N. Shanks, *Brute Science: Dilemmas of Animal Experimentation* (London: Routledge, 1996); McKinlay, J. B., and S. McKinlay, The Questionable Contribution of Medical Measures to the Decline of Mortality in the United States in the Twentieth Century, *Health and Society* 55 (1977): 405–428.

HUGH LaFOLLETTE AND NIALL SHANKS

V

VEAL CALVES

For a dairy cow to produce milk, milk production must be initiated by her giving birth to a calf on a yearly basis. Her female calves are often kept on the dairy farm and raised to be possible replacements for old or low-producing cows. The male calves are of no use to most dairy farmers, so they are sold at local auctions or to calf dealers. Some of the calves may be slaughtered; most, however, are used for either "formula-fed" veal or are raised in groups, go into feed lots, and are then used for "dairy" beef. Most formula-fed calves are raised in 24-inch-wide crates their entire lives, 16 to 18 weeks, and are fed a liquid diet twice a day.

The iron intake of formula-fed veal calves is very closely regulated. When the calves first arrive at veal barns, they are often given iron supplements. However, prior to slaughter iron intake is restricted to below normal levels and the calves are made anemic. Anemia is necessary so that a pale or white color of the meat can be achieved. The paleness of a carcass is the most important factor in grading the meat and the price paid to the producer. Iron intake is easy to restrict as long as the calves do not have access to the normal sources of iron that a calf on pasture would have, for example, grass or dirt. In fact, one of the reasons for using wooden crates is to prevent the possibility of calves ingesting iron from a metal crate.

Numerous studies have found that the confinement of veal calves is stressful to the calves. Studies have also shown that calves housed in crates or on slatted floors have an increased motivation to exercise and that the thwarting of drives may be one of the contributing factors that make rearing veal calves in confinement difficult.

Veal growers are starting with very young calves who may not have received adequate colostrum shortly after birth, and the calves have gone through a marketing system in which some calves may have been sold as many as five times before reaching the veal grower. Up through the 1980s, most veal rations were heavily medicated with antibiotics to stimulate growth and prevent high death losses. Due to public concern over the feeding of such large amounts of antibiotics, manufacturers of veal rations have stopped mixing antibiotics in their diets, leaving the administration of antibiotics up the individual vealer. At slaughter, U.S. Department of Agriculture (USDA) inspectors spot-check calf carcasses for residues of several commonly used drugs. By self-policing to ensure that vealers stop using the drugs for which the USDA tests in time for the residue of the drugs in the calves to be reduced, the veal industry has succeeded in reducing its drug-residue violation rate.

The state of the present U.S. veal industry is a good example of the impact public opinion can have on a segment of animal agriculture. Formula-fed veal production started in the United States in the 1960s. It rapidly grew through the 1970s and early 1980s until 1985, when 3.4 million veal calves were slaughtered. Because of the low fat content of veal, it appeared that the demand for the product would continue to increase. However, public concern over the methods used to raise veal calves, fueled by the activities of a number of animal welfare* and animal rights* groups, grew dramatically during the mid-1980s, resulting in a decrease in the demand for formula-fed veal. Also, the first studies in the United States that addressed some of the welfare issues of confinement veal production, sponsored by the USDA, were published in 1985. These studies were consistent with earlier European studies and the general knowledge in the dairy industry that raising calves in crates or in groups on slatted floors was associated with increased health problems. Virtually all subsequent studies published in reputable peer-reviewed journals have been critical of confinement veal production. The production of and demand for formula-fed veal have dropped precipitously since 1985, but now have stabilized at approximately 800,000 calves per year, a decrease of 425%.

Rather than look for compromise and modify its production practices, the American Veal Association (AVA) took a hard-line stand and attempted to suppress any public discussion of the issue. The consumer, however, spoke through the marketplace. What was once perceived as a delicacy served at banquets and dinner parties is now shunned by many people.

Selected Bibliography. Dellmeier, G. R., T. H. Friend, and E. E. Gbur, Comparison of Four Methods of Calf Confinement. II. Behavior, *Journal of Animal Science* 60 (1985): 1102–1109; Du Vernay, Alan, The Ten Year History of the Vealer, *Vealer* 10 (1988): 6; Friend, T. H., and G. R. Dellmeier, Common Practices and Problems Related to Artificially Rearing Calves: An Ethological Analysis, *Applied Animal Behaviour Science* 20 (1988): 47–62; Friend, T. H., G. R. Dellmeier, and E. E. Gbur, Comparison of Four Methods of Calf Confinement. I. Physiology, *Journal of Animal*

Science 60 (1985): 1095–1101; Le Neindre, P., Evaluating Housing Systems for Veal Calves, *Journal of Animal Science* 71 (1993): 1345–1354.

TED FRIEND

VEGETARIANISM

Paul Amato and Sonia Partridge offer the following useful classification of vegetarianism: *lacto-ovo vegetarians* eat eggs and dairy products but no meat; those who eat dairy products but no eggs or meat are *lacto-vegetarians*; those who eat eggs but no dairy products or meat are *ovo-vegetarians; vegans* consume no meat, dairy products, or eggs; *macrobiotic vegetarians* live on whole grains, sea and land vegetables, beans, and miso; *natural hygienists* eat plant foods, combine foods in certain ways, and believe in periodic fasting; *raw foodists* eat only uncooked nonmeat foods; *fruitarians* eat fruits but also nuts, seeds, and certain vegetables; *semivegetarians* are those who include small amounts of fish and/or chicken in their diet.

Arguments for vegetarianism can be categorized as follows:

1. *Health.* Whether a vegetarian diet is as healthy as or healthier than one including meat is a source of much debate. It may seem that good health is a matter of one's own long-term self-interest, but some philosophers (e.g., Immanuel Kant*) have argued that we have duties to ourselves, others (e.g., Aristotle) that we must always strive to attain the virtuous (or morally decent) life. On both views, health (and thus a sound diet) is a precondition of being able to carry out these obligations and is therefore a matter of moral concern in the larger sense. Persons to whom we have responsibilities likewise have a stake in our health, as does society, which has an interest in our being productive, nonburdensome members. If a vegetarian diet were healthier, then it would be the one we should choose.

2. *Animal suffering and death.* There is no method for rearing food animals without pain* and suffering.* Whatever the method used, death is the final outcome. Confinement, transportation, and slaughtering (*see* TRANSPORTATION AND SLAUGHTER) are the main sources of pain and suffering. Modern "factory farming"* maximizes the problems, and its cruelties are well documented. Utilitarians (*see* UTILITARIANISM) are typically concerned with promoting pleasure and other interests of sentient beings, and with reducing or eliminating pain, suffering, and other conditions that frustrate welfare. They argue for vegetarianism as a way of helping reach this general end of morality. Animal rights* theorists see many nonhuman animals as irreplaceable individuals who have morally significant interests and hence rights, including the right to live and not be caused suffering. On the rights view, even totally painless meat production that gave great pleasure to human consumers would still be unacceptable.

3. *Impartiality and moral well-being.* An impartial person who is well in-

formed about animals understands that they have morally significant interests, such as life and well-being* (health and contentment), which can only be respected if we refrain from eating them. Using animals instrumentally for food violates the condition of impartiality and demonstrates speciesism.*

4. *Environmental concerns*. Large-scale meat production by agribusiness causes great environmental depletion and degradation, including excrement in waterways, loss of topsoil, deforestation, and wild-habitat destruction. Vegetarianism is seen as a way to lessen or eliminate such abuses.

5. *World hunger and social justice*. Food-animal production that relies on feedlots rather than natural foraging is extremely wasteful, yielding far less protein output than the protein input required to fuel it. Vegetarianism would aid in freeing up resources to feed the world's hungry by undermining the artificially created economy of scarcity.

6. *Interconnected forms of oppression*. Some ecofeminists* (*see* ANIMAL RIGHTS, Ecofeminists' Perspectives) have argued that various forms of domination, oppression, and exploitation are causally and conceptually connected. Those who are more powerful than others tend to exercise power over them, to see them as inferior, and to treat them as merely serving their own interests. A vegetarian way of life can contribute to breaking out of this traditional pattern.

7. *Universal compassion and kinship*. Evolutionary considerations of biological kinship reinforce the idea that humans should exercise compassion toward other animals. Vegetarianism accords with a compassionate approach to life.

8. *Universal nonviolence (ahimsa)*. Mohandas Gandhi* taught that violence begets more violence, that nonviolence (or *ahimsa*) is a superior moral force, and that humans have a duty to avoid or minimize the harm they cause all sentient beings. A vegetarian diet minimizes harm to other sentient beings.

9. *Religious arguments (see also* RELIGION AND ANIMALS). Some religions, notably Jainism, Hinduism, and the Pythagorean cult in ancient Greece, share a belief in reincarnation and in the ensoulment of humans and nonhuman animals. The Pythagoreans held that animals may contain the souls of former humans and thus should not be eaten. Many Hindus, Jains, and Buddhists refrain from eating animals out of respect for kindred beings with souls. Vegetarianism is sometimes advocated for the benefit of abstinence or spiritual purification. Some Christian and Jewish thinkers have taught that God granted humans stewardship rather than dominion over nature. Islam has also been presented as a stewardship religion, with the stronger proviso that causing grievous harm to nature is a direct offense to Allah. Vegetarianism may be seen as required to carry out the task of stewardship. Finally, the wisdom traditions of Indigenous Peoples teach that a spiritual identity or unity binds together all living things. Although this most often entails killing animals only out of necessity, reverently and wasting

nothing, it sometimes issues in a prescription for a vegetarian or semivege-tarian diet.

Taken together, these arguments have considerable persuasive force. Veg-etarianism, finally, can be seen as a means of focusing our attention not only on the human-animal or human-nature relationship, but also on the choice of a way of life that is morally and ecologically preferable.

Selected Bibliography. Akers, Keith, *A Vegetarian Sourcebook* (Arlington, VA: Vegetarian Press, 1983): Amato, Paul R., and Sonia A. Partridge, *The New Vegetar-ians: Promoting Health and Protecting Life* (New York and London: Plenum, 1989); Brown, Les, *Cruelty to Animals: The Moral Debt* (Houndmills, Basingstoke, Hamp-shire: Macmillan, 1988); Robbins, John, *Diet for a New America* (Walpole, NH: Still-point, 1987); Rosen, Steven, *Food for the Spirit: Vegetarianism and the World Religions* (New York: Bala, 1987); Wynne-Tyson, Jon, *Food for a Future: The Complete Case for Vegetarianism* (New York: Universe; Fontwell, Sussex: Centaur, 1979).

MICHAEL ALLEN FOX

Vegetarian Diets: Ethics and Health

Increasingly, people are adopting vegetarian diets for reasons of health or ethics. Vegetarian diets vary greatly, however, and different varieties of veg-etarianism might be endorsed by people with different moral commitments.

Nutritionists commonly recognize the following varieties of vegetarian: *vegans*, *lacto-ovo vegetarians*, *pesco-vegetarians*, and *semivegetarians*. People who are vegetarians on moral grounds can consistently use any of these diets, depending on what specific moral reasons they have for becoming vegetar-ians.

Many people have become vegetarians out of concern for human starva-tion. In *Diet for a Small Planet* Francis Moore Lappé argued that a lacto-ovo vegetarian diet would feed the world's human population more efficiently because a cow must eat many pounds of vegetable matter to grow a pound of meat, and much of that vegetable matter could have been used to feed humans. However, as large areas of the world that are not suited to farming could nevertheless support grazing animals, a semi-vegetarian diet could also be inspired by concern about human starvation.

Especially since the 1970s, many have become vegetarians out of concern for the well-being of farm animals (*see* FARM-ANIMAL WELFARE). Many have become lacto-ovo vegetarians, consuming only products that can be obtained without slaughtering the animals in question. Additional concern over the day-to-day confinement and handling of farm animals has led others to become vegans. In particular, the tight confinement of most laying hens today (*see* CHICKENS) has led some to avoid eggs. Also, it has been pointed out that because modern milking cows are impregnated yearly and spend an average of only three to four years in production, the dairy industry is closely tied both to the veal industry (*see* VEAL CALVES) and to beef slaughter (*see* TRANSPORTATION AND SLAUGHTER) in general.

With vegetarian diets as diverse as the moral reasons for adopting them, it is not surprising that estimates of the number of vegetarians vary widely (from 2% or 3% up to 10% or more for the United States), but clearly more and more people are becoming vegetarians for moral and/or health-related reasons. While nutritionists increasingly acknowledge the health benefits of these less meat-based diets, controversy remains regarding the safety of the more restrictive diets like veganism.

Some nutritionists claim that people with high metabolic needs, like pregnant or lactating women and children, face significantly higher risks of nutritional deficiency if they exclude both meats and dairy products from their diets. They claim, for instance, that (1) it is difficult for vegan women to get enough iron because iron from nonmeat sources is less efficiently absorbed than the iron available in meat (iron deficiency is a problem for women because menstruation removes iron from their systems monthly), (2) vegans cannot get enough vitamin B_{12} (deficiencies of which cause severe neurological damage) because the vitamin is produced by microorganisms in the digestive tracts of animals, and (3) it is particularly difficult for women to get enough calcium from a dairy-free diet (osteoporosis, a condition characterized by brittle bones, is a serious problem for postmenopausal women).

Other nutritionists claim that with some planning and variety, even a strict vegan diet is not significantly more risky than an average diet. For instance, they respond to the claims in the foregoing paragraph by arguing that (1) the efficacy of iron supplements is acknowledged, (2) vegans can get enough vitamin B_{12} from fermented vegetable products like tempeh or from microorganisms in their own digestive tracts, and (3) the high calcium intakes suggested for women today are only necessary in high-protein, largely meat-based diets. Nutritionist Colin Campbell has gone further, claiming that his long-term study of dietary habits in mainland China shows that a low-fat (10–20% of total calories), plant-based diet could significantly reduce the incidence of a variety of chronic degenerative diseases such as cancers and cardiovascular diseases in Western countries.

Selected Bibliography. Campbell, T. Colin, and J. S. Chen, Diet and Chronic Degenerative Diseases: Perspectives from China, *American Journal of Clinical Nutrition* 59 (1994): 1153–61; Comstock, Gary (Ed.), Might Morality Require Veganism? [Special edition], *Journal of Agricultural and Environmental Ethics* 7(1) (1994), including essays on both the morality of vegetarianism and the nutritional adequacy of vegetarian diets; Lappé, Francis Moore, *Diet for a Small Planet*, rev. ed. (New York: Ballantine Books, 1975); Singer, Peter, *Animal Liberation* (New York: NY Review of Books, 1990); Singer, Peter, *Practical Ethics*, 2nd ed. (New York: Cambridge University Press, 1993).

GARY VARNER

VETERINARIANS

Veterinarians' expertise and skill are vital for any animal-protection organization's work, and yet these two parties have not always had harmonious relations. The veterinary community has often come under criticism from animal-protection organizations. Disagreements between the veterinary and animal-protection communities have roots in the philosophy, practice, and economic reality of veterinary medicine.

Veterinarians practice in a societal context in which most animals are the legal property of individual owners (*see* LAW AND ANIMALS). Theirs is a business dealing largely with animals who are owned by people other than the veterinarian, who pay the bills and make decisions about their animals' lives. Rarely does a veterinarian have sole authority over how an animal is to be treated.

Veterinarians daily see the many ways in which animals' interests and human interests compete. Pet practitioners have clients who cannot or will not pay for needed medical care, who refuse to neuter their pets, who elect to have a dog's* ears trimmed or tail docked (*see* DOCKING), or who will choose to euthanize (*see* EUTHANASIA) a healthy animal.

In 1981, the American Veterinary Medical Association (AVMA) formed an Animal Welfare Committee to study the issues and make recommendations for AVMA position statements. The committee's intent is to focus on the scientific aspects of animal welfare.* Prior to formation of the Animal Welfare Committee, most AVMA ethical principles defined how veterinarians should treat their human clients and other veterinarians. Since 1981, the AVMA has developed positions on how veterinarians should treat their animal patients. No veterinarian is bound to abide by any of these AVMA policies. Following are a few positions of concern to people in the animal welfare and animal rights* movements.

Confinement rearing of livestock and poultry: Although confinement rearing is scorned by animal protectionists as factory farming,* the AVMA sees many opportunities to enhance the health and welfare of food animals by protecting them from weather and predators, assuring food and water supplies, and allowing the farmer (or producer) to carefully observe animals. The wide disparity between protectionists and veterinarians lies in differing definitions of animal welfare and its assessment. While protectionists focus on freedom of movement and animal behavior, veterinarians focus on physical health and disease control, especially the control of animal epidemics (epizootics).

Ear cropping: The AVMA is opposed to trimming dog ears for cosmetic and show reasons; it is a medically unnecessary procedure. The AVMA has called on the American Kennel Club and other breed associations to ban dogs with cropped ears from dog shows. Some veterinarians and other animal protectionists believe that the AVMA should adopt the position that it is unethical for veterinarians to perform this procedure, but the AVMA has disagreed. As long as trimmed ears are allowed or encouraged by breed standards, the AVMA believes that veterinarians should be the people performing the surgery, with good sterility practices and use of anesthetics and painkilling drugs.

Steel-jaw leghold traps (see also HUNTING; TRAPPING): The AVMA has changed its position on steel-jaw leghold traps over the course of a decade. The current policy statement no longer highlights the device's usefulness and states quite simply that the steel-jaw leghold trap is inhumane.

Low-cost spay/neuter clinics: Perhaps no issue has caused greater division between the veterinary community and the animal welfare community than the establishment of low-cost community facilities to surgically sterilize animals. Though veterinarians have expressed concern that such large-volume clinics might not maintain acceptable standards of sterility, anesthesia, and surgical expertise, financial and business disputes have resulted in controversy and lawsuits. Much of the mistrust and misunderstanding of the 1970s and 1980s has given way to a variety of programs, including animal shelter* clinics staffed by local veterinarians or their own staff veterinarians, and a variety of government and privately subsidized voucher systems to provide animals access to veterinary care who might not otherwise receive it.

Despite these several controversial issues, the veterinary and humane communities have agreed on several points, such as calling for full funding for enforcement of the Laboratory Animal Welfare Act,* opposing the use of performance-altering drugs in racehorses, and condemning the sports of dogfighting and cockfighting.

Selected Bibliography. American Veterinary Medical Association, *The Veterinarian's Role in Animal Welfare* (Schaumburg, IL: American Veterinary Medical Association, 1995); Maggitti, P., Veterinarians: For or against Animal Rights? *Animals' Agenda*, February 1989, 12–23; 48; Rollin, B. E., Veterinary Ethics and Animal Rights, *California Veterinarian* 37 (1) (1983): 9–13, 98; Tannenbaum, J., *Veterinary Ethics*, 2nd ed. (St. Louis: Mosby, 1995); Wilson, J. F., B. E. Rollin, and J. A. L. Garbe, *Law and Ethics of the Veterinary Profession* (Yardley, PA: Priority Press, 1988).

LARRY CARBONE

Veterinary Ethics

Veterinary medical ethics is a branch of professional ethics, a field that includes human medical ethics, legal ethics, research ethics—indeed, the ethics of all fields performing a specialized function in society and demanding specialized knowledge and special privileges. All professions face a variety of "pulls" growing out of separate and often conflicting moral obligations. Four obligations are common to all professions: obligations to society, obligations to clients, obligations to peers and the profession, and obligations to self and family. The fifth, obligations to animals, is unique to veterinary medicine.

Society has only recently begun to take seriously the question of our moral obligations to other animals. In 1978, Bernard Rollin identified this issue as the "fundamental question of veterinary medical ethics," that is, do veterinarians have primary obligation to animals rather than animal owners, in a manner similar to pediatricians, or do they have primary obligation to owners, as a garage mechanic does?

All of these varying obligations can and do conflict. For example, one's obligation to clients involves confidentiality. But suppose that client confidentiality conflicts with obligations to society, as when a client is using an unauthorized growth promoter in food animals. In the same way, obligations to animals frequently conflict with obligations to owners, as when an owner demands euthanasia* of a healthy young dog because the owner is moving. A third conflict grows out of the increasing social moral concern for animals mentioned earlier. Society expects veterinarians to be strong animal advocates. For example, a 1985 federal law charges veterinarians with assuring that the pain* and suffering* of research animals are controlled. Yet, at the same time, veterinary medicine has traditionally not led in animal welfare,* tending to defer to established practices in animal use and abuse because of the fact that clients pay the bill.

Before 1980, veterinary ethics focused almost exclusively on professional conduct among peers, and therefore largely on matters of etiquette. For example, the Code of Ethics of the American Veterinary Medical Association had numerous entries on advertising and no entries on convenience euthanasia. This has slowly changed. The first required course ever given at a veterinary school anywhere in the world on genuine ethical issues in veterinary medicine was developed in 1978 by Bernard Rollin at Colorado State University, and since then courses, lectures, and discussions on veterinary ethics have increasingly appeared in veterinary-school curricula. Social concern has galvanized veterinarians' attention to such animal issues as animal research and confinement agriculture. Ethical issues are now discussed at professional meetings and in professional journals. Veterinarians have begun to shoulder responsibility for moral issues involving animals because society expects and demands it of them, because most enter the profession out of strong moral concern for animals, and because they realize that elevating the moral status of animals in society also elevates the compensation and social status of those who care for animals.

Selected Bibliography. Kesel, M. Lynne, Veterinary Ethics, in *Encyclopedia of Bioethics*, rev. ed. (New York: Macmillan, 1995), 2520–2525; Rollin, Bernard E., Updating Veterinary Medical Ethics, *Journal of the American Veterinary Medical Association* 173(8) (1978): 1015–1018; Rollin, Bernard E., Veterinary and Animal Ethics, in J. F. Wilson, B. E. Rollin, and J. A. L. Garbe (Eds.), *Law and Ethics of the Veterinary Profession* (Yardley, PA: Priority Press, 1988), 24–48; Rollin, Bernard E., Veterinary Medical Ethics, monthly column in *Canadian Veterinary Journal* (1991–) analyzing real ethically problematic cases sent to the journal by veterinarians; Tannenbaum, Jerrold, *Veterinary Ethics*, 2nd ed. (Baltimore, MD: Williams and Wilkins, 1995).

BERNARD E. ROLLIN

VICTIMIZATION OF ANIMALS

Victimization is tied to the idea of rights. Human beings have rights and are victimized when those rights are violated. An assumption behind this discussion is that nonhuman animals have rights too, and that nonhuman animals are victimized when humans exploit, harm, or kill them for human gain. The question remains, however, whether or not an animal who is harmed or killed by another animal has been victimized. It is valuable to consider the processes that determine precisely when an animal becomes a "victim."

Humans victimize animals most fundamentally in the reasons we find for removing them from ethical concern. Rationalizations can be religious, moral, or scientific. By circular reasoning, God's world would be unbearably cruel if animal suffering mattered. In the seventeenth century, René Descartes* advanced this line of thought by asserting on weak grounds that animals feel no pain* at all. Descartes's notion of the beast-machine sustained the victimization of animals over the Enlightenment period when the sphere of rights and entitlements was expanding. The belief that animals lack consciousness, even if they do feel pain, is still occasionally claimed in scientific literature.

Until the early 20th century, animals in European cultures could be tried, convicted, and punished for crimes. Church officials sometimes subjected animals to torture to get confessions from them. Animals found guilty could be burned at the stake.

A survey of cultural representations of animal victims reveals many contradictions. Among domestic animals, pets arouse emotional and moral anxiety, while, until recently, livestock have received little consideration. One notes, for example, the utility of victimizing pets in horror and suspense movies. Often, the first victim of violence is the family pet, as in the films *Straw Dogs* and *Fatal Attraction*. Suspense mounts as the killers move up the scale to human victims.

Often, too, pathos surrounds the animal victim. The death of Bambi's mother in the Disney film is a case in point, as are the deaths of Redruff the partridge and the mother rabbit Molly Cottontail in stories from *Wild Animals I Have Known* by Ernest Thompson Seton. Pathos relies upon the humanization of the animal victim, giving the animal victim a name and a personality.

Individuation alone can generate compassion. In Fred Bodsworth's *The Last of the Curlews* and Allan W. Eckert's *The Great Auk*, the misfortunes of the lone surviving member of the Eskimo curlew and great auk species, respectively, tug at the heartstrings of the reader. When an animal is viewed as one member of a group, victimization is easier. Thus, if a domestic animal (*see* DOMESTICATION) normally destined for slaughter (*see* TRANSPORTATION AND SLAUGHTER) is singled out in a work of the imagination, that animal becomes the subject of moral and emotional concern. Examples are the recent movie *Babe* and Beat Sterchi's novel *Cow*, in which an individual pig and an individual cow are the focus of narrative interest and sympathy.

Wild animals have been demonized to justify assaults upon them. Wolves, bears, snakes, and sharks are foremost among animals claimed to have evil designs upon humans and other, innocent animals. Predatory animals in general excite human moralization. In a presumably animal-sympathetic film like *Benji the Hunted*, for instance, the primary enemy is a wolf, and audiences are meant to feel relief when Benji, a small, civilized dog, tricks the wolf into running off a cliff to his death. By the application of human moral values, any strong animal can become an enemy deserving assault and death. Jack London employs this device in *White Fang*, in which the hero sled dog Buck is tormented by another dog, Spitz: eventually the whole pack turns on Spitz and kills him.

In some literary works, the killing of animals proves or restores the virility of males. D. H. Lawrence's *The Fox* illustrates this theme when a man shoots a particularly mesmerizing fox and ultimately gains a similarly hypnotic power over the female protagonist. Even the appreciation of animal death* can prove manhood, as the *aficion* of Ernest Hemingway's castrated hero Jake for the bullfight demonstrates in *The Sun Also Rises*.

Selected Bibliography. Arluke, Arnold, and Boria Sax, Understanding Nazi Animal Protection and the Holocaust, *Anthrozoös* 5 (1992): 6–31; Cavalieri, Paola, and Peter Singer (Eds.), *The Great Ape Project: Equality beyond Humanity* (London: Fourth Estate, 1993); Evans, E. P., *The Criminal Prosecution and Capital Punishment of Animals* (New York: E. P. Dutton, 1906); Passmore, John, The Treatment of Animals, *Journal of the History of Ideas* 36 (April–May 1975): 195–218; Scholtmeijer, Marian, *Animal Victims in Modern Fiction: From Sanctity to Sacrifice* (Toronto: University of Toronto Press, 1993); Spiegel, Marjorie, *The Dreaded Comparison: Human and Animal Slavery* (Philadelphia: New Society Publishers, 1988).

MARIAN SCHOLTMEIJER

VIRTUE ETHICS

The term "virtue" generally refers to any one of many desirable traits, habits, skills, abilities, dispositions, excellences, and so on. For example, knowledge, literacy, compassion, humility, moderation, strength, courage, wealth, and beauty are virtues. This usage is somewhat different from popular language, where the term "virtuous" often means "pure" or "untainted."

Virtue ethics is very different from the two leading modern moral theories, which are utilitarianism* (or consequentialism) and rights (or nonconsequentialism, including duties). Utilitarianism and rights are concerned primarily with *actions*: which actions are right and which are wrong. Virtue ethics looks at the *character* of the agent. Virtue ethics is concerned with what kind of person would kill a pet (*see* COMPANION ANIMALS AND PETS) merely because the pet is old and no longer frisky, questioning whether or not that is the kind of person one ought to be.

Three issues are of interest to the theory of virtue ethics. First is the question of the nature or essence of virtue. Second is the unity of the virtues. We want to know how the virtues are connected, whether there is one virtue such that the many virtues are aspects of it, and whether the virtues have a ranking, or hierarchy, where one is premier or central and the others subordinate or peripheral.

The third issue is especially important for the question of whether animals have virtues. Here we want to know the relationship of knowledge to virtue. To what extent, for a virtue to be virtuous, must it be known, understood, consciously valued, chosen, or nurtured by its possessor? We also want to know whether virtues can be taught, and whether, if the trait or habit is innate, instinctive, biological, or environmentally conditioned, it is a virtue.

Contemporary theories of animal ethics have been developed largely within the utilitarian and rights perspectives (by Peter Singer and Tom Regan, respectively). Equally important for virtue ethics are the positions of Stephen R. L. Clark, Mary Midgley, and Bernard Rollin, who present Aristotelian perspectives on animals. The basic thesis is that animals should be given, or permitted to have, a life according to their kind, a notion roughly similar to Aristotle's concept of natural end (*telos*).

Selected Bibliography. Aquinas, Thomas, *Summa Theologiae*, Ia-Iae, q55–89; IIa-IIae, q.1–170 (1266–1273); Aristotle, *Nichomachean Ethics*, books 1–3; Clark, Stephen R. L., *The Moral Status of Animals* (New York: Oxford University Press, 1984); Hume, David, *An Enquiry Concerning the Principles of Morals* (1751), Kruschwitz, Robert B., and Robert C. Roberts (Eds.), *The Virtues: Contemporary Essays on Moral Character* (Belmont, CA: Wadsworth, 1987); MacIntyre, Alasdair, *After Virtue*, 2nd ed. (Notre Dame, IN: University of Notre Dame Press, 1984).

JACK WEIR

VIVISECTION. *See* ANTIVIVISECTIONISM, EDUCATION AND THE USE OF ANIMALS.

WELL-BEING OF ANIMALS

The least controversial component of animal well-being is *experiential well-being*. Sometimes called "quality of life" in other contexts, experiential well-being is more accurately understood as quality of experiences or feelings.* One is experientially well off to the extent that one has such feelings as pleasure, enjoyment, and satisfaction; one is experientially poorly off to the extent that one has such feelings as pain,* distress,* and suffering.*

It is debatable whether animal well-being consists of anything other than experiential well-being. One point of controversy is whether animals have an interest in life—remaining alive—or, equivalently, whether death* harms an animal who dies. The issue is put in focus by asking whether an animal who is painlessly killed while sleeping (so that his or her dying involves no unpleasant experiences) is harmed. It is possible that different answers are appropriate for different sorts of animals. First, it is reasonable to hold that animals who have no feelings at all are not harmed by death (or anything else). Moreover, even if fish* and elephants (*see* CIRCUSES AND CIRCUS ELEPHANTS) both have feelings and therefore an experiential well-being, it might be argued that elephants, but not fish, have an interest in life due to the more complex consciousness and rich, long-term social relationships that characterize elephant lives.

Addressing such issues in any detail requires a theory of well-being. One leading possibility is to conceptualize death as a harm that consists in lost opportunities—for pleasure, enjoyment, and satisfaction, but perhaps also for features of lives whose value is independent of experimental well-being.

A broad theory of well-being will also shed light on other aspects of animal well-being. For example, we can ask about the relationship between well-being and freedom (*see* ANIMAL WELFARE, Freedom) or liberty, understood as the absence of external constraints on movement. Suppose that an animal is given a drug that causes him or her not to mind life in a small cage (and has no unpleasant side effects). Then if confinement causes the animal any harm, this is probably because the animal has an interest in freedom that is independent of experiential well-being, in which case animal well-being is not simply a matter of the quality of experiences.

A similar issue arises with respect to functioning. Functioning, both mental and physical, is clearly important to well-being. Typically, for example, dogs* who are brain damaged are vulnerable in unique ways. But consider a dog who becomes brain damaged (either naturally or by surgery) in a way that cuts his intelligence in half, so that his practical problem solving is much poorer, but also changes his emotional life such that he is equally contented on the whole. If the brain damage has harmed the dog, whose experiential well-being is no worse, it would seem that dogs have an interest in functioning that is not reducible to its protection of experiential well-being.

However animal well-being is understood, trade-offs are imaginable. For example, a whale might be caused some distress today by veterinarians* giving her needed medical treatment but benefit in the long run from improved health. Some sorts of trade-off are difficult to view confidently due to unresolved theoretical questions about the nature of animal well-being. Determining what is best for animals depends on a detailed understanding of animal well-being.

Selected Bibliography. DeGrazia, David, D., *Taking Animals Seriously: Mental Life and Moral Status* (Cambridge: Cambridge University Press, 1996); Regan, Tom, *The Case for Animal Rights* (Berkeley: University of California Press, 1983); Rollin, B. E., *Animal Rights and Human Morality*, 2nd ed. (Buffalo, NY: Prometheus Books, 1992); Sapontzis, S. F. *Morals, Reason, and Animals* (Philadelphia: Temple University Press, 1987); Singer, Peter, *Animal Liberation* (New York: New York Review of Books, 1990).

DAVID D. DeGRAZIA

Assessment of Quality of Animal Life

For the past 30 years, animal welfare* scientists have tried to assess the quality of life of animals in order to identify which housing and management systems are better, or even best, for farm, laboratory, and zoo* animals. This has proven, however, to be a difficult task. Animal (or human) quality of life is determined by the total impact of a large variety of factors affecting the individual. The difficulty in applying such a holistic approach, however, lies not so much in identifying potentially relevant factors as in combining all factors into a single decision about the relative merit of housing or husbandry systems.

Cost-benefit analysis, as borrowed from the field of economics, has been widely used to weigh the advantages and disadvantages of various alternatives, but these have traditionally been assessed numerically. There is, however, another method, cost-benefit dominance (CBD), which requires only that we can determine, for each factor, whether one alternative is preferable to the other. This method has been used in human personality assessment and in evaluating human quality of life.

As an example, let us assume that we wish to compare two housing systems for pregnant pigs.* We first list all factors affecting the pigs' quality of life: freedom from disease, comfortable ambient temperature, adequate food intake, and so on. Then we decide (based on current scientific evidence) which system is preferable for the pigs with respect to each of the factors on our list. If, at the end of the assessment, system A is preferable for all factors, or preferable for some factors and equal for all others, then we conclude that it dominates system B. More likely, however, system A will dominate for some factors and system B for others. In such situations, CBD offers four strategies to incorporate information about the relationship between factors, for although we cannot numerically weight them, we do know that some are more important (or differently important) than others.

In summary, qualitative cost-benefit analysis offers an assessment alternative that makes the logic of the assessment process more clear, yet does not require quantification of individual factors or their relationships. In these ways, CBD provides a useful framework for an orderly comparison of housing and management systems with regard to the quality of animal life.

Selected Bibliography. Hurnik, J. F., and H. Lehman, The Philosophy of Farm Animal Welfare: A Contribution to the Assessment of Farm Animal Well-being, in R. M. Wegner (Ed.), *Proceedings of the Second European Symposium on Poultry Welfare* (Celle, Federal Republic of Germany, June 10–13, 1985), 256–266; Michalos, A. C., *North American Social Report: A Comparative Study of the Quality of Life in Canada and the USA from 1964 to 1974*, vol. 1, *Foundations, Population, and Health* (Dordrecht: D. Reidel Publishing Co., 1980); Taylor, A. A., Theoretical and Practical Aspects of Assessing the Quality of Life of Laying Hens in Alternative Housing Systems (Ph.D. thesis, Department of Animal and Poultry Science, University of Guelph, Ontario, 1994); Taylor, A. A., J. F. Hurnik, and H. Lehman, The Application of Cost-Benefit Dominance Analysis to the Assessment of Farm Animal Quality of Life, *Social Indicators Research* 35 (1995): 313–329.

ALLISON A. TAYLOR

WESLEY, JOHN

A fellow of Lincoln College, Oxford, John Wesley (1704–1791) began the spiritual revival later known as Methodism. He was one of the few English reformers to advocate ethical care for animals. His sermon in defense of animal immortality has become a classic: "But what does it answer to dwell upon this subject [the future life of animals] which we so imperfectly understand? . . . It may enlarge our hearts towards these poor creatures, to reflect that, vile as they may appear in our eyes, not one of them is forgotten in the sight of our Father which is in heaven" ("The General Deliverance," 285).

Selected Bibliography. Telford, John, *The Life of John Wesley* (London: Epworth Press, 1947); Wesley, John, The General Deliverance, in *Sermons on Several Occasions*, biographical note by John Beecham, vol. 2 (London: Wesleyan Conference Office, 1874), no. 60; Wesley, John, *Journal*, standard ed., vol. 4 (London: Charles H. Kelley, 1909).

ANDREW LINZEY

WHITE, CAROLINE EARLE

Caroline Earle White (1833–1916) was the founder of all three of Pennsylvania's most significant animal organizations and a founding member of the American Humane Association. The daughter of a well-known Quaker abolitionist, White was the mainspring of the Pennsylvania Society for the Prevention of Cruelty to Animals (SPCA), founded in 1867. The gender politics of the era led her to reorganize the Women's Auxiliary as an independent organization, and under her guidance, the Women's Pennsylvania SPCA began operation of the nation's first animal shelter,* an alternative to the dog pound where unwanted animals were brutally killed by uncaring municipal employees. She was one of the earliest promoters of animal adoption in America, as well as one of the first animal advocates to struggle with vivisectionists (*see* ANTIVIVISECTIONISM) over the use of pound and shelter animals in research. After a personal meeting with Frances Power Cobbe,* White founded the first antivivisection society in the United States, the American Anti-Vivisection Society, in 1883.

Selected Bibliography. Coleman, Sydney Haines, *Humane Society Leaders in America* (New York: American Humane Association, 1924); Lovell, Mary F., *Outline of the History of the Women's Pennsylvania Society for the Prevention of Cruelty to Animals from Its Foundation April 14, 1869, to December 31, 1899* (Philadelphia: Women's Pennsylvania Society for the Prevention of Cruelty to Animals, 1900); Mrs. C. E. White, Humanitarian, Dies, *Philadelphia Inquirer*, September 7, 1916; White, Caroline Earle, The History of the Anti-Vivisection Movement, in *Proceedings of the International Anti-Vivisection and Animal Protection Congress* (New York: Tudor Press, 1914), 25–35.

BERNARD UNTI

WILD ANIMALS, DUTIES TO

The question of duties to wild animals is often disputed among ethicists. Leading issues surround hunting* and trapping,* animal suffering,* appropriate levels of management by humans, poisoning, habitat degradation, feral* animals, restoration, and endangered species.*

Duties to wild animals, if they involve care, also involve noninterference, sometimes called hands-off management. In February 1983 a bison fell through the ice into the Yellowstone River and struggled to get out. Snowmobilers looped a rope around the animal's horns and attempted a rescue. They failed, and the park authorities ordered them to let the animal die and refused even to mercy-kill it. "Let nature take its course" is the park ethic. In 1981–1982, bighorn sheep in Yellowstone caught pinkeye (conjunctivitis). Partial blindness often proves fatal on craggy slopes. More than 300 bighorns died, over 60% of the herd. Wildlife veterinarians* might have treated the disease, as in any domestic herd, but the Yellowstone ethicists claimed that the disease should be left to run its natural course as a part of natural selection.

Some respond that human nature urges compassion for suffering, and we should let human nature take its course. But compassion is not the only consideration, and in environmental ethics it plays a different role than in humanist ethics. Animals live in the wild, subject to natural selection, and the integrity of the species is a result of these selective pressures. To intervene artificially is not to produce any benefit for the good of the kind, although it would benefit an individual bison or whale. Human beings, by contrast, live in culture, where the forces of natural selection are relaxed, and a different ethic is appropriate.

Wild animals are often affected by human-introduced changes, and this can change the ethic. Colorado wildlife veterinarians have made extensive efforts to rid the Colorado bighorns of a lungworm disease. Arguments were that some think that the lungworm parasite was contracted from imported domestic sheep, or that even if it was a native parasite, the bighorns' natural resistance is weakened because human settlements in the foothills deprive sheep of their winter forage and force them to winter at higher elevations. There, undernourished, they contract the lungworm first and later die of pneumonia.

The ethic changes again when an endangered species is involved. In the spring of 1984 a sow grizzly and her three cubs walked across the ice of Yellowstone Lake to Frank Island, two miles from shore. They stayed several days to feed on two elk carcasses, while the ice bridge melted. Soon afterward, they were starving on an island too small to support them. This time park authorities rescued the mother and her cubs and released them on the mainland.

Despite the protests of some in the ranching community, wolves have recently been reintroduced to Yellowstone National Park, having been exterminated there early in this century. Such restoration arises, according to most supporters, from a duty to the wolf as a species, coupled with the fact that the wolf was historically, and ought to be again, the top predator in the

Yellowstone ecosystem. Conservationists also realize that problem wolves will have to be relocated and often killed, and believe that this is an acceptable killing of individuals in order to have the wolf species present.

Duties to animals can conflict with concern for endangered animal or plant species. In a 1996 case, the U.S. Fish and Wildlife Service moved to poison 6,000 gulls at Monomoy National Wildlife Refuge off Cape Cod in order to save 35 piping plovers, an endangered species. A U.S. District Court rejected an appeal by the Humane Society of the United States* to stop the killing.

San Clemente Island, off the coast of California, has both native plant species and a population of feral goats, introduced by Spanish sailors two centuries ago. To protect plants numbering in the few hundreds, the Fish and Wildlife Service and the U.S. Navy have shot tens of thousands of feral goats. The Fund for Animals protested that it is inhumane to count a few plant species more than many mammal lives, but again the ethic of species triumphed.

Selected Bibliography. Hargrove, Eugene C. (Ed.), *The Animal Rights/Environmental Ethics Debate* (Albany: State University of New York Press, 1992); Rolston, Holmes, III, Ethical Responsibilities toward Wildlife, *Journal of the American Veterinary Medical Association* 200 (1992): 618–622; Rolston, Holmes, III, Higher Animals: Duties to Sentient Life, chapter 2 in *Environmental Ethics* (Philadelphia: Temple University Press, 1988).

HOLMES ROLSTON III

WILD BIRD CONSERVATION ACT

The federal Wild Bird Conservation Act was passed by Congress in 1992. A major incentive to congressional action was the cruel treatment of the birds by the big bird dealers who caused tens of thousands of deaths by cramming transport crates so tightly that only the hardiest survived the trip to pet stores in the United States, Europe, or Japan. Many magnificent species were depleted and sank from threatened to endangered status as the trade, combined with logging of forests and other habitat destruction, eliminated them from areas where they had thrived for countless years. Macaws and smaller members of the parrot family Psittacidae were especially pressed by the trade because of their beauty, intelligence, and capacity for companionship with human beings.

As the public outrage at the cruel mistreatment of the wild-caught birds increased, the airlines who had been carrying them for the international pet industry dropped out one by one. A hundred airlines had refused wild bird shipments by the time the Wild Bird Conservation Act was being considered by Congress. Lufthansa, a major carrier of animals, was the first to refuse all wild bird shipments.

Before the Wild Bird Conservation Act was finally passed, concentrated efforts to pass state laws were repeatedly made by humane organizations and conservation groups. New York State passed the Wild Bird Law in August 1984. The pet trade mounted a legal challenge charging that the law was unconstitutional and that pet shops would be put out of business. The law withstood the pet industry's attack, and far from being put out of business, the pet stores increased their profits. In 1991, the state of New Jersey passed a law banning importation and sale of wild-caught birds. But attempts to pass similar laws in other states all failed due to highly organized opposition by the Pet Industry Joint Advisory Council (PIJAC). Finally, PIJAC approached the World Wildlife Fund with a proposal to phase out wild bird imports in five years and write a bill with conservation and animal-protection groups, which never happened.

Two bills were presented to Congress, and the bill brought to the floor of the U.S. House of Representatives by Gerry Studds (Democrat of Massachusetts) was a compromise between the across-the-board import bans of New York State and New Jersey and the regulation proposed by PIJAC allies. The Studds bill passed both houses of Congress and was signed into law by President Bush.

The act immediately banned the import of the ten species whose survival was most severely threatened by capture for the pet trade. A year later, it prohibited importation of all species of birds listed by the Convention on International Trade in Endangered Species (CITES). The U.S. Department of Interior (USDI) duly prohibited the birds in CITES' Appendix I (endangered) and Appendix II (threatened), but continued to allow birds in Appendix III (protected species in individual countries) to enter the United States. Bird-protection groups sued the department on behalf of these birds and won. Judge Charles Richey ruled that the species of birds listed by their native countries as receiving protection are included in America's Wild Bird Conservation Act.

The Wild Bird Conservation Act has substantially reduced the volume of wild-caught birds exported to the United States. The numbers grow lower year by year. The bird trade has largely adjusted to sale of domestically bred exotic birds, and this business is thriving.

Hostility to the act still exists among those bird dealers who profited from the immense markups that characterized the wild-caught bird trade. A trapper in the rainforest would receive a pittance from the middleman who, in turn, shipped the birds out of the country for the big international dealers, who might increase the price as much as 1,000%. Because of this excessive profiteering, smuggling was rife, and the U.S. Fish and Wildlife Service's "Operation Renegade" uncovered evidence and took the perpetrators to court, where many were convicted of violating the act.

CHRISTINE STEVENS

WOOLMAN, JOHN

John Woolman (1720–1772) was an English Quaker divine and ardent abolitionist who expressed a reverence-for-life philosophy in his writings and personal practices. In his *Journal* (1772, 178–179), Woolman recorded that he was especially disturbed by the suffering of barnyard fowl carried for food on the ship on which he made his journey to England. Earlier, he recorded his conviction that "true religion" consisted in exercising "true justice and goodness not only towards all men but also towards the brute creatures" (*Journal*, 1720–1742, 28). Woolman declined to use stagecoaches and would not even send letters by couriers, finding the horses badly abused by their owners' habits of running them to death in an effort to maintain reputations for speed and efficiency. By practicing vegetarianism,* Woolman complemented his boycott against cotton, sugar, and indigo dye produced by slave labor with a conscious witness against animal exploitation. The significance of Woolman's witness was not lost on later Quakers, as they became the first Christian sect to oppose blood sports and incorporate kindness to animals as an article of faith (see *Advices and Queries*, 1928, 1964, Query 19).

Selected Bibliography. *Advices and Queries* (London: Society of Friends, 1928 and 1964); Lawson, Chris, *Some Quaker Thoughts about Animal Welfare* (London: Quaker Social Responsibility and Education, 1985); Woolman, John, *Journal* (London: Moulton, 1720–1742); Woolman, John, *Journal* (London: Moulton, 1772).

ANDREW LINZEY AND BERNARD UNTI

X

XENOGRAFT

The demand for transplantable tissues and organs is much greater than the supply. Many people die every year on transplant waiting lists. Physicians and medical researchers have long been fascinated by the idea that nonhuman animals might become an appropriate source for organs, and that xenografts (organs or tissues transplanted between animals of different species) could even solve the organ scarcity problem. Supporters of this idea have imagined setting up "farms" in which animals would be kept at the ready for human beings who need new hearts, livers, kidneys, lungs, or other body parts.

The idea that no one need die waiting for an organ is an attractive one, but there are many obstacles, both technical and ethical, in the way of xenograft's becoming the solution to this problem. Technically, organs from nonhumans have not yet been shown to be feasible for use in humans. In fact, every effort of this kind, from the implantation of a chimpanzee* heart into a 68-year-old man in 1964, through the transplantation of a baboon's heart into the infant "Baby Fae" twenty years later, to the 1994 attempt to transplant a pig's* liver into a 26-year-old woman, has ended dismally. In every case, the patient died shortly after receiving the xenograft.

Yet even should the technical problems someday be solved, the moral problems would remain. The central ethical challenge to xenograft concerns whether taking organs from healthy animals for use in human beings can be justified.

A number of serious moral arguments conclude that animals may not be treated in this way, even if doing so would offer a human being a considerable chance of living longer. For example, Tom Regan's claim that many animals (including those who might become attractive organ sources for humans) "have a distinctive kind of value in their own right, if we do; therefore, they too have a right not to be treated in ways that fail to respect this value" would, if correct, imply that xenografting is immoral. An allied view, based on the argument from marginal cases,* would also condemn xenograft unless we were willing to regard the mentally handicapped or other "marginal" members of our species as potential sources of transplant organs as well.

Those who favor trying to develop xenografting as a reliable method of obtaining organs often point out that we take animal lives for many less serious reasons than obtaining organs for people who will die without them. For example, we eat and wear animal products when there is no real life-or-death need to do so. Further, xenograft is just a particularly visible way in which animals are used in medical research, education,* and therapy: a great deal of what happens to any patient in very many medical encounters involves the suffering* and death* of animals, on whom drugs were tested and physicians and surgeons studied. Finally, there is great interest among those who are involved in xenograft research in using pigs rather than primates as sources of organs. Whereas primates are scarce, expensive, and disturbingly humanlike, pigs are breakfast food; if it is morally legitimate to raise pigs in confinement settings and then eat sausage, why is it not morally legitimate to genetically engineer (*see* GENETIC ENGINEERING) pigs in laboratories and then use their organs for people who may die without them? The answer to this question may simply be that it is not morally legitimate to use animals for food and clothing, even though people commonly do, and not defensible to use animals as we have done in medical research, testing, and education.

Selected Bibliography. Caplan, Arthur, Is Xenografting Morally Wrong? *Transplantation Proceedings* 24 (1992): 722–727; Discussion, *Transplantation Proceedings* 2 (1970); Kushner, Thomasine, and Raymond Belliotti, Baby Fae: A Beastly Business, *Journal of Medical Ethics* 11 (1985); Nelson, James Lindemann, Animals as a Source of Human Transplant Organs, in J. Humber and R. Almeder (Eds.), *Biomedical Ethics Reviews 1987* (Towtowa, NJ: Humana Press, 1988); Nelson, James Lindemann, Transplantation through a Glass Darkly, *Hastings Center Report* 22 (1992): 6–8; Regan, Tom, *The Case for Animal Rights* (Berkeley: University of California Press, 1983).

JAMES LINDEMANN NELSON

Z

ZOOS

History of Zoos

With few exceptions, the earliest collections of captive wild animals were privately held menageries that were symbols of wealth and power. Ancient Egyptians are thought to have been the first people to keep collections of wild animals. Animals of religious significance were kept as representatives of gods. In 1490 B.C., the Egyptian queen Hatshepsut directed an animal-collecting trip through Africa to fill her royal menagerie and to trade with neighboring countries. Chinese emperor Wen Wang, of the Chou dynasty, kept a variety of plants and animals in a 1,500-acre "Intelligence Park" around 1100 B.C. Like the menageries in Egypt, it was intended primarily to show the wealth of the empire. By the third century B.C., private Greek collections of animals were used for study, experimentation, and pets. Alexander the Great opened the first public menagerie in Alexandria in Egypt. Wealthy Romans kept small menageries and aviaries in villas. By the second and first centuries B.C., most captive animals were kept on exhibit in public menageries until they were sent into the arena or killed for food.

In the 1200s, Kublai Khan's collection in Asia held elephants, monkeys, fish,* hawks, and other species found in his vast empire. In 1519, conquistador Hernando Cortés visited a large menagerie held by the Aztec king Montezuma in Mexico that was staffed by 300 keepers. The collection included exhibits featuring American animals as well as human dwarfs and slaves. Like many of today's exhibits, the animals were exhibited in barless, moated enclosures.

By the 1600s, foreign conquests, trade, and the spread of agriculture and industry into undeveloped lands brought tales of great beasts and occasionally living specimens to Western nations. Because collections were still mostly private, the demand for animals that could be seen by the public in traveling menageries grew.

The first "modern" zoos were European zoological collections like Tierpark Schönbrunn in Austria, which opened in 1765, Menagerie du Jardin des Plantes in Paris, which opened in 1793, and the London Zoological Garden, which opened in 1828. Animal exhibits were surrounded by exotic plants in a gardenlike setting. These combined zoos and gardens (hence the term "zoological garden") differed from earlier menageries in that closely related species were exhibited near each other. They were established for scientific studies and education.

The first true European-style zoo in the United States was the Philadelphia Zoo (opened in 1874), which was modeled after the London Zoological Garden. Animals were housed in permanent ornate buildings, and the zoo was supported by a zoological society and managed by a director knowledgeable about wildlife. Soon there was a competition among zoos to have as many different kinds of animals as possible represented. The emphasis was on a great variety of species. Expeditions were organized to trap and transport great numbers of wild animals to the zoo. Animal mortality during capture and transport and at the zoo was high. Since little was known about animal care, many exhibits were small and barren. Exhibits were barred cages for the safety of the visitors and the animals, and to allow visitors to see the animals as close as possible. Animal buildings were designed for the pleasure of the visitor.

Around 1907, some zoos began to take advantage of the "Hagenbeck Revolution." At his zoo, Carl Hagenbeck Tierpark, animal supplier Carl Hagenbeck designed concrete moats around exhibits that kept animals in and visitors out and eliminated the need for bars. His exhibits were re-creations of nature as he saw it during his world travels. Exhibit illusions such as a lion sharing space with antelope were created by a moat separating the two animals that was hidden from the visitor's view.

The major purposes of zoos that have persisted over the years remain unchanged. The National Zoo in Washington, D.C., was established in 1891 "for the advancement of science and the instruction and recreation of the people." Some zoos, such as the New York Zoological Park, made conservation a priority by breeding and reintroducing native species like the American bison that were nearly extinct in the wild.

As the sciences of zoo biology, animal behavior, veterinary medicine, and animal nutrition grew in the 20th century, animal management improved, more species bred in captivity, and emphasis was no longer on large collections of many species, but on fewer species exhibited in larger, more naturalistic enclosures. There were more mixed-species exhibits and exhibits with

History of Zoos: Carl Hagenbeck Tierpark (*left*) introduced zoos to "naturalistic" enclosures for animals. This type of exhibit was a major influence on the New York Zoological Park (*right*). *Source*: Postcards c. 1910.

social groups of one species. Animals could be exhibited by themes like species relatedness, geographic zone, or habitat. With the recognition that many species of animals were becoming threatened with extinction due to human activities, zoos of the 1980s and 1990s became major centers of conservation and public education.

Selected Bibliography. Fisher, J., *Zoos of the World: The Story of Animals in Captivity* (Garden City, NY: Natural History Press, 1967); Hoage, R. J., and W. A. Deiss (Eds.), *New Worlds, New Animals: From Menagerie to Zoological Park in the Nineteenth Century* (Baltimore: Johns Hopkins University Press, 1996); Maier, F., and J. Page, *Zoo: The Modern Ark* (New York: Facts on File, 1990); Mann, W. M., *Wild Animals in and out of the Zoo* (New York: Smithsonian Institution Series, 1930); Mullan, B., and G. Marvin, *Zoo Culture* (London: Weidenfeld and Nicolson, 1987).

MICHAEL D. KREGER

Roles of Zoos

If animals have a right to freedom (*see* ANIMAL WELFARE, Freedom), zoos seem to infringe on it and to be questionable on welfare grounds also. Today's thousands of zoos, attracting millions of visitors worldwide, vary

The "Pollenarium" is an exhibit at the National Zoological Park in Washington, D.C., where the visitor walks through a botanic garden habitat filled with uncaged butterflies and hummingbirds. This theme exhibit highlights animal adaptations and living communities rather than a specific scientific group of related animals. Photo by M. Kreger, 1996.

enormously from so-called roadside zoos (condemned outright by reputable ones) to zoological parks whose animals, many of them in large, naturalistic, and/or behaviorally enriched enclosures (*see* ENRICHMENT FOR ANIMALS), often give every indication of being in a state of well-being.*

The question remains whether it is still misguided, as some feel, to maintain wild animals, however well cared for, outside their natural habitats, to which millions of years of evolution have adapted them. Zoos and their critics agree now that wild species must be protected and reputable zoos now take very few animals—especially mammals—from the wild (though they need to do this occasionally for serious conservational reasons). If it is acceptable to keep domesticated animals (*see* DOMESTICATION), perhaps it is not wrong to keep what can only be relatively wild animals in zoos. Indeed, some of them could be argued to be slightly domesticated because of their individual adjustment to zoo conditions or because of some perhaps unavoidable selective breeding. It is true that many domesticated animals— intensively reared hens (*see* CHICKENS) and pigs*—are kept in appalling conditions, but this is because of economic greed, not because they cannot be kept humanely. Zoo animals' captive environments can similarly be vastly improved by study of their behavioral requirements.

The degree to which animals show their natural behavior is a main criterion for judging their well-being or otherwise, as well as a guide to how their facilities may be improved. Other criteria include their degree of physical health, their readiness to breed, and the degree to which they show (or do not show) abnormal behavior such as the stereotyped "weaving" of some captive polar bears (*see* STEREOTYPIES IN ANIMALS).

If animals in zoos are only relatively wild or even slightly domesticated, this makes keeping them more acceptable, but at the same time it casts doubt on zoos' claim to maintain truly wild animals (and on whether these or their descendants could successfully be reintroduced to the wild). This is one of many real problems for zoos, and some critics deny their ability to save animals who are wild in any meaningful sense. On the other hand, zoos now have elaborate conservational arrangements to help to maintain their animals' wildness, at least genetically. These include studbooks for many endangered species and computerized, linked animal records (part of ISIS, the International Species Information System, started twenty years ago) to assist in the management of zoo animals as members of total captive populations with minimal inbreeding and maximal genetic diversity, as in a wild population. Enlightened zoo conditions help to maintain behavioral wildness also. Successful reintroductions have already occurred, such as the reintroduction of the Arabian oryx. However, just how successful some reintroductions have been, for example, the golden lion tamarin, is arguable. Thus zoos' ability to save, or at least reintroduce, many wild species remains unproven. However, threats face many wild species (from the hunting* of rhinos and tigers to the threats to almost all wild habitats from the exploding human population), and zoos can help considerably. Again, some critics see a concentration on captive breeding as a dangerous distraction from the primary conservational task of protecting actual wild habitats. But zoos see their captive breeding as merely complementing this, and some zoo scientists assist greatly in the protective management of actual wild populations. Many more zoos help to educate the public about threats to wild habitats. Zoos' conservational roles also bring their own moral problems, such as whether saving endangered species* can justify killing surplus animals, for example, nearly eighty hybrid orangutans in American zoos who are unsuitable for reintroductions.

Serious zoos are in many ways allies of all those who care about animals as individuals and about their survival as species. Apart from their conservational captive breeding, zoos constitute a kind of powerhouse of ordinary people's fondness and concern for animals. Though zoo critics tend to see zoos as demonstrations of domination over nonhumans, many of the millions who visit zoos probably do it because of animals' appeal to them. Such people are potentially a huge body of support for conservation and animal protection. A first step here is the introduction of legislation to regulate zoos (such legislation exists in Britain, for example, but not in the United States).

Selected Bibliography. Bostock, S. St. C., *Zoos and Animal Rights: The Ethics of Keeping Animals* (London and New York: Routledge, 1993); Broom, D. M., and K. G. Johnson, *Stress and Animal Welfare* (London: Chapman and Hall, 1993); International Union of Directors of Zoological Gardens/Conservation and Breeding Specialist Group (International Union for the Conservation of Nature and Natural Resources [now World Conservation Union]/Species Survival Commission), *The World Zoo Conservation Strategy: The Role of the Zoos and Aquaria of the World in Global Conservation* (Brookfield, IL: Chicago Zoological Society, 1993); Norton, B. G., M. Hutchins, E. F. Stevens, and T. L. Maple (Eds.), *Ethics on the Ark: Zoos, Animal Welfare, and Wildlife Conservation* (Washington, DC: Smithsonian Institution Press, 1995); Tudge, C., *Last Animals at the Zoo* (Oxford: Oxford University Press, 1992).

STEPHEN ST. C. BOSTOCK

Zoos and Ethical Animal Care

Ethics deals with what is good and bad and with moral duty and obligation. Assuming that zoos will always exist in some form, what factors decide if a zoo treats its animals well or poorly?

The survival of an animal in captivity is totally dependent on its human caretakers. Animals, like most humans, have an interest in avoiding pain* and perhaps in experiencing pleasure. Because humans have chosen to put zoo animals into a restricted area, humans are obligated to provide them with a good quality of life (*see* WELL-BEING OF ANIMALS). Zoos cannot and would not try to duplicate nature completely because animals in the wild are eaten by predators, suffer from disease or parasites, and experience droughts and starvation.

An important aspect of ethical animal care is maintaining good health. Professionally managed zoos have strong veterinary care programs designed to prevent, monitor, and treat illness, disease, and parasites. Zoo veterinarians, pathologists, and nutritionists seek to improve animal health. Nutritionists make sure that each animal gets a healthy (and tasty) diet that considers the animal's species, age, gender, and physical condition (such as pregnancy). As important as these types of services are to animal care, other types of services are important in providing for the animal's psychological health. For example, some zoos provide animals with interesting and stimulating environments.

Zoos with environmental enrichment* programs provide stimulation and challenges that are not life-threatening but give the animal some control over its activities. For example, wild bears spend much of their day searching for food. To simulate that behavior in captivity, some animal keepers hide food throughout the exhibit—under logs, in crevices in rocks, or frozen in blocks of ice. The bear must work to find and then remove the food to eat. If zoo bears are fed all their food in one large serving at the same time of the day every day, some bears may spend the rest of the day pacing or inactive.

Although some people argue that a zoo animal's life should be completely stress free, stress* occurs in nature, and some stress is actually good. Many behavioral enhancements at the zoo can be stressful to the animal (for example, the first time an animal is introduced to a new group member). A stress-free life, however, may lead to boredom (see ANIMAL BOREDOM) and abnormal behaviors such as overgrooming or repetitive pacing that sometimes occurs to fill the animal's free time (see STEREOTYPIES IN ANIMALS).

Professional and ethical animal care includes providing appropriate shade, lighting, humidity, temperature, and flooring material such as soil for animals that dig or pools of water for animals that swim or bathe. Keeping social animals in appropriately sized groups is another source of stimulation. Shelters protect animals from inclement weather and allow isolation for breeding or nesting animals. There should also be enough complexity in cage and holding-area furniture (perches, shelves, retreats) that an animal is not forced to interact with other animals or visitors should it choose not to do so.

Animal care is regulated by law (such as the Laboratory Animal Welfare Act* in the United States) and by professional guidelines (such as those of the American Zoo and Aquarium Association [AZA]). Many zoos have strict policies on when animals can be used in educational demonstrations or animal rides. The policies include which animals can be used, under what conditions, and for how long. In addition, zoos often provide training to animal care staff about species biology, care, handling, and other management techniques.

While the role that professionally managed or accredited zoos play in educating people about animals and in conserving animals and their habitats has much public support, other ethical issues resulting from these goals have caused criticism. For example, with limited space and resources available for breeding and exhibiting certain "priority" species, ethical decisions must be made about what should be done with the other animals whose space in the collection is needed.

One solution is "strategic collection planning," which involves developing a plan for a zoo's entire collection of animals. It is based on the principle that every animal in the zoo is there to promote the conservation of its species or habitat in the wild. This may lead to a trend toward fewer species in larger, more appropriate naturalistic "habitats." These new enclosures are better designed to meet the species' physical and behavioral needs as well as to educate the visiting public.

Selected Bibliography. Hutchins, M., and N. Fascione, Ethical Issues Facing Modern Zoos, *Proceedings of the American Association of Zoo Veterinarians* (1992): 56–64; Kreger, M. D., M. Hutchins, and N. Fascione, Context, Ethics, and Environmental Enrichment in Zoos, in D. Shepherdson, J. Mellen, and M. Hutchins (Eds.), *Second Nature: Environmental Enrichment for Captive Animals* (Washington, DC:

Smithsonian Institution Press, 1998); Mench, J. A., and M. D. Kreger, Ethical and Welfare Issues Associated with Keeping Wild Mammals in Captivity, in D. Kleiman, M. Allen, K. Thompson, S. Lumpkin, and H. Harris (Eds.), *Wild Mammals in Captivity: Principles and Techniques* (Chicago: University of Chicago Press, 1996), 5–15; Norton, B. G., M. Hutchins, E. Stevens, and T. Maple (Eds.), *Ethics on the Ark* (Washington, DC: Smithsonian Institution Press, 1995); Rowan, A. N. (Ed.), *Wildlife Conservation, Zoos, and Animal Protection: A Strategic Analysis* (North Grafton, MA: Tufts Center for Animals and Public Policy, 1996).

MICHAEL D. KREGER AND MICHAEL HUTCHINS

Zoos And Animal Welfare

In recent years there has been a great deal of discussion about the welfare of animals who are raised for food, used in research, or confined in zoos. This has led to discussion of what welfare consists of, attempts at "behavioral enrichment" (*see* ENRICHMENT FOR ANIMALS), and debate about whether adequate levels of animal welfare can ever be secured in zoos, laboratories, and slaughterhouses.

In addition to these concerns about welfare, another critique has developed that appeals to a wide range of interests that animals may have. Some critics have argued that keeping animals in zoos and laboratories is unjust; that animals may suffer in these institutions is only part of what makes them unjust. What is wrong with zoos, in this view, is not just that they cause animal suffering,* but that they violate a whole range of interests that are central to the lives of many animals.

This second critique can only have moral force among people who already believe that animals have significant moral standing.* Once this is granted, zoos become morally problematic, for virtually all creatures with moral status have an interest in directing their own lives. If animals are to be confined in zoos, then the moral claim in favor of respecting this interest will have to be overcome.

Some, like Tom Regan, argue that this moral claim cannot be overcome. Humans and many nonhumans enjoy equal moral status that manifests in rights. Fundamental rights can almost never be infringed. Zoos infringe on the rights of many of these animals; thus they are morally indefensible.

Others, like Dale Jamieson, believe that in principle this presumption could be overcome if there were weighty-enough reasons for keeping animals in captivity. In recent years education and conservation have been used most frequently to justify zoos. But even if we grant that zoos are successful in educating the public in some positive way, given the technological resources that are now coming on line, it is far from clear that holding animals in captivity is necessary for delivering positive educational results. For example, Zoo Atlanta is now piloting a virtual-reality exhibit that allows people to take the perspective of a gorilla in interacting with the social and natural environment.

Conservation is the justification most often appealed to by scientists in the zoo community. There are variations on the theme. Some want to use zoos as bases for captive breeding and reintroduction. Others want to use the economic and political power of zoos to protect habitat. Still others would be satisfied if zoos could be constituted as genetic libraries for animals who no longer exist in viable populations. However, most zoos have no habitat-conservation programs, and among those that do, it is rare that more than 1–2% of the budget is spent on them.

Reintroduction has been a mixed success. Benjamin Beck, chair of the American Zoo and Aquarium Association's Reintroduction Advisory Group, writes, "We must acknowledge frankly at this point that there is not over-whelming evidence that reintroduction is successful" (157). David Hancocks, executive director of the Arizona-Sonora Desert Museum, writes that "[t]here is a commonly held misconception that zoos are not only saving wild animals from extinction but also reintroducing them to their wild habitats" (181).

Whatever the role of captive breeding and reintroduction in species preservation may be, an inconsistency arises when it is enlisted as a justification for zoos. Zoos are places where people can see animals. They are places to take children on Sunday afternoons. They are amenities (like football and baseball teams) that can be boasted about by city boosters. Increasingly they are even the sites of rock concerts and fund-raisers. But the best institutions for captive breeding and reintroduction would not play these roles. They would remove animals from excessive contact with people, give them relatively large ranges, and prepare them for reintroduction in ways that zoo visitors might find shocking (e.g., by developing their competence as predators).

Selected Bibliography. Beck, Benjamin, Reintroduction, Zoos, Conservation, and Animal Welfare, in B. Norton, M. Hutchins, E. Stevens, and T. Maple (Eds.), *Ethics on the Ark: Zoos, Animal Welfare, and Wildlife Conservation* (Washington, DC: Smithsonian Institution Press, 1995), 155–163; Hancocks, David, An Introduction to Reintroduction, in B. Norton, M. Hutchins, E. Stevens, and T. Maple (Eds.), *Ethics on the Ark: Zoos, Animal Welfare, and Wildlife Conservation* (Washington, DC: Smithsonian Institution Press, 1995), 181–183; Jamieson, Dale, Against Zoos, in Peter Singer (Ed.), *In Defence of Animals* (New York: Basil Blackwell, 1985), 108–117; Jamieson, Dale, Zoos Revisited, in B. Norton, M. Hutchins, E. Stevens, and T. Maple (Eds.), *Ethics on the Ark: Zoos, Animal Welfare, and Wildlife Conservation* (Washington, DC: Smithsonian Institution Press, 1995), 52–66: Norton, B., M. Hutchins, E. Stevens, and T. Maple (Eds.), *Ethics on the Ark: Zoos, Animal Welfare, and Wildlife Conservation* (Washington, DC: Smithsonian Institution Press, 1995); Regan, T., Are Zoos Morally Defensible? in B. Norton, M. Hutchins, E. Stevens, and T. Maple (Eds.), *Ethics on the Ark: Zoos, Animal Welfare, and Wildlife Conservation* (Washington, DC: Smithsonian Institution Press, 1995), 38–51; Regan, T., *The Case for Animal Rights* (Berkeley: University of California Press, 1983).

DALE JAMIESON

Zoo Visitor–Animal Interactions

Surveys have shown that one reason people visit zoos is to touch or get close to animals. With a growing urban population worldwide, zoos have become islands where people can interact with species they may otherwise see only in television documentaries. One of the ways that zoos attempt to provide a recreational visit with their education and conservation roles is through visitor–animal interactions such as animal demonstrations, rides, feeding, and children's zoos.

The study of these types of interactions is important from an animal welfare* perspective. Research has shown that simply the presence of the visitors in front of zoo exhibits can disrupt behavior of social animals like primates. Unpredictable or loud noises may also have negative effects on the health and behavior of animals. Cotton-top tamarins, for example, learn vocalizations like alarm calls from other animals in their group. They must hear and separate important vocalizations from background noise made by visitors.

Visitors like to see animal movement. Unfortunately, this sometimes leads to visitors pounding on exhibit windows, throwing objects, or taunting animals to get some kind of response. Another way that visitors can see movement and get a direct personal response from an animal is to feed it. However, because such feeding causes sickness in animals, it is outlawed in British zoos.

Since the average visitor spends about 30 seconds to 2 minutes at a typical exhibit and only reads some labels, an animal show or demonstration is a way for visitors to get close to animals while receiving an educational message. Some studies suggest that messages about conservation are better remembered during demonstrations when live animals are used, particularly if they are handled. So as not to send messages of human domination over animals while handling animals during demonstrations, zoo educators include statements against keeping wild animals as pets, use fewer animals, or use biofacts (furs, feathers, and the like) while discussing issues like habitat, animal adaptations, and biodiversity.

Many zoos make a serious effort to safeguard the welfare of animals against stressful conditions during demonstrations. The Laboratory Animals Welfare Act* regulates how mammals are treated during presentations. It specifies that animal handling should not cause unnecessary discomfort, behavioral stress,* or physical harm to the animal or the handler. An animal that the public may contact can only be displayed for periods of time under conditions that keep it healthy and comfortable. Performing animals must be allowed a rest period between performances equal to the time for one performance.

In the United States all facilities that keep marine mammals (for example, whales, dolphins, polar bears, or seals) or apply for a permit to capture or

Zoo Visitor–Animal Interaction: Chimpanzee "Baldy" wearing a keeper's uniform at the New York Zoological Park c. 1910. Such entertaining practices are now seen by most zoos as uneducational and demeaning to species. *Source*: Postcard c. 1910. Photo by M. Kreger.

import a marine mammal for public display must offer an education program based on professionally recognized standards. An education/conservation message must be part of all marine mammal demonstrations.

Another form of visitor–animal interaction is the animal ride. Some include a conservation message or education about the ride animals. Revenue is used to fund conservation programs or zoo operating expenses. Fewer zoos are offering animal rides due to their high insurance liability, because they send the wrong message, and because rides such as trams and monorails generate more income.

Children's zoos allow children to get close to and touch animals. They aim to foster animal appreciation. Children's zoo themes may include farm animals, animal habitats, or adaptations. Many include "contact areas" where visitors can touch the animals. Handling animals in contact areas or educational demonstrations may or may not be stressful. Animals who do not receive frequent handling may view people as predators.

A 1995 Roper poll showed that 69% of Americans are concerned about zoo, aquarium, and animal-park treatment of captive animals (although most are supportive of zoo missions). To address these concerns, zoos must balance the visitor's desire to interact with animals with the method (most humane way of interacting) and context (educational or recreational value) of the interactions.

Selected Bibliography. Eagles, P. F. J., and S. Muffitt, An Analysis of Children's Attitudes toward Animals, *Journal of Environmental Education* 21(3) (1990): 41–44; Kreger, M. D., and J. A. Mench, Visitor-Animal Interactions at the Zoo, *Anthrozoös* 8(3) (1995): 143–158; Kreger, M. D., and J. A. Mench, Visitor-Animal Interactions at the Zoo: Animal Welfare, in *AZA Annual Conference Proceedings 1995*, 1995, 310–315; Sherwood, K. P., S. F. Rallis, and J. Stone, Effects of Live Animals vs. Preserved Specimens on Student Learning, *Zoo Biology* 8 (1989): 99–104; Tunnicliffe, S. D., Why Do Teachers Visit Zoos with Their Pupils? *International Zoo News* 41/5(254) (1994): 4–13.

MICHAEL D. KREGER

Zoos and Environmental Enrichment

The term "environmental enrichment" refers to the modifications that can be made to animal enclosures that increase the complexity and diversity of an animal's surroundings (*see* ENRICHMENT FOR ANIMALS). Animals in zoos are expected to live long lives in good health and, especially for endangered species,* to reproduce naturally in captivity. Zoos also strive to educate the public about the natural behavior and adaptations of animal species. The public and zoo professionals alike assess the psychological well-being of zoo animals by the resemblance of their behavior to that of their wild counterparts. In general, environmental enrichment improves psychological welfare by allowing the animal to perform behavior that it is naturally motivated to perform, such as seeking food or a mate, demarcating a territory, building a nest, maintaining its physical condition, escaping conspecifics or hiding itself, or interacting with a mate or social partner. In so doing, environmental enrichment may reduce stress,* relieve boredom (*see* ANIMAL BOREDOM), increase activity and alertness, and decrease abnormal behaviors. Enrichment of an animal's surroundings in a manner that stimulates it to behave as it would in the wild is, therefore, a major goal of the modern zoo, both for public education* and for successful captive propagation.

Heini Hediger was one of the first biologists to write about the importance of providing environments for zoo animals that allow them to express species-specific, natural behaviors that enable them to breed. Hediger's emphasis in 1950 that "one of the most urgent problems in the biology of zoological gardens arises from the lack of occupation of captive animals" was acted upon in the early 1960s by Desmond Morris, then curator of mammals at the London Zoo. He presented some of the first papers describing methods of providing "occupational therapy" for zoo and laboratory animals to prevent abnormal behavior, boredom, and laziness.

Environmental enrichment in zoos can be categorized into four general types based on behavioral improvements:

1. Environmental complexity can be enhanced by providing structures that increase surface area over which an animal can move and that make use

of the vertical space of a cage or enclosure. Crisscrossing grids or walkways or natural tree branches can be used to create a multidimensional network of pathways. Soft, natural substrate materials allow animals to dig, burrow, bury food items, dustbathe, or search for insects.

2. Feeding enrichment is perhaps the most important type of enrichment in terms of providing occupation. In the wild, many species spend most of their waking hours looking for, pursuing, gathering, handling, or hiding food. Gorillas, for example, spend up to 70% of their day foraging and feeding, and black bears, 75%. In the vast majority of captive situations animals are fed one or several daily meals by human caretakers. No effort is expended to acquire food, and it is consumed in a short time. Methods of feeding enrichment involve presenting food in a manner in which the animal must search for and/or gather its own food or spend more time handling it. One method is to scatter small food items such as grain or mealworms in a substrate such as hay, woodchips, or brush. Releasing live insects into an enclosure will stimulate extended periods of complex foraging and capture behaviors. Many simple feeding devices can be devised: plastic bottles or pipes with holes hanging on a rope and containing food, mazes in which peanuts or other snacks must be moved with a finger through holes to a goal area where they can be claimed, freezing food in ice blocks that need to be chipped away before the food can be consumed, hollowed-out logs and pipes filled with honey, peanut butter, or other foods and then re-plugged, and so on. Providing browse, whole carcasses or meat on bones, or whole fruits and vegetables is a common method of increasing food-handling time.

3. Novel objects, odors, sounds, and events provide substrates for investigation, manipulation, and play when care is taken to ensure that they elicit species-appropriate activity. For example, spraying cologne on tree stumps and branches stimulates investigation and rubbing in canids; many carnivores will attack and "kill" cardboard boxes (with staples removed); and beer kegs and oil drums floating in water will be used in inventive ways by great apes and large carnivores.

4. Keeping animals in appropriate social groupings is an extremely important means of enriching environments. Social partners are an infinite source of stimulation, as well as essential to normal rearing and development. Many primate species in the past never reproduced in captivity until they were kept in larger social groups instead of in pairs. The relationship of an animal with its keeper can also be a source of social enrichment. Fear of the keeper or unpredictable keeper behavior may lead to animal stress, ill health, or disturbance of maternal, parental, and other social behaviors. Rewards from keepers in the form of snack feedings appear to be favorable in a number of species. Training is also a way of managing animals, particularly large species, that reduces fear of people and increases predictability of human actions for the animal.

Selected Bibliography. Carlstead, K., and D. Shepherdson, Effects of Environmental Enrichment on Reproduction, *Zoo Biology* 13 (1994): 447–458; Chamove, A. S., Environmental Enrichment: A Review, *Animal Technology* 40(3) (1989): 155–178; Hutchins, M., D. Hancocks, and C. Crockett, Naturalistic Solutions to the Behavioral Problems of Captive Animals, *Zoologische Garten*, N. F. (Jena) 54 (1984): 28–42; Markowitz, H., Engineering Environments for Behavioral Opportunities in the Zoo, *Behavior Analyst* (Spring 1978); Morris, D., Occupational Therapy for Captive Animals, in *Laboratory Animals Center: Collected Papers*, vol. 2 (Carshalton, Surrey, UK: M. R. C. Laboratories, 1962), 37–42.

KATHY CARLSTEAD

Appendix: Resources on Animal Welfare and Humane Education

This is a representative list[1] of organizations that provide humane education materials directly pertaining to animals or that have information materials related to animal welfare available, either for the asking or for a fee. Space does not allow a complete listing of organizations; extensive lists of international organizations are available from many of the organizations listed here. Nearly all of the curricular and activity materials listed here are sold, even if they are underwritten by a nonprofit organization. Humane education is considered a part of environmental education, and environmental education part of global or peace education; consequently, a few organizations pertaining to these broader concepts are also included. Addresses, names, and telephone and fax numbers in this list are, of course, subject to change. Websites on animal rights and related matters are updated at the following website: http://www.liv.ac.uk/~srlclark/animal.html. Information is also updated at Animal Rights Updates, P.O. Box 51, Yellow Springs, Ohio 45387 (please send self-addressed stamped envelope). A selection of Canadian and British organizations has also been included. Many national, regional, and local organizations promote or provide humane and environmental education, and the local phone book may reveal one closer or better than any listed here.

AMERICAN ANTI-VIVISECTION SOCIETY (AAVS)

Noble Plaza, 801 Old York Road, Suite 204
Jenkintown, Pennsylvania 19046-1685
Phone: 215-887-0816; 800-SAY-AAVS (orders)
Fax: 215-887-2088
URL: http://www.aavs.org/
Contact: Tina Nelson, Executive Director

[1]Compiled, with revisions and additions by the editors, by David C. Anderson, Information Specialist at the University of California Center for Animal Alternatives, Davis, California, and Editor of the *Interactions Bibliography* (*Humans and Other Species*) of Rockydell Resources.

An advocate for the abolition of animal experimentation, AAVS conducts public outreach programs, research, and lobbying. It publishes the *AV Magazine* and pamphlets (e.g., *Why We Oppose Vivisection; Point/Counterpoint: Responses to Typical Pro-Vivisection Arguments*). Through its scientific arm, the Alternatives Research and Development Foundation (ARDF), AAVS awards grants to researchers for development of alternatives to traditional animal use in research.

AMERICAN HUMANE ASSOCIATION (AHA)

63 Inverness Drive East
Englewood, Colorado 80112
Phone: 303-792-9900; 800-227-4645
Fax: 303-792-5333
URL: http:///www.amerhumane.org
Contact: Michael E. Kaufmann, Humane Education Coordinator

AHA's Animal Division offers its members, principally public and private humane societies, animal shelters, and animal control offices, materials for use in their own programs. AHA's annual two-day workshops cover trends in the field (e.g., the link between child and animal abuse; age-specific humane education). AHA trains local people in the business of humane education. Its *Operational Guide: Humane Education* describes strategies for teaching humane education at all grade levels and includes suggestions on handling controversial issues and on building awareness for animal programs. Lesson plans, activity packets, and teachers' guides are available, including *Favorite Lessons by Humane Educators; The Animal Shelter, a Home Away from Home*; and *Pet Responsibility: Citizenship Lessons for Elementary Students*.

AMERICAN HUMANE EDUCATION SOCIETY (AHES)

350 South Huntington Avenue
Boston, Massachusetts 02130
Phone: 617-541-5095
Fax: 617-983-5449
Contact: Judith A. Golden

AHES facilitates Operation OutReach–U.S.A., a national humane education and literacy program for elementary schools. Operation OutReach–U.S.A. provides training for teachers at the local level. It also provides teachers with classroom materials, lesson plans, and free books for students to encourage literacy and the responsible treatment of all living things.

AMERICAN SOCIETY FOR THE PREVENTION OF CRUELTY TO ANIMALS (ASPCA)

424 East 92nd Street
New York, New York 10128
Phone: 212-876-7700
Fax: 212-348-3031
Adopt-a-School phone: 800-427-7228
Contact: Stephen Zawistowski, Humane Education Department

The ASPCA "Extend the Web" program offers a wide variety of low-cost educational materials, curricula, videos, books, and flyers for educators, children, and parents. The Web of Life consists of classroom lessons that engage children in hands-on, minds-on role-playing simulations involving humane concepts.

ANIMAL PROTECTION INSTITUTE OF AMERICA (API)

P.O. Box 22505
Sacramento, California 95820
Phone: 916-731-5521; 800-348-7387
Fax: 916-731-4467
Contact: Fran Stricker, Coordinator, Educational Services

A.P.E. News, API's animal-protection education newsletter, is available at no charge to educators across the United States. It includes ideas for use in the classroom, ideas for introducing children to animal issues, excerpts from recent books, reviews, and educational programs.

ANIMAL RIGHTS LAW CENTER, RUTGERS UNIVERSITY

15 Washington Street
Newark, New Jersey 07102
Phone: 201-648-5989
Contacts: Anna Charlton, Gary Francione

ANIMAL WELFARE INFORMATION CENTER (AWIC)

National Agricultural Library
10301 Baltimore Avenue
Beltsville, Maryland 20705
Phone: 301-504-6212 (direct line M–F, 8:00 A.M. to 4:30 P.M., Eastern Time);
    301-504-5704 (ATS)
Fax: 301-504-7125
E-mail: awic@nal.usda.gov
Contact: Jean Larson, Coordinator

Many AWIC bibliographies are available at the URL site: http://netvet.wustl.edu/awic.htm.

ANIMAL WELFARE INSTITUTE (AWI)

P.O. Box 3650, Georgetown Station
Washington, District of Columbia 20007
Phone: 202-337-2332
Fax: 202-338-9478

AWI publications produced for teachers include *Factory Farming, the Experiment That Failed*; *Facts about Furs*; and *First Aid and Care of Small Animals*.

THE ARK TRUST, INC.

Gretchen Wyler, Founder and President
5551 Balboa Boulevard
Encino, California 91316
Phone: 818-501-2275
Fax: 818-501-2226
E-mail: genesis@arktrust.org

The Ark Trust primarily focuses on promoting positive coverage of animal issues in the media. It presents the Genesis Awards, which honor people in the major media and the entertainment industry for works that have helped sensitize the public to the physical and psychological needs of animals.

ASSOCIATION FRANÇAISE D'INFORMATION ET DE RECHERCHE SUR L'ANIMAL DE COMPAGNIE (AFIRAC)

7, rue du Pasteur Wagner
75011 Paris, France
Phone: 49 29 12 00
Fax: 49 06 55 65
Contact: Anelyne Alanvert

Amies pour la Vie is a French-language educational package for elementary school children.

ASSOCIATION OF VETERINARIANS FOR ANIMAL RIGHTS (AVAR)

Nedim C. Buyukmihci, V.M.D., President
Association of Veterinarians for Animal Rights
P.O. Box 208
Davis, California 95617-0208
Phone: 916-759-8106
Fax: 916-759-8116
E-mail: AVAR@igc.apc.org
URL: http://www.envirolink.org.arrs/avar/avar__www.htm
Contact: Teri Barnato, National Director

The AVAR Alternatives in Education Database, a stand-alone database for DOS-based personal computers, cites adjunct and supplemental teaching tools for use from grade school through medical or veterinary school. Additional software is not required for this stand-alone database. It is available for $5.00 on either 3.5- or 5.5-inch diskettes. It is also available from the URL site in either a Windows or MS-DOS version.

BORN FREE FOUNDATION

Cherry Tree Cottage
Coldharbour Darking
Surrey, RH5 6JA, England
Phone: 01306 712091/13431
Fax: 01306 713350

BUNNY HUGGERS' GAZETTE

P.O. Box 601
Temple, Texas 76503

Once a year, the *Bunny Huggers' Gazette* devotes an issue to the current addresses of many North American and some international animal-protection organizations. The list is indexed by nation, in the United States by state, and by special interest or focus.

CANADIAN ENVIRONMENTAL NETWORK/RESEAU CANADIEN DE L'ENVIRONNEMENT (CEN/RCE)

251 Laurier Avenue West, Suite 1004
P.O. Box 1289 Station B
Ottawa, Ontario K1P 5R3
Phone: 613-563-2078
Fax: 613-563-7236
E-mail: cen@web.apc.org
Contact: Eva Schacherl, Executive Director

CEN/RCE is a nongovernmental, nonprofit network of over 1,800 environmental organizations, providing a cooperative forum for its groups to share knowledge and expertise. Its concerns include clean air, energy, environmental assessment, wilderness, forests, education, and international affairs. It publishes *Bulletin of the Canadian Environmental Network (BCEN)* and *The Green List.* There are a number of regional networks.

CANADIAN FEDERATION OF HUMANE SOCIETIES (CFHS)

Suite 102, 30 Concourse Gate
Nepean, Ontario K2E 7V7, Canada
Phone: 613-224-8072 (9:00 A.M. to 4:30 P.M. weekdays)
Fax: 613-723-0252

The Canadian Federation of Humane Societies is active in all areas of animal protection. Its education program works to integrate humane education studies into the Canadian school system to promote a more humane attitude toward animals, people, and the environment. Among its resources are the videos *Pet Pals* (level K–5) and *Animal Crackers* (level 3-8) and manuals from its Humane Education Workshops.

CENTER FOR COMPASSIONATE LIVING (CCL)

P.O. Box 1209
Blue Hill, Maine 04614
Phone: 207-374-8808
Fax: 207-374-8851
E-mail: ccl@downeast.net
Contacts: Zoe Weil, Rae Sikora

CCL offers training, consulting, workshops, and outdoor experiences for people who want to help the planet and all its inhabitants. Programs are designed for adults and young people for animal protection, environmental and social justice groups, humane educators, activists, business and civic groups, and students of all ages. CCL workshops are intended to provide tools and information to fully live a vision of a healthy life for people, the planet, and other beings. CCL has inaugurated a humane education certification program as an off-campus correspondence program, with one to two weeks of on-site training annually. The workshops are recommended for those who would like to apply for certification. *Sowing Seeds: A Humane Education Workbook*, designed for educators, provides specific suggestions for presentations, as well as guidelines for communicating and stimulating critical thinking.

CENTER FOR ENVIRONMENTAL EDUCATION

400 Columbus Avenue
Valhalla, New York 10595
Phone: 914-747-8200
Fax: 914-727-8299
E-mail: cee@earthspirit.org
Contact: Robert Zuber, Executive Director

The Green School Program of the Center for Environmental Education is a four-part high-school supplementary curricular program that uses existing environmental education materials: Peer Partners in Environmental Education (grade 9); School Organic Garden Program (grade 10); Student/School Greening Partnership (grade 11); and Student/Business Greening Partnership (grade 12). It publishes a newsletter, *Grapevine*.

CONNECTICUT UNITED FOR RESEARCH EXCELLENCE, INC. (CURE)

P.O. Box 5048
Wallingford, Connecticut 06492
Phone: 203-294-3521

CURE is a nonprofit coalition of more than 50 Connecticut universities, research institutes, health-related professional societies and corporations, hospitals, and volunteer health organizations. The occasional publication *BioRAP: Biomedical Research for Animals and People* is designated for classroom use with teachers' guides and is distributed nationally.

CONSUMERS FOR HEALTHY OPTIONS IN CHILDREN'S EDUCATION (CHOICE)

P.O. Box 30654
Bethesda, Maryland 20824
Phone: 800-470-3275

CHOICE is a program of the Farm Animal Reform Movement (FARM). CHOICE recommends and supplies *What Are We Feeding Our Kids?* (Workman, 1994); *Healthy School Lunch Action Guide* (Earth-Save); and *How on Earth!*, a quarterly magazine.

COUNCIL FOR ENVIRONMENTAL EDUCATION (CEE)

University of Reading
London Road
Reading, Berkshire RG1 5AQ, UK
Phone: (01734) 76-60-61
Fax: (01734) 76-62-64
Contact: Christine Midgley, Head of Information

CEE encourages increasing understanding of the role of environmental education nationwide. CEE publishes the *Annual Review of Environmental Education* and a *Newssheet*, which includes details on resources and events.

CRUELTY FREE INVESTING

Cynthia Kessler
7700 Wisconsin Avenue
Suite 300
Bethesda, Maryland 20814-3522
Phone: 800-311-4212 (message center; enter: 301-404-1245)

EARTHKIND

Humane Education Centre
Bounds Green Road
London N22 4EU, UK
Phone: (+44-181) 889-1595
Fax: (+44-181) 881-7662
URL: http://www.zynet.co.uk/beacon/earthkind/anchor.html
Contact: Cindy Milburn, Chief Executive

EarthKind is a dynamic partnership of people working to improve the well-being of animals and our environment. EarthKind's wildlife rescue ship, *Ocean Defender*, was launched in 1994. Members receive the magazine *The Living World* and the *Ocean Defender Newslog*.

ETHICAL SCIENCE EDUCATION COALITION (ESEC)

167 Milk Street #423
Boston, Massachusetts 02109-4315
Phone: 617-367-9142

The ESEC Resource Garden provides dissection alternatives (software, models, videotapes, and other resources) to instructors on a temporary loan basis. ESEC also prepared the catalog *Beyond Dissection: Innovative Teaching Tools for Biology Education* for the New England Anti-Vivisection Society (NEAVS) (see the listing for that organization).

EUROPEAN NETWORK OF INDIVIDUALS AND CAMPAIGNS FOR HUMANE EDUCATION (EURONICHE)

Nick Jukes, Coordinator
11 Beckingham Road
Leicester LE2 1HB, UK
Phone/Fax: (+44-116) 255 3223
E-mail: lynx@gn.apc.org

Alternative Contact:
Ursula Zinko
Klockartorget c1
96232 Jokkmokk, Sweden
Phone: (+46-971) 12455
E-mail: euroniche.alts@jokkmokk.mail.tolia.com

EuroNICHE offers information and support to students, lecturers, and campaigners across Europe. It strives "for the right to freedom of conscience and to promote alternative teaching methods to replace animals in undergraduate medical, biological and veterinary science."

FAY SPRING CENTER

534 Red Bud Road
Winchester, Virginia 22603
Phone: 540-665-2827
Fax: 304-728-7315
URL: http://members.aol.com/FaySpring/FaySpring.html

The Fay Spring Center coordinates and distributes Focus on Animals, a humane education program. It produces and distributes videotape documentaries, creates teaching guides for use with its own tapes, networks with producers and consumers of audiovisuals, assists producers with original footage and resource materials, works with teachers to encourage a more compassionate youth, and works with the media nationwide.

FEMINISTS FOR ANIMAL RIGHTS (FAR)

P.O. Box 16425
Chapel Hill, North Carolina 27516
Phone/Fax: 919-286-7333
E-mail: finla001@mc.duke.edu

Editorial Office
P.O. Box 694, Cathedral Station
New York, New York 10025
Phone/Fax: 212-866-6422
E-mail: BatyaB@aol.com

Dedicated to ending all forms of abuse against women and animals, FAR believes that the exploitation of animals and women "derives from the same patriarchal mentality" and that the feminist movements' neglect of animal rights has "done a great disservice to women and animals." It publishes *FAR Newsletter*.

FOOD AND NUTRITION INFORMATION CENTER

The Food and Nutrition Information Center prepares and revises bibliographies and source lists on nutrition, for example, *Sources of Free or Low-cost Food and Nutrition Materials* (which lists nutrition organizations, both national and local, food-related associations, and food companies), and *Nutrition Education Materials and Audiovisuals for Grades 7 through 12* (which lists curricula, lesson plans, learning activities, audiovisuals, and resources for adults). To obtain copies, send a request and a self-addressed mailing label to Reference Division, National Agricultural Library, 10301 Baltimore Boulevard, Beltsville, Maryland 20705-2351; 301-504-5755.

FOUNDATION FOR BIOMEDICAL RESEARCH (FBR)

818 Connecticut Avenue N.W., Suite 303
Washington, District of Columbia 20006
Phone: 202-457-0654
Fax: 202-457-0659
E-mail: nabr-fbr@access.digex.com
Contact: Frankie L. Trull, President

FBR and NABR (National Association for Biomedical Research) are sister organizations representing the scientific community on the issues of humane care and treatment of research animals. FBR serves as the public information and education program and works to educate the public on the importance of animal research for the diagnosis and treatment of human disease. FBR considers itself "a formal opposition to animal rights activists who formerly went unchallenged" and maintains a speakers' bureau and public relations programs. FBR publishes booklets (e.g., *Caring for Laboratory Animals; Health Benefits of Animal Research*), videos (e.g., *Caring for Life*), and a *Directory of Animal Rights/Animal Welfare Organizations*. Write for a current list of publications and a speakers' kit.

FUND FOR ANIMALS

Companion Animals Education Office
808 Alamo Drive, Suite 306
Vacaville, California 95688
Phone: 707-451-1306

National Office
200 West 57th Street
New York, New York 10019

Animal Crusaders, subtitled the *Newsletter for Teachers and Students Who Want to Help*, concentrates on getting students involved in correcting the abuse and injustice to which animals are subjected. This eight-page quarterly includes activity suggestions, learning sheets, and resources. It is available at no charge to classroom teachers and humane educators; $25 a year to others.

THE GREEN BRICK ROAD (GBR)

c/o 8 Dumas Court
Don Mills, Ontario M3A 2N2, Canada
Phone: 416-465-1597; 800-477-BOOK
URL: http://gbr.org/home.htm

GBR is a nonprofit organization that specializes in resources and information for teachers and students of global and environmental education.

HOW ON EARTH! HOE!

P.O. Box 3347
West Chester, Pennsylvania 19381
Phone: 717-529-8638

How on Earth! is a quarterly for and by youth who support compassionate, ecologically sound living. It covers a variety of environmental, animal, and social justice issues and encourages activism and empowerment among youth who are concerned about the earth and all beings. *HOE!* holds that being a vegetarian is an essential component of compassionate, sustainable living, so vegetarian recipes, nutrition advice, and lifestyle information are important features.

HUMANE SOCIETY OF THE UNITED STATES (HSUS)

2100 L Street, N.W.
Washington, District of Columbia 20037
Phone: 202-452-1100
Fax: 202-778-6132
URL: http://www.hsus.org
Contact: Jonathan Balcombe, Associate Director for Education, Animal Issues
Phone: 301-258-3046
Fax: 301-258-3082
E-mail: hsuslab@ix.netcom.com

The nation's largest animal-protection organization is "not opposed to the legitimate and appropriate utilization of animals" for human needs and further believes that humans have "neither the right nor the license to exploit or abuse any animals in the process." Its educational arm is the National Association for Humane and Environmental Education (NAHEE). HSUS promotes public education to foster respect, understanding, and compassion for all creatures. It publishes the magazines *HSUS News* and *Animal Activist Alert* and numerous brochures ("Companion Animals"; "Fur Seals"; "Factory Farming"; "The Living Science: A Humane Approach to the Study of Animals in Elementary and Secondary School Biology").

INTERNATIONAL INSTITUTE FOR GLOBAL EDUCATION

Faculty of Education, University of Toronto
371 Bloor Street West
Toronto, Ontario M5S 2R7, Canada
Phone: 416-978-1863
Fax: 416-978-4612
Contacts: Graham Pike, David Selby, codirectors
E-mail: david_selby@tednet.feut.utoronto.ca;
   graham_pike@tednet.feut.utoronto.ca

The institute aims to contribute to the growth of global education in Ontario, Canada, and internationally through teaching programs, curriculum development, research, and networking.

INTERNATIONAL SOCIETY FOR ANIMAL RIGHTS (ISAR)

4212 South Summit Street
Clarks Summit, Pennsylvania 18411
Phone: 717-586-2200; 800-543-ISAR
Fax: 717-685-9580
Contact: Helen E. Jones, President

ISAR seeks to enlighten the public about the exploitation and suffering of animals. It publishes a quarterly *ISAR Report* and numerous pamphlets (e.g., *Experimental Psychology; Cosmetic Tests on Animals*).

IOWA STATE UNIVERSITY BIOETHICS INSTITUTE

425 Catt Hall
Iowa State University
Ames, Iowa 50011
URL: http://www.public.iastate.edu/~grad_college/bioethics/

The Iowa State University Bioethics Institute is a nationally recognized faculty-development workshop for nonmedical life scientists. Funded in part by a major grant from the National Science Foundation, the institute has its roots in a program begun in 1991 at Iowa State University. These institutes offer a creative and unique approach to solving the problem: they improve the quality of undergraduate and graduate education in the life sciences by expanding partnership ventures among academic disciplines such as biochemistry and philosophy and they lead to long-term relationships among life scientists, humanists, and others. A particularly interesting and somewhat controversial feature of the institute is that it provides only vegetarian lunches to participants. The project director is Gary Comstock, Bioethics Program, Iowa State University, 403 Ross Hall, Ames, Iowa 50011. Its newsletter, *Ag Ethics Bioethics*, is available from the editor.

JANE GOODALL INSTITUTE FOR WILDLIFE RESEARCH, EDUCATION, AND CONSERVATION

P.O. Box 14890
Silver Spring, Maryland 20911-4890
Phone: 301-565-0086
Fax: 301-565-3188

Dilys Vass, Executive Director
15 Clarendon Park
Lymington, Hants, SO41 8AX, England
Phone: (+44-1590) 671188
Fax: (+44-1590) 670887

JEWS FOR ANIMAL RIGHTS (JAR)

255 Humphrey Street
Marblehead, Massachusetts 01945
Phone: 781-631-7601
E-mail: micah@micahbooks.com
URL: http://www.micahbooks.com
Contact: Roberta Kalechofsky

JAR, a nonmembership organization, promotes animal rights and the alleviation of animal suffering. JAR believes that "the earth and all life is sacred because God created it." JAR encourages vegetarianism, preventive medicine, and alternatives to animals in research. It provides materials on celebrating bar/bat mitzvahs, confirmations, and other holidays in a manner consistent with JAR's goals. Micah Publications is its publishing arm (e.g., the *JAR Newsletter* and books such as *Autobiography of a Revolutionary: Essays on Animal and Human Rights; The Dark Face of Science; In Pity and in Anger;* and *Judaism and Animal Rights: Classical and Contemporary Responses*).

JOHNS HOPKINS CENTER FOR ALTERNATIVES TO ANIMAL TESTING (CAAT)

111 Market Place, Suite 840
Baltimore, Maryland 21202-6709
Phone: 410-955-3343
Fax: 410-955-0258

E-mail: CAAT@jhuhyg.sph.jhu.edu
URL: http://www.jhu.edu/caat
Contacts: Alan E. Goldberg, Joanne Zurlo, Deborah Rudacille

Individuals and corporations united to develop in vitro alternatives to the use of whole animals in evaluating and testing commercial and medical products founded CAAT. CAAT validates alternative testing methods, encourages their use, and conducts education and research programs. Besides its newsletter, CAAT publishes a newsletter for middle schools, the *CAATalyst*, on alternatives in product safety testing.

THE LATHAM FOUNDATION

Latham Plaza Building, Clement & Schiller
Alameda, California 94501
Phone: 510-521-0920
Fax: 510-521-9861
URL: http://www.latham.org/home.html
Contact: Hugh H. Tebault, President

Latham promotes the ideas of interdependence of all living things, justice, kindness, and compassion for all life and broadcasts a children's radio program and a weekly television series. Its publications include books (e.g., *Dynamic Relationships: Animals in the Helping Professions; Universal Kinship: The Bond between All Living Things*) and the quarterly *Latham Letter*.

LIVINGEARTH LEARNING PROJECT

P.O. Box 2160
Boston, Massachusetts 02106
Phone: 617-367-8687

The LivingEarth Learning Project, the humane education arm of the New England Anti-Vivisection Society (NEAVS) (see the listing for that organization), offers a series of educational programs about animal and environmental issues for grades 3 through college in New England and parts of New York. The classroom presentations are interactive and flexible in length and format. LivingEarth also has a Video Loan Library, provides speakers for teacher in-service training and conferences, and publishes lesson plans, classroom activity materials, and other resource materials.

MEDICAL RESEARCH MODERNIZATION COMMITTEE

P.O. Box 2751 Grand Central Station
New York, New York 10163-2751
Phone: 212-832-3904

The Medical Research Modernization Committee publishes newsletter reports and books devoted mainly to the use and abuse of animals in medical experimentation.

MONITOR: THE CONSERVATION, ENVIRONMENT, AND ANIMAL WELFARE CONSORTIUM

Craig van Note, Executive Vice-President
1506 19th Street N.W.
Washington, District of Columbia 20036
Phone: 202-234-6576
Fax: 202-234-6577

NATIONAL ALLIANCE FOR ANIMALS

P.O. Box 77196
Washington, District of Columbia 20013-7196
Phone: 703-810-1085

NATIONAL ANTI-VIVISECTION SOCIETY (NAVS)

53 West Jackson Boulevard, Suite 1552
Chicago, Illinois 60604-3795
Phone: 312-427-6065; 800-888-NAVS (6287)
Fax: 312-427-6524
Dissection hot line: 800-922-FROG (6734)
Contacts: Mary Margaret Cunniff, Executive Director; Linda M. Petty, Dissection
   Alternatives Program Director

The National Anti-Vivisection Society was founded in 1929 and has over 50,000 members. For nonanimal alternatives to dissection, NAVS offers three-dimensional models of the frog and fetal pig on loan to educators, students, and concerned individuals. These state-of-the-art models are hand painted, anatomically accurate replicas of an adult female bullfrog and a fetal pig. Instructors are encouraged to examine the effectiveness of these models on a firsthand basis. Contact Linda Petty; a credit card or check deposit is required to assure return of the model. The NAVS dissection hot line provides additional information on nonanimal alternatives to dissection and manuals for students (*Saying No to Dissection: Elementary; Objecting to Dissection: High School; Objecting to Dissection: College*).

NATIONAL ASSOCIATION FOR HUMANE AND ENVIRONMENTAL EDUCATION (NAHEE)

Norma Terris Humane Education Center
67 Salem Road, P.O. Box 362
East Haddam, Connecticut 06423-0362
Phone: 203-434-8666
Fax: 203-434-9579
Contact: Dorothy Waller, Director of Education Outreach

NAHEE is the Youth Education Division of the Humane Society of the United States (HSUS) and seeks to improve humane and environmental education programs nationwide. It provides consultation to school systems, educational organizations, and humane societies interested in incorporating humane concepts into their educational master plan. NAHEE's programs include the Adopt-a-Teacher Program, in which a teacher receives *KIND News* (in bundles of 32 copies a month), *KIND Teacher*, a teaching guide for *KIND News*, classroom posters, and KIND Club membership cards. Adopt-a-Teacher Programs are available to organizations or individuals and are provided at no cost to the teacher or school district. *KIND News*, written for elementary-school children, is published at three reading levels. The *Student Network News* and *Student Action Guide* are intended for middle and high schools.

NATIONAL CATTLEMEN'S BEEF ASSOCIATION (NCBA)

Education Department
444 North Michigan Avenue
Chicago, Illinois 60611
Phone: 312-467-5520
Fax: 312-467-9729
URL: CowTown America: http://www.cowtown.org/
Contact: Barbara Selover, Executive Director of Education
E-mail: selover@meatboard.org

NCBA was recently formed from the National Cattlemen's Association and the National Live Stock and Meat Board. It conducts research, information, education, and legislative programs for the beef industry. The association develops science-based school materials for grades K–12 and provides a wealth of nutrition education materials. It has two kits including information on animal care: *Things We Can Learn from a Cow and a Worm*, a poster with teacher's guide and student activities for science curriculums in grades 5-6, and *Caretakers All*, a study kit with teacher's guide and student activities for grades 3-4.

NATIONAL CONSORTIUM FOR ENVIRONMENTAL EDUCATION AND TRAINING (NCEET)

c/o School of Natural Resources and Environment
University of Michigan
Ann Arbor, Michigan 48109
Phone: 313-998-6727
Gopher server: telnet nceet.snre.umich.edu; logon: eelink
E-mail: eelink@eelink.umich.edu
Contact: Paul Nowak, Jr., Project Manager, EE-Link
E-mail: cappaert@umich.edu

NCEET helps educators explore the environment and investigate current issues with students and is building resources to support K–12 environmental education, including lists of media specialists, in-service providers, nature-center staff, and curriculum developers. It is a partner in the Environmental Education Training Partnership.

NATIONAL FFA ORGANIZATION (NFFAO)

National FFA Center
5632 Mount Vernon Memorial Highway
Box 15160
Alexandria, Virginia 22309-0160
Phone: 703-360-3600
Fax: 703-360-5524
Contact: Dr. Larry Case, CEO

Animal Welfare Instructional Materials (Alexandria, VA: National Council for Agricultural Education, 1995), one volume (looseleaf), is distributed by the National FFA Foundation, PO Box 45205, Madison, Wisconsin 53744-5205; fax: 608-829-3195; 608-829-3105. It is made available through the National Council for Agricultural Education as a special project of the National FFA Foundation (David M. Coffey, Project Director). The project involved 17 sponsors, including the National Pork Producers Council.

NATIONAL 4-H COUNCIL (N4-HC)

7100 Connecticut Avenue
Chevy Chase, Maryland 20815-4999
Phone: 301-961-2820
Fax: 301-961-2937
E-mail: sturm@fourhcouncil.edu
Contact: Richard J. Sauer, President
National 4-H Supply Service contact information:
Phone: 301-961-2934
Fax: 301-961-2937
E-mail: 4hsupply@fourhcouncil.edu

For a current sourcebook, contact 4H Supply. Individual state 4-H Curriculum Committee catalogs are available. For example, the California 4-H Curriculum Committee catalog is available from county agricultural extension offices or the University of California Cooperative Extension (ANR Publications, University of California, 6701 San Pablo Ave., Oakland CA 94608-1239; 510-642-2431; fax: 510-643-5470; e-mail: anrpubs@ucdavis.edu). Additional resources listed in the catalog, *California 4-H Publications, 1996-1997*, must be ordered from other sources (e.g., 4-H Oak Tree Project Video and Project Manual from Calaveras County UCCE, 891 Mountain Ranch Road, San Andreas, California 95249; 209-754-6477).

NATIONAL HUMANE EDUCATION SOCIETY (NHES)

521-A East Market Street
Leesburg, Virginia 22705
Phone: 703-771-8319
Fax: 703-771-4048

NHES publishes a variety of materials, including a *Quarterly Journal* and *Because We Love Them: A Handbook for Animal Lovers* by Anna C. Briggs.

NATIONAL RIFLE ASSOCIATION (NRA)

Hunter Services Division
11250 Waples Mill Road
Fairfax, Virginia 22030
Phone: 703-267-7100; 800-368-5714

NRA lobbies to protect the right of the individual citizen to own and use firearms. NRA argues that hunting is a vital part of wildlife conservation and publishes the magazines *American Hunter* and *American Rifleman*, as well as brochures (e.g., "Improving Access to Private Land").

NATIONAL WILDLIFE FEDERATION (NWF)

1400 Sixteenth Street N.W.
Washington, District of Columbia 20036-2266
Phone: 800-222-9919; 800-245-5485 (Conservation Education Department)

The NWF encourages the intelligent management and appreciation of our natural resources. It operates Ranger Rick's Wildlife Camp, sponsors National Wildlife Week, and produces daily and weekly radio programs. NWF manages a large library of conservation-related publications and publishes *Ranger Rick's Nature Magazine* and the *National Wildlife Magazine*.

NETWORK OF INDIVIDUALS AND CAMPAIGNS FOR HUMANE EDUCATION (NICHE)

Department of Psychology
University of Stirling
Stirling, Scotland FK9 4LA, UK
Phone: (01786) 73171, extension 2077
Contact: Francine Dolins, Secretary/Treasurer

The *NICHE Newsletter* is distributed to members by EarthKind, the Humane Education Centre.

NEW ENGLAND ANTI-VIVISECTION SOCIETY (NEAVS)

333 Washington Street, Suite 850
Boston, Massachusetts 02108
Phone: 617-523-6020; TDD/TTY 617-523-0181

NEAVS opposes vivisection and product safety testing on animals. Its humane education arm is the LivingEarth Learning Project. NEAVS's Library Project offers kits to school librarians, which may be requested at no charge on school-library stationery. *Making a Difference: Action Guide for Students Who Love Animals*, available in both high-school and college versions, is intended for the beginning student group or the group looking for new ideas. It includes steps for getting started, effective communication skills, action ideas for the group and the individual, and information designed to keep the group going.

The catalog *Beyond Dissection: Innovative Teaching Tools for Biology Education*, edited by Sandra Larson (Boston: NEAVS, 1995), is also available. It is a comprehensive printed catalog of nonanimal alternatives to dissection, covering nearly 400 product listings. Listings cover all major whole-animal dissections, human and comparative anatomy, organ or system anatomy and physiology, embryology, and genetics. Product listings are suitable for all grade levels, elementary through college, and are available in all price ranges.

NORTH AMERICAN ASSOCIATION FOR ENVIRONMENTAL EDUCATION (NAAEE)

1255 Twenty-Third Street, Suite 400
Washington, District of Columbia 20037-1199
Phone: 202-884-8912
Fax: 202-884-8701
Contact: Edward McCrea, Executive Director

NAAEE is a multinational organization of individuals and environmental organizations, with students in environmental education and studies as associates. Its objectives are to promote environmental education programs at all levels, coordinate environmental educational activities among programs and educational institutions, disseminate information about environmental educational activities appropriate for its members, assist educational institutions in beginning or developing programs and serve as a resource to them, and foster research and evaluation in connection with environmental education.

NORWEGIAN INVENTORY OF AUDIOVISUAL ALTERNATIVES (NORINA)

Karina Smith
Laboratory Animal Unit
Norwegian College of Veterinary Medicine
P.O. Box 8146 Dep.
N-0033 Oslo, Norway
Fax: +47 22 96 45 35
URL: http://oslovet.veths.no; http://www.bio.mq.eu.au.NORINA
E-mail: karina.smith@veths.no
URL: http://www.vetsh.no/norina/fullversion; consists of 81K
http://www.veths.no/norina/state_here.html
http://www.bio.mq.edu.au/norina (mirror site at MacQuarrie University, Australia)

PEOPLE FOR THE ETHICAL TREATMENT OF ANIMALS (PETA)

501 Front Street
Norfolk, Virginia 23510
Phone: 757-622-PETA (7382); Student line: ext. 691
Fax: 757-622-0457
URL: http://envirolink.org/arrs/peta/

PETA is an educational and activist organization that works to stop animal abuse and animal research. It advocates vegetarianism and the use of cruelty-free products. PETA's publications for elementary-school teachers and students include the Lifetime Learning Systems' Share the World, a humane education curriculum unit for grades 3-5. It includes a teacher's guide and activity packets. A noncopyrighted reproducible coloring book (We're All Animals Coloring Book) is also available. PETA offers elementary-school teachers Kids Can Save The Animals! 101 Easy Things to Do, by Ingrid Newkirk (New York: Warner, 1991), and, to secondary-school and college teachers, Save the Animals! 101 Easy Things You Can Do, by Ingrid Newkirk (New York: Warner, 1990). Teacher packets include Bringing Animal Issues into Elementary and Middle School Classrooms and Bringing Animal Issues into High School and College Classrooms. PETA offers students refusing dissection a dissection pack and teachers the video Their Future Is in Your Hands (Tonbridge, Kent: Animal Aid, 1992). PETA's student magazine is Grrr! The 'Zine That Bites Back.

PERFORMING ANIMAL WELFARE SOCIETY (PAWS)

P.O. Box 849
Galt, California 96532

PSYCHOLOGISTS FOR THE ETHICAL TREATMENT OF ANIMALS (PsyETA)

P.O. Box 1297
Washington Grove, Maryland 20880-1297
Phone: 301-963-4751
Fax: 301-963-4751
URL: http://www.psyeta.org
E-mail: kshapiro@capaccess.org
Contact: Kenneth J. Shapiro, Executive Director

PsyETA, an organization of psychologists, graduate students, institutions, animal rights organizations, and interested individuals, seeks to ensure the proper treatment of animals used in behavioral research and education. It urges revision of curricula to include ethical issues in the treatment of animals. PsyETA has a speakers' bureau, tips on how to organize, and sample student rights policies and supports students who are discriminated against in animal behavioral laboratories. PsyETA published the annual notebook Humane Innovations and Alternatives from 1987 through 1994 and produces brochures (e.g., "The Student Rights Option: A Student Guide to Objecting to Psychology Animal Labs"). Portions of its newsletter and other publications can also be found at the Web site. It also publishes Society and Animals and Journal of Applied Animal Welfare Science.

RESEARCH DEFENCE SOCIETY

Grosvenor Gardens House
Grosvenor Gardens, London SW1W 0BS, UK

The Research Defence Society "takes the view that we must first educate the public, and particularly its younger members, about medical progress and the research which underlies it before they will be able to fully appreciate why animal-based research needs to be done. This will clearly be a long-term task requiring the help of many of those involved in biological research and teaching."

ROYAL SOCIETY FOR THE PREVENTION OF CRUELTY TO ANIMALS (RSPCA)

The Causeway
Horsham, West Sussex RH12 1HG, UK
Phone: (+44-1403) 26 41 81
Fax: (+44-1403) 24 10 48

The RSPCA is a multinational organization of individuals and organizations concerned about the well-being of wild and domestic animals in the United Kingdom. It opposes unnecessary animal experimentation, habitat destruction, factory farming, and blood sports and promotes attitudes and behaviors supporting the dignity and rights of all animals, spaying and neutering of pets, and proper treatment of pets. The RSPCA publishes a pamphlet series, RSPCA Information (e.g., "Ethical Concerns for Animals"; "Guide to Products Not Tested on Animals"; Alternatives to Animal Experiments"). The series RSPCA Campaigns targets issues (e.g., "Bullfighting—Ban the Business"; "Thinking of Buying a Parrot?").

SCIENTISTS CENTER FOR ANIMAL WELFARE (SCAW)

Golden Triangle Building One
7833 Walker Drive, Suite 340
Greenbelt, Maryland 20814
Phone: 301-345-3500
Fax: 301-345-3503
Contact: Lee Krulisch, Executive Director

SCAW, organized in 1978, is a nonprofit organization concerned about animal welfare. It supports the responsible and humane treatment of research animals. SCAW sponsors seminars and conferences and publishes conference proceedings and other educational materials.

STUDENT ACTION CORPS FOR ANIMALS (SACA)

P.O. Box 15588
Washington, District of Columbia 20003
Phone: 202-543-8983
Contact: Rosa Feldman, Cofounder

Members of SACA are primarily high-school and college students. SACA coordinates a Stop-Dissection Campaign throughout the United States. A counseling group on issues of students' rights and empowerment, SACA assists students in saying no to dissection and saying yes to vegetarianism. SACA offers slide shows and speakers and the publication *101 Non-Animal Biology Lab Methods*.

SWEDISH SOCIETY AGAINST PAINFUL EXPERIMENTS ON ANIMALS

P.O. Box 2005, S-125 02
Älvsjö, Sweden
Phone: +46 8 749 20 40
Fax: +46 8 749 20 02

The society publishes an extensive international list of organizations that are concerned with various animal welfare issues, including vegetarianism.

TUFTS CENTER FOR ANIMALS AND PUBLIC POLICY

Tufts University School of Veterinary Medicine
200 Westboro Road
North Grafton, Massachusetts 01536
Phone: 508-839-7991
Fax: 508-839-2953

The center publishes two newsletters, *Animals and Public Policy* and *The Alternatives Report*, and proceedings of meetings it sponsors on various animal-related policy issues (e.g., *The Animal Research Controversy; Zoos and Wildlife Conservation*).

UNITED POULTRY CONCERNS (UPC)

P.O. Box 59367
Potomac, Maryland 20859
Phone: 301-948-2406
URL: http://www.envirolink.org/arrs/upc
Contact: Karen Davis

UPC produces the quarterly *Poultry Press*, books, and videos, as well as fact sheets and handouts. It provides *Replacing School Hatching Projects: Alternative Resources and How to Order Them*, which discusses the issues and lists books, a videodisc, videos, overhead transparencies, a model, and hands-on ecology projects.

UNIVERSITIES FEDERATION FOR ANIMAL WELFARE (UFAW)

The Old School
Brewshouse Hill
Wheathampstead
Hertfordshire AL4 8AN, UK
Phone: +44-1582 831838
Fax: +44-1582 831414
E-mail: ufaw@ufaw.org.uk
URL: http://www.users.dircon.co.uk/~ufaw3/

The Universities Federation for Animal Welfare (UFAW) is a scientific and educational charity that was founded to promote humane behavior toward the animals used and managed by humans. UFAW is essentially a research, development, and education group working in the general field of animal welfare, enlists the support of university graduates, students, and professional men and women, and obtains and disseminates relevant knowledge. UFAW cooperates with Parliament, government departments, industry, the scientific community, learned societies, and other appropriate organizations. UFAW holds symposia and workshops and publishes the proceedings of these meetings; it carries out and sponsors scientific research and field investigations into many aspects of the biology and welfare of farmed, companion, wild, zoo, and laboratory animals; it produces standard texts on animal care and management; and it publishes a newsletter, a publications list, technical reports, and the quarterly refereed journal *Animal Welfare*.

UNIVERSITY OF CALIFORNIA CENTER FOR ANIMAL ALTERNATIVES

School of Veterinary Medicine
University of California, Davis
Davis, California 95616-8684
Phone: 916-752-1800
Fax: 916-752-8391
Contacts: Lynette A. Hart, Director (916-752-7722); R. Lee Zasloff, Associate Director

The University of California Center For Animal Alternatives publishes and continually updates a set of information resource guides on animal welfare and alternatives: *Bibliographies, Ethical Use of Animals, Internet Resources, Higher Education, Recommended Journals, Organizations*, and *PreCollege Science Education*. This last guide is an introduction to resources on alternatives to animal use in the classroom, appropriate husbandry of animals, dissection (advocacy, opposition, and alternatives), and the improvement of precollege science education. These guides are available at the World Wide Web site, http://www.vetmed.ucdavis.edu/Animal_Alternatives/main.htm. The center also produces an occasional newsletter, *UC Alert*.

VEGETARIAN RESOURCE GROUP

P.O. Box 1463
Baltimore, Maryland 21203
Phone: 410-366-VEGE
URL: http://www.envirolink.org/arrs/vrg/home.html

The Vegetarian Resource Group maintains an active publishing program, which includes the monthly *Vegetarian Journal* and books (e.g., *The Vegetarian Software Game, an IBM Compatible Program; Guide to Natural Food Restaurants in the United States and Canada*) and brochures ("Guide to Non-Leather Shoes"). The group also supports the establishment of local vegetarian groups ("Hints for Starting a Vegetarian/Environmental/Animal Rights Group at Your School or College").

VIRTUAL FROG DISSECTION KIT

URL: http://george.lbl.gov/vfrog/
Contact: David Roberston, owner, at dwrobertson@lbl.gov

The Virtual Frog Dissection Kit was developed and placed on the Web by the Imaging and Distributed Computing Group of Lawrence Livermore Laboratory. The kit allows interactive dissection. Available in a number of languages (e.g., French, Czech), it contains an overview, a tutorial, and the Virtual Frog Builder Game to test the viewer's knowledge of frog anatomy. It needs a browser that supports forms and sensitive images that are generated "on the fly."

WARDS, INC.

8150 Leesburg Pike, #512
Vienna, Virginia 22812-1655

An eighteen-page *Directory of Animal Protection Organizations* is available at no charge. WARDS publishes the newsletters *Our Animal Wards* and *Science and Animal Welfare*.

Sources

This bibliography contains general source material covering all sides of the important issues presented in this encyclopedia (see also the Appendix). The list is not meant to be exhaustive (see the individual entries for many more references) but rather includes books and articles that are frequently cited and journals that often publish essays that are concerned with animal rights and animal welfare.

BOOKS, BOOK CHAPTERS, JOURNAL ARTICLES, NEWSLETTERS, AND REPORTS

Achor, A. B. *Animal Rights: A Beginner's Guide*. Yellow Springs, OH: WriteWare, 1996.

Adams, C. J. 1994. *Neither Man nor Beast: Feminism and the Defense of Animals*. New York: Continuum.

Allen, C., and M. Bekoff. 1997. *Species of Mind: The Philosophy and Biology of Cognitive Ethology*. Cambridge, MA: MIT Press.

Animal Welfare Institute. 1990. *Animals and Their Legal Rights: A Survey of American Laws from 1641 to 1990*. 4th ed. Washington, DC: Animal Welfare Institute.

Appleby, M. C., and B. O. Hughes (Eds.). 1997. *Animal Welfare*. New York: CAB International.

Arluke, A., and C. R. Sanders. 1996. *Regarding Animals*. Philadelphia: Temple University Press.

Baker, R. M., G. Jenkin, and D. J. Mellor (Eds.). 1994. *Improving the Well-Being of Animals in the Research Environment*. Glen Osmond, South Australia: Australian and New Zealand Council for the Care of Animals Used in Research and Teaching.

Bayne, K.A.L., and M. D. Kreger (Eds.). 1995. *Wildlife Mammals as Research Models: In the Laboratory and Field*. Greenbelt, MD: Scientists Center for Animal Welfare.

Beck, A., and A. Katcher. 1996. *Between Pets and People: The Importance of Animal Companionship*. Rev. ed. West Lafayette, IN: Purdue University Press (contains list of World Wide Web sites as of 1996).

Bekoff, M. 1994. Cognitive Ethology and the Treatment of Nonhuman Animals: How Matters of Mind Inform Matters of Welfare. *Animal Welfare* 3:75–96.

Bekoff, M. 1998. Cognitive Ethology, Deep Ethology, and the Great Ape Project: Expanding the Community of Equals. In J. Gluck and B. Orlans (Eds.), *Applied Ethics in Animal Research*. West Lafayette, IN: Purdue University Press.

Bekoff, M. 1998. Deep Ethology. *AV Magazine* (a publication of the American Anti-Vivisection Society) Winter: 10–18.

Bekoff, M., and D. Jamieson. 1991. Reflective Ethology, Applied Philosophy, and the Moral Status of Animals. *Perspectives in Ethology* 9:1–47.

Birke, L. 1994. *Feminism, Animals, and Science: The Naming of the Shrew*. Philadelphia: Open University Press.

Blum, D. 1994. *The Monkey Wars*. New York: Oxford University Press.

Bostock, S. St. C. 1993. *Zoos and Animal Rights*. London: Routledge.

Brestrup, C. 1997. *Disposable Animals: Ending the Tragedy of Throwaway Pets*. Leander, TX: Camino Bay Books.

Broom, D. M., and K. G. Johnson. *Stress and Animal Welfare*. New York: Chapman and Hall.

Carson, G. 1972. *Men, Beasts, and Gods: A History of Cruelty and Kindness to Animals*. New York: Scribner's.

Cavalieri, P., and P. Singer (Eds.). 1993. *The Great Ape Project: Equality beyond Humanity*. London: Fourth Estate.

Clark, S.R.L. 1984. *The Moral Status of Animals*. New York: Oxford University Press.

Cohen, D. 1993. *Animal Rights: A Handbook for Young Adults*. Brookfield, CT: Millbrook Press.

Cohen, H. 1995. Federal Animal Protection Statutes. *Animal Law* 1:143–161.

Cooper, N., and R.J.C. Carling (Eds.). 1996. *Ecologists and Ethical Judgements*. London: Chapman and Hall.

Davis, H., and D. Balfour (Eds.). 1992. *The Inevitable Bond: Examining Scientist-Animal Interactions*. New York: Cambridge University Press.

Dawkins, M. S. 1980. *Animal Suffering: The Science of Animal Welfare*. New York: Chapman and Hall.

Dawkins, M. S. 1993. *Through Our Eyes Only: The Search for Animal Consciousness*. San Francisco: W. H. Freeman.

DeGrazia, D. 1996. *Taking Animals Seriously: Mental Life and Moral Status*. New York: Cambridge University Press.

Dewsbury, D. A. 1990. Early Interactions between Animal Psychologists and Animal Activists and the Founding of the APA Committee on Precautions in Animal Experimentation. *American Psychologist* 45:315–327.

Dol, M., S. Kasamoentalib, S. Lijmbach, E. Rivas, and R. van den Bos (Eds.). 1997. *Animal Consciousness and Animal Ethics*. Assen: Van Gorcum.

Fadali, M. A. 1996. *Animal Experimentation: A Harvest of Shame*. Los Angeles: Hidden Springs Press.

Finsen, L., and S. Finsen. 1994. *The Animal Rights Movement in America: From Compassion to Respect*. New York: Twayne.

Forsman, B. 1993. *Research Ethics in Practice: The Animal Ethics Committees in Sweden, 1979–1989*, Studies in Research Ethics no. 4. Göteborg: Royal Society of Arts and Sciences in Gothenburg, Centre for Research Ethics.

Fouts, R., and S. Mills. 1997. *Next of Kin*. New York: William Morrow and Co.

Fox, M. W. 1990. *Inhumane Society: The American Way of Exploiting Animals*. New York: St. Martin's Press.

Francione, G. L. 1995. *Animals, Property, and the Law*. Philadelphia: Temple University Press.

Francione, G. L. 1996. *Rain without Thunder: The Ideology of the Animal Rights Movement*. Philadelphia: Temple University Press.

Francione, G. L., and A. E. Charlton. 1992. *Vivisection and Dissection in the Classroom: A Guide to Conscientious Objection*. Jenkintown, PA: American Anti-Vivisection Society.

Frey, R. G. 1980. *Interests and Rights: The Case against Animals*. New York: Oxford University Press.

Garner, R. 1993. *Animals, Politics, and Morality*. Manchester, England: Manchester University Press.

Gluck, J., and Orlans, F. B. (Eds.). 1998. *Applied Ethics in Animal Research*. West Lafayette, IN: Purdue University Press.

Godlovitch, S., R. Godlovitch, and J. Harris (Eds.). 1972. *Animals, Men, and Morals: An Enquiry into the Maltreatment of Non-Humans*. New York: Taplinger.

Guillermo, K. S. 1993. *Monkey Business: The Disturbing Case That Launched the Animal Rights Movement*. Washington DC: National Press Books.

Harnack, A. (Ed.). 1996. *Animal Rights: Opposing Viewpoints*. San Diego, CA: Greenhaven Press.

Harrison, R. 1964. *Animal Machines*. London: Vincent Stuart.

Hart, L. (Ed.). 1998. *Responsible Conduct of Research in Animal Behavior*. New York: Oxford University Press.

Hastings Center Report. 1990. Animals Science, and Ethics [Special supplement], Vol. 20.

Hoage, R. J. (Ed.). 1989. *Perceptions of Animals in American Culture*. Washington, DC: Smithsonian Institution Press.

Kellert, S. R., and E. O. Wilson (Eds.). 1993. *The Biophilia Hypothesis*. Washington, DC: Island Press.

LaFollette, H., and N. Shanks. 1996. *Brute Science: Dilemmas of Animal Experimentation*. London: Routledge.

Langley, G. R. (Ed.). 1989. *Animal Experimentation: The Consensus Changes*. Basingstoke, England: Macmillan.

Lawrence, A. B., and J. Rushen (Eds.). 1993. *Stereotypic Animal Behaviour: Fundamentals and Applications to Welfare*. Wallingford, Oxon, England: CAB International.

Lawrence, E. A. 1984. *Rodeo: An Anthropologist Looks at the Wild and the Tame*. Chicago: University of Chicago Press.

Lawrence, E. A. 1991. Animals in War: History and Implications for the Future. *Anthrozoös* 4:145–153.

Leahy, T. 1991. *Against Liberation*. New York: Routledge.

Linzey, A. 1976. *Animal Rights*. London: SCM Press.

Linzey, A. 1987. *Christianity and the Rights of Animals*. New York: Crossroad.

Linzey, A. 1995. *Animal Theology*. Urbana: University of Illinois Press.

Linzey, A., and D. Cohn-Sherbok. 1997. *After Noah: Celebrating Animals in Judaism and Christianity*. London: Cassell.

Mack, A. (Ed.). 1995. In the Company of Animals. *Social Research* 62:415–838.

Magel, C. R. 1989. *Keyguide to Information Sources in Animal Rights*. Jefferson, NC: McFarland.

Masson, J. M., and S. McCarthy. 1995. *When Elephants Weep: The Emotional Lives of Animals*. New York: Delacorte Press.

Midgley, M. 1983. *Animals and Why They Matter*. Athens: University of Georgia Press.

Miller, H. B., and W. H. Williams (Ed.). 1983. *Ethics and Animals*. Clifton, NJ: Humana Press.

Newkirk, I. 1992. *Free the Animals: The Untold Story of the U.S. Animal Liberation Front and Its Founder, "Valerie."* Chicago: Noble Press.

Norton, B. G., M. Hutchins, E. F. Stevens, and T. L. Maple (Eds.). 1995. *Ethics on the Ark: Zoos, Animal Welfare, and Wildlife Conservation*. Washington, DC: Smithsonian Institution Press.

Noske, B. 1997. *Beyond Boundaries: Humans and Animals*. Montreal: Black Rose.

Orlans, F. B. 1993. *In the Name of Science: Issues in Responsible Animal Experimentation*. New York: Oxford University Press.

Orlans, F. B., et al. (Eds.). 1998. *The Human Use of Animals: Case Studies in Ethical Choice*. New York: Oxford University Press.

Pimple, K. D., F. B. Orlans, and J. P. Gluck (Eds.). 1997. Ethical Issues in the Use of Animals in Research. *Ethics and Behavior* 7(2).

Pluhar, E. B. 1995. *Beyond Prejudice: The Moral Significance of Human and Nonhuman Animals*. Durham, NC: Duke University Press.

Preece, R., and L. Chamberlain. 1993. *Animal Welfare and Human Values*. Waterloo, Ontario, Canada: Wilfrid Laurier University Press.

Rachels, J. 1990. *Created from Animals: The Moral Implications of Darwinism*. New York: Oxford University Press.

Regan, T. 1983. *The Case for Animal Rights*. Berkeley: University of California Press.

Regan, T., and P. Singer (Eds.). 1989. *Animal Rights and Human Obligations*. 2nd ed. Englewood Cliffs, NJ: Prentice-Hall.

Regenstein, L. G. 1991. *Replenish the Earth: A History of Organized Religion's Treatment of Animals and Nature—Including the Bible's Message of Conservation and Kindness toward Animals*. New York: Crossroad.

Ritvo, H. 1987. *The Animal Estate: The English and Other Creatures in the Victorian Age*. Cambridge, MA: Harvard University Press.

Rodd, R. 1990. *Biology, Ethics, and Animals*. New York: Oxford University Press.

Rollin, B. E. 1989. *The Unheeded Cry: Animal Consciousness, Animal Pain, and Science*. New York: Oxford University Press.

Rollin, B. E. 1992. *Animal Rights and Human Morality*. Rev. ed. Buffalo, NY: Prometheus Books.

Rollin, B. E. 1995. *Farm Animal Welfare: Social, Bioethical, and Research Issues*. Ames: Iowa State University Press.

Rollin, B. E., and M. L. Kesel (Eds.). 1990, 1993, 1995. *The Experimental Animal in Biomedical Research*. Vols. 1–3. Boca Raton, FL: CRC Press.

Rowan, A. 1984. *Of Mice, Models, and Men: A Critical Evaluation of Animal Research.* Albany: State University of New York Press.

Rowan, A. N., F. M. Loew, and J. C. Weer. 1995. *The Animal Research Controversy: Protest, Process, and Public Policy—An Analysis of Strategic Issues.* Medford, MA: Tufts University School of Veterinary Medicine.

Russell, W.M.S., and R. L. Burch. 1992. *The Principles of Humane Experimental Technique.* Potters Bar, Herts, UK: Universities Federation for Animal Welfare.

Ryder, R. 1975. *Victims of Science: The Use of Animals in Research.* London: Davis-Poynter.

Ryder, R. D. 1989. *Animal Revolution: Changing Attitudes towards Speciesism.* Oxford: Blackwell.

Salisbury, J. E. 1994. *The Beast Within: Animals in the Middle Ages.* New York: Routledge.

Sapontzis, S. F. 1987. *Morals, Reason, and Animals.* Philadelphia: Temple University Press.

Serpell, J. 1996. *In the Company of Animals: A Study of Human-Animal Relationships.* New York: Cambridge University Press.

Sharpe, R. 1988. *The Cruel Deception.* Wellingborough, Northants, UK: Thorsons.

Shepherson, D. J., J. D. Mellen, and M. Hutchins (Eds.). 1998. *Second Nature: Environmental Enrichment for Captive Animals.* Washington, DC: Smithsonian Institution Press.

Sherry, C. J. 1995. *Animal Rights.* Santa Barbara, CA: ABC-Clio.

Singer, P. 1990. *Animal Liberation.* 2nd ed. New York: New York Review of Books.

Singer, P. (Ed.). 1986. *In Defense of Animals.* New York: Harper and Row.

Smith, J. A., and K. M. Boyd (Eds.). 1991. *Lives in the Balance: The Ethics of Using Animals in Biomedical Research.* New York: Oxford University Press.

Sperling, S. 1988. *Animal Liberators: Research and Morality.* Berkeley: University of California Press.

Tannenbaum, J. 1995. *Veterinary Ethics.* 2nd ed. St. Louis: Mosby.

Tobias, M., and K. Solisti (Eds.). 1998. *Kinship with the Animals.* Portland, OR: Beyond Words Publishers.

van Zutphen, L.F.M., and M. Balls (Eds.). 1997. *Animal Alternatives, Welfare and Ethics.* New York: Elsevier.

Waal, F. de. 1996. *Good Natured: The Origins of Right and Wrong in Humans and Other Animals.* Cambridge, MA: Harvard University Press.

Webster, J. 1995. *Animal Welfare: A Cool Eye towards Eden.* Oxford: Blackwell.

Wynne-Tyson, J. (Ed.). 1988. *The Extended Circle: A Commonplace Book of Animal Rights.* New York: Paragon House.

Zinko U., N. Jones, and C. Gericke. 1997. From Guinea Pig to Computer Mouse: Alternative Methods for a Humane Education. European Network of Individuals and Campaigns for Humane Education (EuroNICHE).

Zurlo, J., D. Rudacille, and A. M. Goldberg. 1994. *Animals and Alternatives in Testing: History, Science, and Ethics.* New York: Mary Ann Liebert.

JOURNALS AND MAGAZINES

Acta Agriculturae Scandinavica, Section A, Animal Sciences
Agriculture and Human Values
Alternatives to Laboratory Animals
American Journal of Primatology
American Psychologist
Animal Activist Alert
Animal Behaviour
Animal Biotechnology
Animal Issues
Animal Law
Animal People
Animal Policy Report (Tufts University)
Animal Welfare
Animal Welfare Information Center Newsletter
Animals' Agenda
Anthrozoös: A Multidisciplinary Journal of the Interactions of People and Animals
Applied Animal Behaviour Science
Between the Species: A Journal of Ethics
Biodiversity and Conservation
Biological Conservation
Biology and Philosophy
British Poultry Science
Bunny Huggers' Gazette
Canadian Journal of Animal Science
Canadian Journal of Veterinary Research
Conservation Biology
Directions (Association of Veterinarians for Animal Rights [AVAR])
E: The Environmental Magazine
Environmental Conservation
Environmental Ethics
Environmental Law
Environmental Values
Ethics
Ethics and Behavior
Etica & Animali
Hastings Center Report

Humane Society of the United States (HSUS) News
InterActions Bibliography (continued as *Humans and Other Species* [1997–   ])
International Society for Environmental Ethics
Johns Hopkins Center for Alternatives to Animal Testing Newsletter
Journal of Agricultural and Environmental Ethics
Journal of Animal Science
Journal of Applied Animal Research
Journal of Applied Animal Welfare Science
Journal of Dairy Science
Journal of the American Veterinary Medical Association
Journal of Zoo and Wildlife Medicine
Lab Animal
Laboratory Animal Science
Laboratory Animals
Laboratory Primate Newsletter
Medical Research Modernization Committee Report
Our Animal Wards
Planet 2000 Newsletter
Politics and the Life Sciences
Poultry Science
Public Affairs Quarterly
Royal Society for the Prevention of Cruelty to Animals Science Review
Satya
Shelter Sense (continued as *Animal Sheltering* [1996–   ])
Social Justice
Social Research
Society and Animals
Wildlife Society Bulletin
Zoo Biology

Index

About the Editors and Contributors

COLIN ALLEN is associate professor of philosophy at Texas A&M University.

DAVID C. ANDERSON publishes the quarterly *Humans & Other Species*. He was formerly information specialist at the University of California Center for Animal Alternatives.

ARNOLD ARLUKE is a professor of sociology and anthropology at Northeastern University and a senior fellow at Tufts University Center for Animals and Public Policy.

MARTHA ARMSTRONG is vice president for companion animals at the Humane Society of the United States and an adjunct professor at Tufts University School of Veterinary Medicine.

JONATHAN BALCOMBE is associate director for education, Animal Research Issues, with the Humane Society of the United States.

TOM L. BEAUCHAMP is professor of philosophy and senior research scholar at Georgetown University.

ALAN M. BECK is a professor in the School of Veterinary Medicine at Purdue University in Indiana.

ANNE C. BEKOFF is a professor of biology at the University of Colorado, Boulder.

MARC BEKOFF is professor of biology at the University of Colorado, Boulder, a Fellow of the Animal Behavior Society, and a Guggenheim Fellow.

MARJORIE BEKOFF has served as an unaffiliated member of an animal care and use committee.

PIERS BEIRNE is a professor of criminology at the University of Southern Maine.

BETH BENNETT is a research associate at the Institute for Behavioral Genetics, investigating a genetic basis for alcoholism.

LYNDA BIRKE is at the Centre for the Study of Women and Gender at the University of Warwick.

STEVEN J. BISSELL is head of education for the Colorado Division of Wildlife.

STEPHEN ST. C. BOSTOCK is education officer at Glasgow Zoo and honorary research fellow in philosophy at the University of Glasgow.

JOHN P. BROIDA is an associate professor of psychology at the University of Southern Maine.

DONALD M. BROOM has been professor of animal welfare at Cambridge University Veterinary School since 1986.

JOSEPH BRUCHAC is an Abenaki storyteller and writer whose most recent book is *Lasting Echoes* (1997), an oral history of Native American people.

GORDON M. BURGHARDT is a professor in the Department of Psychology and Ecology and Evolutionary Biology at the University of Tennessee, Knoxville.

NEDIM C. BUYUKMIHCI is a professor of ophthalmology at the University of California and president of the Association of Veterinarians for Animal Rights.

LARRY CARBONE is a veterinarian and a graduate student of veterinary ethics at Cornell University.

KATHY CARLSTEAD is a researcher at the National Zoological Park, Washington, DC.

MATT CARTMILL is a professor of biological anthropology at Duke University.

ANN S. CAUSEY is in the Department of Philosophy at Auburn University.

PAOLA CAVALIERI is the editor of the international journal *Etica & Animali*.

ANNA E. CHARLTON is clinical staff attorney and co-director of the Rutgers Animal Rights Law Center.

STEPHEN R. L. CLARK is professor of philosophy at the University at Liverpool.

DAN COHN-SHERBOK is professor of Judaism at the University of Wales, Lampeter.

EILEEN CRIST is assistant professor in the Department of Science and Technology Studies at Virginia Polytechnic University.

HANK DAVIS is a professor of psychology at the University of Guelph, Ontario.

DAVID D. DeGRAZIA is associate professor of philosophy at George Washington University.

REBECCA DRESSER is a professor in the School of Law and Center for Biomedical Ethics, School of Medicine, Case Western Reserve University.

IAN J. H. DUNCAN is a professor specializing in animal welfare research and education at the University of Guelph, Ontario.

ANDRZEJ ELZANOWSKI is a professor of vertebrate zoology at the University of Wroclaw, Poland.

SALLY FEKETY is the director of animal sheltering issues for the Humane society of the United States in Washington, DC.

LAWRENCE FINSEN is a professor of philosophy at the University of Redlands in California.

SUSAN FINSEN is professor and chair of the Department of Philosophy at California State University, San Bernadino.

JOHN ANDREW FISHER is professor of philosophy at the University of Colorado at Boulder.

BIRGITTA FORSMAN is an associate professor of philosophy of science and research ethics at Lund University, Sweden.

DEBORAH FOUTS is the co-director of the Chimpanzee and Human Communication Institute at Central Washington University.

ROGER FOUTS is the co-director of the Chimpanzee and Human Communication Institute and a professor of psychology at Central Washington University.

MICHAEL ALLEN FOX is a professor of philosophy at Queen's University, Kingston, Ontario, specializing in environmental ethics and ethics and animals.

MICHAEL W. FOX is senior advisor to the president, the Humane Society of the United States, Washington, DC.

GARY L. FRANCIONE is professor of law at Rutgers University and faculty director of the Rutgers Animal Rights Law Center.

DAVID FRASER is professor of animal welfare at the University of British Columbia.

TED FRIEND is a professor of applied ethology in the Department of Animal Science at Texas A&M University.

R. G. FREY is an applied ethicist in the Philosophy Department at Bowling Green State University.

BENNETT G. GALEF, JR., is professor of psychology and adjunct professor of biology at McMaster University.

MICHAEL GARNER studied mathematics, works as a translator of art literature, and is a member of the international Board of the Great Ape Project.

JOHN P. GLUCK is professor of psychology at the University of New Mexico.

ALAN M. GOLDBERG is professor of toxicology and director of the Center for Alternatives to Animal Testing at the Johns Hopkins School of Public Health.

JANE GOODALL is director of science and research at the Jane Goodall Institute for Wildlife Research, Education and Conservation.

TEMPLE GRANDIN is professor of animal science at Colorado State University.

LORI GRUEN teaches philosophy and is affiliated with the Ethics in Society Program at Stanford University.

JOHN HADIDIAN is the director of the Urban Wildlife Protection Program at the Humane Society of the United States.

LYNETTE A. HART is director of the University of California Center for Animal Alternatives, Davis.

HAROLD A. HERZOG, JR., is a professor of psychology at Western Carolina University.

NED HETTINGER teaches philosophy at the College of Charleston in South Carolina.

MICHAEL HUTCHINS is director of conservation and science at the American Zoo and Aquarium Association.

ROBERT G. JAEGER is John Chance Professor of Biology at the University of Southwestern Louisiana.

DALE JAMIESON is Henry R. Luce Professor in the Human Dimensions of Global Change at Carleton College.

KARIN KARCHER studied philosophy and ethology and is a member of the international Board of the Great Ape Project.

MICHAEL D. KREGER is a technical information specialist at the U.S. Department of Agriculture's Animal Welfare Information Center.

HUGH LaFOLLETTE is professor of philosophy at East Tennessee State University.

GILL LANGLEY is scientific adviser to the Dr. Hadwen Trust, England, and an international expert on animal experimentation issues.

HERBERT LANSDELL is a guest researcher at the National Institute of Neurological Disorders and Stroke in Bethesda, Maryland.

ELIZABETH ATWOOD LAWRENCE, a veterinarian and cultural anthropologist, is a professor at the Tufts University School of Veterinary Medicine.

ANDREW LINZEY is the IFAW senior research fellow of Mansfield College, Oxford, and honorary professor at the University of Birmingham.

CATHY LISS is executive director of the Animal Welfare Institute and senior research associate of the Society for Animal Protective Legislation.

RANDALL LOCKWOOD is vice president for training initiatives at the Humane Society of the United States.

PAULA MacKAY is a community organizer working with ecological and social issues.

HAL MARKOWITZ is a professor of biology at San Francisco State University.

JIM MASON is the author of *An Unnatural Order* and other books.

CARRON A. MEANEY is a research associate at the Denver Museum of Natural History and the University of Colorado Museum.

JOY A. MENCH is a professor in the Department of Animal Science at the University of California, Davis.

MICHAEL MENDL is a lecturer in animal behavior at the University of Bristol, England.

SLAVOLJUB MILEKIC teaches psychology at Hampshire College.

ROBERT W. MITCHELL is an associate professor in the Psychology Department at Eastern Kentucky University.

DAVID B. MORTON is head of the Department of Biomedical Science and Ethics at the University of Birmingham, England.

SAMANTHA MULLEN is director of training resources at the Humane Society of the United States.

JAMES LINDEMANN NELSON is a professor of philosophy at the University of Tennessee, Knoxville.

CHARLES S. NICOLL is a professor in the Department of Integrative Biology, University of California at Berkeley.

BARBARA NOSKE is a research scholar at the Faculty of Environmental Studies, York University, Canada.

PATRICIA OLSON is the director of veterinary affairs and studies at the American Humane Association in Englewood, Colorado.

F. BARBARA ORLANS is a senior research fellow at the Kennedy Institute of Ethics, Georgetown University.

J. BRUCE OVERMIER is a professor of psychology specializing in animal models of human dysfunction.

WAYNE PACELLE is vice president for government affairs and media for the Humane Society of the United States.

ELIZABETH PAUL researches the psychology of human-animal relationships at the University of Bristol, England.

ANDREW J. PETTO is at the National Center for Science Education in Madison, Wisconsin.

EVELYN PLUHAR is professor of philosophy at the Pennsylvania State University, Fayette Campus.

JAMES RACHELS is professor of philosophy at the University of Alabama, Birmingham.

TOM REGAN is professor of philosophy and head of the Department of Philosophy and Religion at North Carolina State University.

VIKTOR REINHARDT is a veterinarian and ethologist specializing in animal welfare issues.

HARRIET RITVO is the Arthur J. Conner Professor of History at MIT.

BERNARD E. ROLLIN is professor of philosophy, professor of physiology, and director of bioethical planning at Colorado State University.

HOLMES ROLSTON III is University Distinguished Professor of Philosophy at Colorado State University.

ANTHONY ROSE is executive director of the Biosynergy Institute in Hermosa Beach, California.

ANDREW N. ROWAN is senior vice president (Research, Education and International Issues) of the Humane Society of the United States.

DEBORAH RUDACILLE is a science writer at Johns Hopkins University.

SHARON M. RUSSELL is professor in the Department of Integrative Biology at the University of California, Berkeley.

LILLY-MARLENE RUSSOW is an associate professor of philosophy and adjunct associate professor of veterinary pathobiology at Purdue University.

RICHARD D. RYDER is director of animal welfare studies for the International Fund for Animal Welfare (IFAW), a trustee of the Royal Society for the Prevention of Cruelty to Animals (RSPCA), and inventor of the concepts "speciesism" and "painism."

JOYCE E. SALISBURY is a professor of history at the University of Wisconsin, Green Bay.

CLINTON R. SANDERS is a professor in the sociology department at the University of Connecticut.

STEVE F. SAPONTZIS is professor of philosophy at California State University, Hayward.

LISA M. SAVAGE is an assistant professor of psychology at the State University of New York, Binghamton.

MARIAN SCHOLTMEIJER teaches university English and researches the cultural representation of nonhuman animals.

JAMES A. SERPELL is professor of animal welfare at the University of Pennsylvania School of Veterinary Medicine.

NIALL SHANKS is an associate professor in the Department of Philosophy and an adjunct professor in the Department of Biological Sciences at East Tennessee State University.

KENNETH J. SHAPIRO is executive director of Psychologists for the Ethical Treatment of Animals, editor of *Society and Animals*, and coeditor of the *Journal of Applied Animal Welfare Science.*

PETER SINGER is a professor of philosophy in the Centre for Human Bioethics at Monash University.

ADRIAN SMITH is a professor at the Norwegian College of Veterinary Medicine, Oslo.

KARINA SMITH is a consultant, registered nurse, and co-compiler of the NORINA database.

MAREK ŠPINKA is a senior researcher at the Research Institute of Animal Production in Prague, Czech Republic.

LEILA STANFIELD is a founder of Biodiversity Associates/Friends of the Bow, Laramie, Wyoming.

CHRISTINE STEVENS is president of the Animal Welfare Institute and secretary of the Society for Animal Protective Legislation.

CYNTHIA STITELY is associate for animal sheltering issues for the Humane Society of the United States, Washington, DC.

DAVID SZTYBEL is in the Department of Philosophy at the University of Toronto.

ALLISON A. TAYLOR is an animal behaviorist specializing in animal welfare issues.

LARRY D. TERRY is a graduate student in the Department of History at Texas A&M University.

MICHAEL TOBIAS is an ecologist, writer, and filmmaker specializing in issues pertaining to interspecies empathy.

BERNARD UNTI is a graduate student in the Department of History at American University.

GARY VARNER is associate professor of philosophy at Texas A&M University, specializing in environmental ethics and animal rights philosophies.

PAUL WALDAU received his Ph.D. from Oxford University and studies relationships between religion and animals.

JACK WEIR is professor of philosophy at Morehead State University in Kentucky.

FRANÇOISE WEMELSFELDER is a research scientist in the field of animal behavior and welfare at the Scottish Agricultural College, Edinburgh.

ANN B. WOLFE is a graduate student in philosophy at the University of Wisconsin, Madison.

R. LEE ZASLOFF is associate director of the University of California Center for Animal Alternatives.

STEPHEN L. ZAWISTOWSKI is senior vice president and science advisor at the American Society for the Prevention of Cruelty to Animals.

JOANNE ZURLO is the associate director of the Johns Hopkins Center for Alternatives to Animal Testing.